An Introduction to Machine Learning

Miroslav Kubat

An Introduction to Machine Learning

Second Edition

 Springer

Miroslav Kubat
Department of Electrical and Computer Engineering
University of Miami
Coral Gables, FL, USA

ISBN 978-3-319-87669-6   ISBN 978-3-319-63913-0   (eBook)
DOI 10.1007/978-3-319-63913-0

Printed on acid-free paper

This Springer imprint is published by Springer Nature
The registered company is Springer International Publishing AG
The registered company address is: Gewerbestrasse 11, 6330 Cham, Switzerland

To my wife, Verunka.

Contents

1   A Simple Machine-Learning Task 1
  1.1   Training Sets and Classifiers 1
  1.2   Minor Digression: Hill-Climbing Search 5
  1.3   Hill Climbing in Machine Learning 8
  1.4   The Induced Classifier's Performance 11
  1.5   Some Difficulties with Available Data 13
  1.6   Summary and Historical Remarks 15
  1.7   Solidify Your Knowledge 16

2   Probabilities: Bayesian Classifiers 19
  2.1   The Single-Attribute Case 19
  2.2   Vectors of Discrete Attributes 22
  2.3   Probabilities of Rare Events: Exploiting the Expert's Intuition 26
  2.4   How to Handle Continuous Attributes 30
  2.5   Gaussian "Bell" Function: A Standard *pdf* 33
  2.6   Approximating PDFs with Sets of Gaussians 34
  2.7   Summary and Historical Remarks 36
  2.8   Solidify Your Knowledge 40

3   Similarities: Nearest-Neighbor Classifiers 43
  3.1   The k-Nearest-Neighbor Rule 43
  3.2   Measuring Similarity 46
  3.3   Irrelevant Attributes and Scaling Problems 49
  3.4   Performance Considerations 52
  3.5   Weighted Nearest Neighbors 55
  3.6   Removing Dangerous Examples 57
  3.7   Removing Redundant Examples 59
  3.8   Summary and Historical Remarks 61
  3.9   Solidify Your Knowledge 62

4   Inter-Class Boundaries: Linear and Polynomial Classifiers 65
  4.1   The Essence 65
  4.2   The Additive Rule: Perceptron Learning 69
  4.3   The Multiplicative Rule: WINNOW 73
  4.4   Domains with More Than Two Classes 76
  4.5   Polynomial Classifiers 79
  4.6   Specific Aspects of Polynomial Classifiers 81
  4.7   Numerical Domains and Support Vector Machines 84
  4.8   Summary and Historical Remarks 86
  4.9   Solidify Your Knowledge 87

5   Artificial Neural Networks 91
  5.1   Multilayer Perceptrons as Classifiers 91
  5.2   Neural Network's Error 95
  5.3   Backpropagation of Error 97
  5.4   Special Aspects of Multilayer Perceptrons 100
  5.5   Architectural Issues 104
  5.6   Radial-Basis Function Networks 106
  5.7   Summary and Historical Remarks 109
  5.8   Solidify Your Knowledge 110

6   Decision Trees 113
  6.1   Decision Trees as Classifiers 113
  6.2   Induction of Decision Trees 117
  6.3   How Much Information Does an Attribute Convey? 119
  6.4   Binary Split of a Numeric Attribute 122
  6.5   Pruning 126
  6.6   Converting the Decision Tree into Rules 130
  6.7   Summary and Historical Remarks 132
  6.8   Solidify Your Knowledge 133

7   Computational Learning Theory 137
  7.1   PAC Learning 137
  7.2   Examples of PAC Learnability 141
  7.3   Some Practical and Theoretical Consequences 143
  7.4   VC-Dimension and Learnability 145
  7.5   Summary and Historical Remarks 148
  7.6   Exercises and Thought Experiments 149

8   A Few Instructive Applications 151
  8.1   Character Recognition 151
  8.2   Oil-Spill Recognition 155
  8.3   Sleep Classification 158
  8.4   Brain–Computer Interface 161
  8.5   Medical Diagnosis 165

8.6   Text Classification 167
8.7   Summary and Historical Remarks 169
8.8   Exercises and Thought Experiments 170

9   Induction of Voting Assemblies 173
9.1   Bagging 173
9.2   Schapire's Boosting 176
9.3   Adaboost: Practical Version of Boosting 179
9.4   Variations on the Boosting Theme 183
9.5   Cost-Saving Benefits of the Approach 185
9.6   Summary and Historical Remarks 187
9.7   Solidify Your Knowledge 188

10   Some Practical Aspects to Know About 191
10.1   A Learner's Bias 191
10.2   Imbalanced Training Sets 194
10.3   Context-Dependent Domains 199
10.4   Unknown Attribute Values 202
10.5   Attribute Selection 204
10.6   Miscellaneous 206
10.7   Summary and Historical Remarks 208
10.8   Solidify Your Knowledge 208

11   Performance Evaluation 211
11.1   Basic Performance Criteria 211
11.2   Precision and Recall 214
11.3   Other Ways to Measure Performance 219
11.4   Learning Curves and Computational Costs 222
11.5   Methodologies of Experimental Evaluation 224
11.6   Summary and Historical Remarks 227
11.7   Solidify Your Knowledge 228

12   Statistical Significance 231
12.1   Sampling a Population 231
12.2   Benefiting from the Normal Distribution 235
12.3   Confidence Intervals 239
12.4   Statistical Evaluation of a Classifier 241
12.5   Another Kind of Statistical Evaluation 244
12.6   Comparing Machine-Learning Techniques 245
12.7   Summary and Historical Remarks 247
12.8   Solidify Your Knowledge 248

13   Induction in Multi-Label Domains 251
13.1   Classical Machine Learning in Multi-Label Domains 251
13.2   Treating Each Class Separately: Binary Relevance 254
13.3   Classifier Chains 256

13.4   Another Possibility: Stacking 258
13.5   A Note on Hierarchically Ordered Classes........................ 260
13.6   Aggregating the Classes 263
13.7   Criteria for Performance Evaluation............................... 265
13.8   Summary and Historical Remarks.................................. 268
13.9   Solidify Your Knowledge 269

14   Unsupervised Learning 273
14.1   Cluster Analysis 273
14.2   A Simple Algorithm: k-Means..................................... 277
14.3   More Advanced Versions of k-Means 281
14.4   Hierarchical Aggregation 283
14.5   Self-Organizing Feature Maps: Introduction...................... 286
14.6   Some Important Details 289
14.7   Why Feature Maps?.. 291
14.8   Summary and Historical Remarks.................................. 293
14.9   Solidify Your Knowledge 294

15   Classifiers in the Form of Rulesets 297
15.1   A Class Described By Rules 297
15.2   Inducing Rulesets by Sequential Covering........................ 300
15.3   Predicates and Recursion 302
15.4   More Advanced Search Operators................................... 305
15.5   Summary and Historical Remarks.................................. 306
15.6   Solidify Your Knowledge 307

16   The Genetic Algorithm 309
16.1   The Baseline Genetic Algorithm 309
16.2   Implementing the Individual Modules 311
16.3   Why It Works.. 314
16.4   The Danger of Premature Degeneration............................ 317
16.5   Other Genetic Operators 319
16.6   Some Advanced Versions 321
16.7   Selections in k-NN Classifiers 324
16.8   Summary and Historical Remarks.................................. 327
16.9   Solidify Your Knowledge 328

17   Reinforcement Learning 331
17.1   How to Choose the Most Rewarding Action........................ 331
17.2   States and Actions in a Game.. 334
17.3   The SARSA Approach 337
17.4   Summary and Historical Remarks.................................. 338
17.5   Solidify Your Knowledge 338

Bibliography 341

Index 347

Introduction

Machine learning has come of age. And just in case you might think this is a mere platitude, let me clarify.

The dream that machines would one day be able to learn is as old as computers themselves, perhaps older still. For a long time, however, it remained just that: a dream. True, Rosenblatt's perceptron did trigger a wave of activity, but in retrospect, the excitement has to be deemed short-lived. As for the attempts that followed, these fared even worse; barely noticed, often ignored, they never made a breakthrough— no software companies, no major follow-up research, and not much support from funding agencies. Machine learning remained an underdog, condemned to live in the shadow of more successful disciplines. The grand ambition lay dormant.

And then it all changed.

A group of visionaries pointed out a weak spot in the knowledge-based systems that were all the rage in the 1970s' artificial intelligence: where was the "knowledge" to come from? The prevailing wisdom of the day insisted that it should take the form of *if-then* rules put together by the joint effort of engineers and field experts. Practical experience, though, was unconvincing. Experts found it difficult to communicate what they knew to engineers. Engineers, in turn, were at a loss as to what questions to ask and what to make of the answers. A few widely publicized success stories notwithstanding, most attempts to create a knowledge base of, say, tens of thousands of such rules proved frustrating.

The proposition made by the visionaries was both simple and audacious. If it is so hard to tell a machine exactly how to go about a certain problem, why not provide the instruction indirectly, conveying the necessary skills by way of examples from which the computer will—yes—*learn*!

Of course, this only makes sense if we can rely on the existence of algorithms to do the learning. This was the main difficulty. As it turned out, neither Rosenblatt's perceptron nor the techniques developed after it were very useful. But the absence of the requisite machine-learning techniques was not an obstacle; rather, it was a challenge that inspired quite a few brilliant minds. The idea of endowing computers with learning skills opened new horizons and created a large amount of excitement. The world was beginning to take notice.

The bombshell exploded in 1983. *Machine Learning: The AI Approach*[1] was a thick volume of research papers which proposed the most diverse ways of addressing the great mystery. Under their influence, a new scientific discipline was born—virtually overnight. Three years later, a follow-up book appeared and then another. A soon-to-become-prestigious scientific journal was founded. Annual conferences of great repute were launched. And dozens, perhaps hundreds, of doctoral dissertations, were submitted and successfully defended.

In this early stage, the question was not only *how* to learn but also *what* to learn and *why*. In retrospect, those were wonderful times, so creative that they deserve to be remembered with nostalgia. It is only to be regretted that so many great thoughts later came to be abandoned. Practical needs of realistic applications got the upper hand, pointing to the most promising avenues for further efforts. After a period of enchantment, concrete research strands crystallized: induction of the *if-then* rules for knowledge-based systems; induction of classifiers, programs capable of improving their skills based on experience; automatic fine-tuning of Prolog programs; and some others. So many were the directions that some leading personalities felt it necessary to try to steer further development by writing monographs, some successful, others less so.

An important watershed was Tom Mitchell's legendary textbook.[2] This summarized the state of the art of the field in a format appropriate for doctoral students and scientists alike. One by one, universities started offering graduate courses that were usually built around this book. Meanwhile, the research methodology became more systematic, too. A rich repository of machine-leaning test beds was created, making it possible to compare the performance or learning algorithms. Statistical methods of evaluation became widespread. Public domain versions of most popular programs were made available. The number of scientists dealing with this discipline grew to thousands, perhaps even more.

Now, we have reached the stage where a great many universities are offering machine learning as an undergraduate class. This is quite a new situation. As a rule, these classes call for a different kind of textbook. Apart from mastering the baseline techniques, future engineers need to develop a good grasp of the strengths and weaknesses of alternative approaches; they should be aware of the peculiarities and idiosyncrasies of different paradigms. Above all, they must understand the circumstances under which some techniques succeed and others fail. Only then will they be able to make the right choices when addressing concrete applications. A textbook that is to provide all of the above should contain less mathematics, but a lot of practical advice.

These then are the considerations that have dictated the size, structure, and style of a teaching text meant to provide the material for a one-semester introductory course.

---

[1] Edited by R. Michalski, J. Carbonell, and T. Mitchell.
[2] T. Mitchell, *Machine Learning*, McGraw-Hill (1997).

The first problem is the choice of material. At a time when high-tech companies are establishing machine-learning groups, universities have to provide the students with such knowledge, skills, and understanding that are relevant to the current needs of the industry. For this reason, preference has been given to Bayesian classifiers, nearest-neighbor classifiers, linear and polynomial classifiers, decision trees, the fundamentals of the neural networks, and the principle of the boosting algorithms. Significant space has been devoted to certain typical aspects of concrete engineering applications. When applied to really difficult tasks, the baseline techniques are known to behave not exactly the same way they do in the toy domains employed by the instructor. One has to know what to expect.

The book consists of 17 chapters, each covering one major topic. The chapters are divided into sections, each devoted to one critical problem. The student is advised to proceed to the next section only after having answered the set of 2–4 "control questions" at the end of the previous section. These questions are here to help the student decide whether he or she has mastered the given material. If not, it is necessary to return to the previous text.

As they say, only practice makes perfect. This is why at the end of each chapter are exercises to encourage the necessary practicing. Deeper insight into the diverse aspects of the material will then be gained by going through the thought experiments that follow. These are more difficult, but it is only through hard work that an engineer develops the right kind of understanding. The acquired knowledge is then further solidified by suggested computer projects. Programming is important, too. Nowadays, everybody is used to downloading the requisite software from the web. This shortcut, however, is not recommended to the student of this book. It is only by being forced to flesh out all the details of a computer program that you learn to appreciate all the subtle points of the machine-learning techniques presented here.

Chapter 1
A Simple Machine-Learning Task

You will find it difficult to describe your mother's face accurately enough for your friend to recognize her in a supermarket. But if you show him a few of her photos, he will immediately spot the tell-tale traits he needs. As they say, a picture—an example—is worth a thousand words.

This is what we want our technology to emulate. Unable to define certain objects or concepts with adequate accuracy, we want to convey them to the machine by way of examples. For this to work, however, the computer has to be able to convert the examples into knowledge. Hence our interest in algorithms and techniques for *machine learning*, the topic of this textbook.

The first chapter formulates the task as a search problem, introducing hill-climbing search not only as our preliminary attempt to address the machine-learning task, but also as a tool that will come handy in a few auxiliary problems to be encountered in later chapters. Having thus established the foundation, we will proceed to such issues as performance criteria, experimental methodology, and certain aspects that make the learning process difficult—and interesting.

1.1 Training Sets and Classifiers

Let us introduce the problem, and certain fundamental concepts that will accompany us throughout the rest of the book.

The Set of Pre-Classified Training Examples Figure 1.1 shows six pies that Johnny likes, and six that he does not. These *positive* and *negative examples* of the underlying concept constitute a *training set* from which the machine is to induce a *classifier*—an algorithm capable of categorizing any future pie into one of the two *classes*: positive and negative.

© Springer International Publishing AG 2017
M. Kubat, *An Introduction to Machine Learning*,
DOI 10.1007/978-3-319-63913-0_1

Fig. 1.1 A simple machine-learning task: induce a classifier capable of labeling future pies as positive and negative instances of "a pie that Johnny likes"

The number of classes can of course be greater. Thus a classifier that decides whether a landscape snapshot was taken in spring, summer, fall, or winter distinguishes four. Software that identifies characters scribbled on an *iPad* needs at least 36 classes: 26 for letters and 10 for digits. And document-categorization systems are capable of identifying hundreds, even thousands of different topics. Our only motivation for choosing a two-class domain is its simplicity.

Table 1.1 The twelve training examples expressed in a matrix form

Example	Shape	Crust		Filling		Class
		Size	Shade	Size	Shade	
ex1	Circle	Thick	Gray	Thick	Dark	pos
ex2	Circle	Thick	White	Thick	Dark	pos
ex3	Triangle	Thick	Dark	Thick	Gray	pos
ex4	Circle	Thin	White	Thin	Dark	pos
ex5	Square	Thick	Dark	Thin	White	pos
ex6	Circle	Thick	White	Thin	Dark	pos
ex7	Circle	Thick	Gray	Thick	White	neg
ex8	Square	Thick	White	Thick	Gray	neg
ex9	Triangle	Thin	Gray	Thin	Dark	neg
ex10	Circle	Thick	Dark	Thick	White	neg
ex11	Square	Thick	White	Thick	Dark	neg
ex12	Triangle	Thick	White	Thick	Gray	neg

Attribute Vectors To be able to communicate the training examples to the machine, we have to describe them in an appropriate way. The most common mechanism relies on so-called *attributes*. In the "pies" domain, five may be suggested: shape (circle, triangle, and square), crust-size (thin or thick), crust-shade (white, gray, or dark), filling-size (thin or thick), and filling-shade (white, gray, or dark). Table 1.1 specifies the values of these attributes for the twelve examples in Fig. 1.1. For instance, the pie in the upper-left corner of the picture (the table calls it ex1) is described by the following conjunction:

```
(shape=circle) AND (crust-size=thick) AND (crust-shade=gray)
AND (filling-size=thick) AND (filling-shade=dark)
```

A Classifier to Be Induced The training set constitutes the input from which we are to induce the classifier. But *what* classifier?

Suppose we want it in the form of a boolean function that is *true* for positive examples and *false* for negative ones. Checking the expression [(shape=circle) AND (filling-shade=dark)] against the training set, we can see that its value is *false* for all negative examples: while it *is* possible to find negative examples that are circular, none of these has a dark filling. As for the positive examples, however, the expression is *true* for four of them and *false* for the remaining two. This means that the classifier makes two errors, a transgression we might refuse to tolerate, suspecting there is a better solution. Indeed, the reader will easily verify that the following expression never goes wrong on the entire training set:

```
[ (shape=circle) AND (filling-shade=dark) ] OR
[ NOT(shape=circle) AND (crust-shade=dark) ]
```

Problems with a Brute-Force Approach How does a machine find a classifier of this kind? Brute force (something that computers are so good at) will not do here. Just consider how many different examples can be distinguished by the given set of attributes in the "pies" domain. For each of the three different shapes, there are two alternative crust-sizes, the number of combinations being $3 \times 2 = 6$. For each of these, the next attribute, crust-shade, can acquire three different values, which brings the number of combinations to $3 \times 2 \times 3 = 18$. Extending this line of reasoning to *all* attributes, we realize that the size of the *instance space* is $3 \times 2 \times 3 \times 2 \times 3 = 108$ different examples.

Each subset of these examples—and there are 2^{108} subsets!—may constitute the list of positive examples of someone's notion of a "good pie." And each such subset can be characterized by at least one boolean expression. Running each of these classifiers through the training set is clearly out of the question.

Manual Approach and Search Uncertain about how to invent a classifier-inducing algorithm, we may try to glean some inspiration from an attempt to create a classifier "manually," by the good old-fashioned pencil-and-paper method. When doing so, we begin with some tentative initial version, say, shape=circular. Having checked it against the training set, we find it to be *true* for four positive examples, but also for two negative ones. Apparently, the classifier needs to be "narrowed" (specialized) so as to exclude the two negative examples. One way to go about the specialization is to add a conjunction, such as when turning shape=circular into [(shape=circular) AND (filling-shade=dark)]. This new expression, while *false* for all negative examples, is still imperfect because it covers only four (ex1, ex2, ex4, and ex6) of the six positive examples. The next step should therefore attempt some generalization, perhaps by adding a disjunction: { [(shape=circular) AND (filling-shade=dark)] OR (crust-size=thick) }. We continue in this way until we find a 100% accurate classifier (if it exists).

The lesson from this little introspection is that the classifier can be created by means of a sequence of specialization and generalization steps which gradually modify a given version of the classifier until it satisfies certain predefined requirements. This is encouraging. Readers with background in Artificial Intelligence will recognize this procedure as a *search* through the space of boolean expressions. And Artificial Intelligence is known to have developed and explored quite a few of search algorithms. It may be an idea to take a look at least at one of them.

What Have You Learned?

To make sure you understand the topic, try to answer the following questions. If needed, return to the appropriate place in the text.

- What is the input and output of the learning problem we have just described?
- How do we describe the training examples? What is *instance space*? Can we calculate its size?

- In the "pies" domain, find a boolean expression that correctly classifies all the training examples from Table 1.1.

1.2   Minor Digression: Hill-Climbing Search

Let us now formalize what we mean by *search*, and then introduce one popular algorithm, the so-called *hill climbing*. Artificial Intelligence defines *search* something like this: starting from an *initial state*, find a sequence of steps which, proceeding through a set of interim *search states*, lead to a predefined *final state*. The individual steps—transitions from one search state to another—are carried out by *search operators* which, too, have been pre-specified by the programmer. The order in which the search operators are applied follows a specific *search strategy* (Fig. 1.2).

Hill Climbing: An Illustration   One popular search strategy is *hill climbing*. Let us illustrate its essence on a well-known brain-teaser, the sliding-tiles puzzle. The board of a trivial version of this game consists of nine squares arranged in three rows, eight covered by numbered tiles (integers from 1 to 8), the last left empty. We convert one search state into another by sliding to the empty square a tile from one of its neighbors. The goal is to achieve a pre-specified arrangement of the tiles.

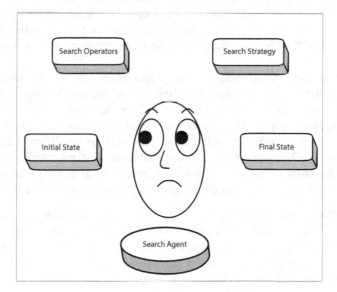

Fig. 1.2 A search problem is characterized by an initial state, final state, search operators, and a search strategy

The flowchart in Fig. 1.3 starts with a concrete initial state, in which we can choose between two operators: "move `tile-6` up" and "move `tile-2` to the left." The choice is guided by an *evaluation function* that estimates for each state its distance from the goal. A simple possibility is to count the squares that the tiles have to traverse before reaching their final destinations. In the initial state, tiles 2, 4, and 5 are already in the right locations; tile 3 has to be moved by four squares; and each of the tiles 1, 6, 7, and 8 have to be moved by two squares. This sums up to distance $d = 4 + 4 \times 2 = 12$.

In Fig. 1.3, each of the two operators applicable to the initial state leads to a state whose distance from the final state is $d = 13$. In the absence of any other guidance, we choose randomly and go to the left, reaching the situation where the empty square is in the middle of the top row. Here, three moves are possible. One of them would only get us back to the initial state, and can thus be ignored; as for the remaining two, one results in a state with $d = 14$, the other in a state with $d = 12$. The latter being the lower value, this is where we go. The next step is trivial because only one move gets us to a state that has not been visited before. After this, we again face the choice between two alternatives ... and this how the search continues until it reaches the final state.

Alternative Termination Criteria and Evaluation Functions Other *termination criteria* can be considered, too. The search can be instructed to stop when the maximum allotted time has elapsed (we do not want the computer to run forever), when the number of visited states has exceeded a certain limit, when something sufficiently close to the final state has been found, when we have realized that all states have already been visited, and so on, the concrete formulation reflecting critical aspects of the given application, sometimes combining two or more criteria in one.

By the way, the evaluation function employed in the sliding-tiles example was fairly simple, barely accomplishing its mission: to let the user convey some notion of his or her understanding of the problem, to provide a hint as to which move a human solver might prefer. To succeed in a realistic application, we would have to come up with a more sophisticated function. Quite often, *many* different alternatives can be devised, each engendering a different sequence of steps. Some will be quick in reaching the solution, others will follow a more circuitous path. The program's performance will then depend on the programmer's ability to pick the right one.

The Algorithm of Hill Combing The algorithm is summarized by the pseudocode in Table 1.2. Details will of course depend on each individual's programming style, but the code will almost always contain a few typical functions. One of them compares two states and returns *true* if they are identical; this is how the program ascertains that the final state has been reached. Another function takes a given search state and applies to it all search operators, thus creating a complete set of "child states." To avoid infinite loops, a third function checks whether a state has already been investigated. A fourth calculates for a given state its distance from the final

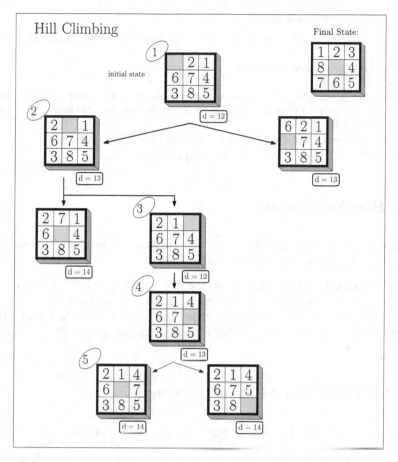

Fig. 1.3 Hill climbing. Circled integers indicate the order in which the search states are visited. d is a state's distance from the final state as calculated by the given evaluation function. Ties are broken randomly

state, and a fifth sorts the "child" states according to the distances thus calculated and places them at the front of the list L. And the last function checks if a termination criterion has been satisfied.[1]

One last observation: at some of the states in Fig. 1.3, no "child" offers any improvement over its "parent," a lower d-value being achieved only after temporary compromises. This is what a mountain climber may experience, too: sometimes, he has to traverse a valley before being able to resume the ascent. The mountain-climbing metaphor, by the way, is what gave this technique its name.

---

[1] For simplicity, the pseudocode ignores termination criteria other than reaching, or failing to reach, the final state.

Table 1.2 Hill-climbing search algorithm

1.	Create two lists, L and L_{seen}. At the beginning, L contains only the initial state, and L_{seen} is empty.
2.	Let n be the first element of L. Compare this state with the final state. If they are identical, stop with success.
3.	Apply to n all available search operators, thus obtaining a set of new states. Discard those states that already exist in L_{seen}. As for the rest, sort them by the evaluation function and place them at the front of L.
4.	Transfer n from L into the list, L_{seen}, of the states that have been investigated.
5.	If $L = \emptyset$, stop and report failure. Otherwise, go to 2.

What Have You Learned?

To make sure you understand the topic, try to answer the following questions. If needed, return to the appropriate place in the text.

- How does Artificial Intelligence define the search problem? What do we understand under the terms, "search space" and "search operators"?
- What is the role of the evaluation function? How does it affect the hill-climbing behavior?

1.3   Hill Climbing in Machine Learning

We are ready to explore the concrete ways of applying hill climbing to the needs of machine learning.

Hill Climbing and Johnny's Pies   Let us begin with the problem of how to decide which pies Johnny likes. The input consists of a set of training examples, each described by the available attributes. The output—the *final state*—is a boolean expression that is *true* for each positive example in the training set, and *false* for each negative example. The expression involves attribute-value pairs, logical operators (conjunction, disjunction, and negation), and such combination of parentheses as may be needed. The evaluation function measures the given expression's error rate on the training set. For the *initial state*, any randomly generated expression can be used. In Fig. 1.4, we chose (shape=circle), on the grounds that more than a half of the training examples are circular.

As for the *search operator*, one possibility is to add a conjunction as illustrated in the upper part of Fig. 1.4: for instance, the root's leftmost child is obtained by replacing (shape=circle) with [(shape=circle) AND (filling-shade=dark)] (in the picture, logical AND is represented by the symbol "∧."). Note how many different expressions this operator generates even in our toy domain. To shape=circle, any other attribute-value pair can be

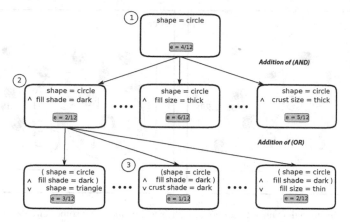

Fig. 1.4 Hill-climbing search in the "pies" domain

"ANDed." Since the remaining four attributes (apart from shape) acquire 2, 3, 2, and 3 different values, respectively, the total number of terms that can be added to (shape=circle) is $2 \times 2 \times 3 = 36$.[2]

Alternatively, we may choose to add a disjunction, as illustrated (in the picture) by the three expansions of the leftmost child. Other operators may "remove a conjunct," "remove a disjunct," "add a negation," "negate a term," various ways of manipulating parentheses, and so on. All in all, hundreds of search operators can be applied to each state, and then again to the resulting states. This can be hard to manage even in this very simple domain.

Numeric Attributes In the "pies" domain, each attribute acquires one out of a few discrete values, but in realistic applications, some attributes will probably be numeric. For instance, each pie has a price, an attribute whose values come from a continuous domain. What will the search look like then?

To keep things simple, suppose there are only two attributes: weight and price. This limitation makes it possible, in Fig. 1.5, to represent each training example by a point in a plane. The reader can see that examples belonging to the same class tend to occupy a specific region, and curves separating individual regions can be defined—expressed mathematically as lines, circles, polynomials. For instance, the right part of Fig. 1.5 shows three different circles, each of which can act as a classifier: examples inside the circle are deemed positive; those outside, negative. Again, some of these classifiers are better than others. How will hill climbing go about finding the best ones? Here is one possibility.

---

[2]Of the 36 new states thus created, Fig. 1.4 shows only three.

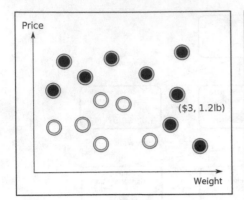 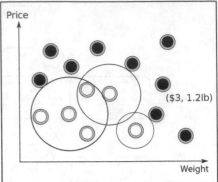

Fig. 1.5 *On the left*: a domain with continuous attributes; *on the right*: some "circular" classifiers

Hill Climbing in a Domain with Numeric Attributes

Initial State A circle is defined by its center and radius. We can identify the initial center with a randomly selected positive example, making the initial radius so small that the circle contains only this single example.

Search Operators Two search operators can be used: one increases the circle's radius, and the other shifts the center from one training example to another. In the former, we also have to determine *how much* the radius should change. One idea is to increase it only so much as to make the circle encompass one additional training example. At the beginning, only one training example is inside. After the first step, there will be two, then three, four, and so on.

Final State The circle may not be an ideal figure to represent the positive region. In this event, a 100% accuracy may not be achievable, and we may prefer to define the final state as, say, a "classifier that correctly classifies 95% of the training examples."

Evaluation Function As before, we choose to minimize the error rate.

What Have You Learned?

To make sure you understand the topic, try to answer the following questions. If needed, return to the appropriate place in the text.

- What aspects of search must be specified before we can employ hill climbing in machine learning?
- What search operators can be used in the "pies" domain and what in the "circles" domain? How can we define the evaluation function, the initial state, and the final state?

1.4   The Induced Classifier's Performance

So far, we have measured the error rate by comparing the training examples' known classes with those recommended by the classifier. Practically speaking, though, our goal is *not* to re-classify objects whose classes we already know; what we really want is to label *future examples*, those of whose classes we are as yet ignorant. The classifier's anticipated performance on these is estimated experimentally. It is important to know how.

Independent Testing Examples   The simplest scenario will divide the available pre-classified examples into two parts: the training set, from which the classifier is induced, and the *testing set*, on which it is evaluated (Fig. 1.6). Thus in the "pies" domain, with its 12 pre-classified examples, the induction may be carried out on randomly selected eight, and the testing on the remaining four. If the classifier then "guesses" correctly the class of three testing examples (while going wrong on one), its performance is estimated as 75%.

Reasonable though this approach may appear, it suffers from a major drawback: a random choice of eight training examples may not be sufficiently representative of the underlying concept—and the same applies to the (even smaller) testing set. If we induce the meaning of a mammal from a training set consisting of a whale, a dolphin, and a platypus, the learner may be led to believe that mammals live in the sea (whale, dolphin), and sometimes lay eggs (platypus), hardly an opinion a biologist will embrace. And yet, another choice of trainingexamples may result in a

Fig. 1.6 Pre-classified examples are divided into the training and testing sets

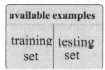

classifier satisfying the highest standards. The point is, a different training/testing set division gives rise to a different classifier—and also to a different estimate of future performance. This is particularly serious if the number of pre-classified examples is small.

Suppose we want to compare two machine learning algorithms in terms of the quality of the products they induce. The problem of non-representative training sets can be mitigated by so-called *random subsampling*.[3] The idea is to repeat the random division into the training and testing sets several times, always inducing a classifier from the i-th training set, and then measuring the error rate, E_i, on the i-th testing set. The algorithm that delivers classifiers with the lower average value of E_i's is deemed better—as far as classification performance is concerned.

---

[3]Later, we will describe some other methodologies.

The Need for Explanations In some applications, establishing the class of each example is not enough. Just as desirable is to know the reasons behind the classification. Thus a patient is unlikely to give consent to amputation if the only argument in support of surgery is, "this is what our computer says." But how to find a better explanation?

In the "pies" domain, a lot can be gleaned from the boolean expression itself. For instance, we may notice that a pie was labeled as negative whenever its shape was square, and its filling white. Combining this observation with alternative sources of knowledge may offer useful insights: the dark shade of the filling may indicate poppy, an ingredient Johnny is known to love; or the crust of circular pies turns out to be more crispy than that of square ones; and so on. The knowledge obtained in this manner can be more desirable than the classification itself.

By contrast, the classifier in the "circles" domain is a mathematical expression that acts as a "black box" which accepts an example's description and returns the class label without telling us anything else. This is not necessarily a shortcoming. In some applications, an explanation is nothing more than a welcome bonus; in others, it is superfluous. Consider a classifier that accepts a digital image of a hand-written character and returns the letter it represents. The user who expects several pages of text to be converted into a Word document will hardly insist on a detailed explanation for each single character.

Existence of Alternative Solutions By the way, we should notice that *many* apparently perfect classifiers can be induced from the given data. In the "pies" domain, the training set contained 12 examples, and the classes of the remaining 96 examples were unknown. Using some simple combinatorics, we realize that there are 2^{96} classifiers that label correctly all training examples but differ in the way they label the unknown 96. One induced classifier may label correctly every single future example—and another will misclassify them all.

What Have You Learned?

To make sure you understand the topic, try to answer the following questions. If needed, return to the appropriate place in the text.

- How can we estimate the error rate on examples that have not been seen during learning?
- Why is error rate usually higher on the testing set than on the training set?
- Give an example of a domain where the classifier also has to explain its action, and an example of a domain where this is unnecessary.
- What do we mean by saying that, "there is a combinatorial number of classifiers that correctly classify all training examples"?

1.5   Some Difficulties with Available Data

In some applications, the training set is created manually: an expert prepares the examples, tags them with class labels, chooses the attributes, and specifies the value of each attribute in each example. In other domains, the process is computerized. For instance, a company may want to be able to anticipate an employee's intention to leave. Their database contains, for each person, the address, gender, marital status, function, salary raises, promotions—as well as the information about whether the person is still with the company or, if not, the day they left. From this, a program can obtain the attribute vectors, labeled as positive if the given person left within a year since the last update of the database record.

Sometimes, the attribute vectors are automatically extracted from a database, and labeled by an expert. Alternatively, some examples can be obtained from a database, and others added manually. Often, two or more databases are combined. The number of such variations is virtually unlimited.

But whatever the source of the examples, they are likely to suffer from imperfections whose essence and consequences the engineer has to understand.

Irrelevant Attributes   To begin with, some attributes are important, while others are not. While Johnny may be truly fond of poppy filling, his preference for a pie will hardly be driven by the cook's shoe size. This is something to be concerned about: *irrelevant* attributes add to computational costs; they can even mislead the learner. Can they be avoided?

Usually not. True, in manually created domains, the expert is supposed to know which attributes really matter, but even here, things are not so simple. Thus the author of the "pies" domain might have done her best to choose those attributes she believed to matter. But unsure about the real reasons behind Johnny's tastes, she may have included attributes whose necessity she suspected—but could not guarantee. Even more often the problems with relevance occur when the examples are extracted from a database. Databases are developed primarily with the intention to provide access to lots of information—of which usually only a tiny part pertains to the learning task. As to which part this is, we usually have no idea.

Missing Attributes   Conversely, some critical attributes can be missing. Mindful of his parents' finances, Johnny may be prejudiced against expensive pies. The absence of attribute `price` will then make it impossible to induce a good classifier: two examples, identical in terms of the available attributes, can differ in the values of the vital "missing" attribute. No wonder that, though identically described, one example is positive, and the other is negative. When this happens, we say that the training set is *inconsistent*. The situation is sometimes difficult to avoid: not only may the expert be ignorant of the relevance of attribute `price`; it may be impossible to provide this attribute's values, and the attribute thus cannot be used anyway.

Redundant Attributes   Somewhat less damaging are attributes that are *redundant* in the sense that their values can be obtained from other attributes. If the database contains a patient's `date-of-birth` as well as `age`, the latter is unnecessary because it can be calculated by subtracting `date-of-birth` from today's date. Fortunately, redundant attributes are less dangerous than irrelevant or missing ones.

Missing Attribute Values In some applications, the user has no problems identifying the right choice of attributes. The problem is, however, that the value of some attributes are not known. For instance, the company analyzing the database of its employees may not know, for each person, the number of children.

Attribute: Value Noise Attribute values and class labels often cannot be trusted on account of unreliable sources of information, poor measurement devices, typos, the user's confusion, and many other reasons. We say that the data suffer from various kinds of *noise*.

Stochastic noise is random. For instance, since our body-weight varies during the day, the reading we get in the morning is different from the one in the evening. A human error can also play a part: lacking the time to take a patient's blood pressure, a negligent nurse simply scribbles down a modification of the previous reading. By contrast, *systematic noise* drags all values in the same direction. For instance, a poorly calibrated thermometer always gives a lower reading than it should. And something different occurs in the case of *arbitrary artifacts*; here, the given value bears no relation to reality such as when an EEG electrode gets loose and, from that moment on, all subsequent readings will be zero.

Class-Label Noise Class labels suffer from similar problems as attributes. The labels recommended by an expert may not have been properly recorded; alternatively, some examples find themselves in a "gray area" between two classes, in which event the correct labels are not certain. Both cases represent stochastic noise, of which the latter may affect negatively only examples from the borderline region between the two classes. However, class-label noise can also be systematic: a physician may be reluctant to diagnose a rare disease unless the evidence is overwhelming—his class labels are then more likely to be negative than positive. Finally, arbitrary artifacts in class labels are encountered in domains where the classes are supplied by an automated process that has gone wrong.

Class-label noise can be more dangerous than attribute-value noise. Thus in the "circles" domain, an example located deep inside the positive region will stay there even if an attribute's value is slightly modified; only the borderline example will suffer from being "sent across the border." By contrast, class-label noise will invalidate *any* example.

What Have You Learned?

To make sure you understand the topic, try to answer the following questions. If needed, return to the appropriate place in the text.

- Explain the following types of attributes: irrelevant, redundant, and missing. Illustrate each of them using the "pies" domain.

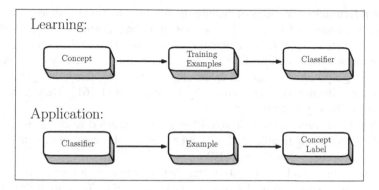

Fig. 1.7 The training examples are used to induce a classifier. The classifier is then employed to classify future examples

- What is meant by "inconsistent training set"? What can be the cause? How can it affect the learning process?
- What kinds of noise do we know? What are their possible sources?

1.6 Summary and Historical Remarks

- Induction from a training set of pre-classified examples is the most deeply studied machine-learning task.
- Historically, the task is cast as search. One can propose a mechanism that exploits the well-established search technique of hill climbing defined by an initial state, final state, interim states, search operators, and evaluation functions.
- Mechanical use of search is not the ultimate solution, though. The rest of the book will explore more useful techniques.
- Classifier performance is estimated with the help of pre-classified testing data. The simplest performance criterion is error rate, the percentage of examples misclassified by the classifier. The baseline scenario is shown in Fig. 1.7.
- Two classifiers that both correctly classify all training examples may differ significantly in their handling of the testing set.
- Apart from low error rate, some applications require that the classifier provides the reasons behind the classification.
- The quality of the induced classifier depends on training examples. The quality of the training examples depends not only on their choice, but also on the attributes used to describe them. Some attributes are relevant, others irrelevant or redundant. Quite often, critical attributes are missing.
- The attribute values and class labels may suffer from stochastic noise, systematic noise, and random artefacts. The value of an attribute in a concrete example may not be known.

Historical Remarks The idea of casting the machine-learning task as search was popular in the 1980s and 1990s. While several "founding fathers" came to see things this way independently of each other, Mitchell [67] is often credited with being the first to promote the search-based approach; just as influential, however, was the family of AQ-algorithms proposed by Michalski [59]. The discipline got a major boost by the collection of papers edited by Michalski et al. [61]. They framed the mindset of a whole generation.

There is much more to search algorithms. The interested reader is referred to textbooks of Artificial Intelligence, of which perhaps the most comprehensive is Russell and Norvig [84] or Coppin [17].

The reader may find it interesting that the question of proper representation of concepts or classes intrigued philosophers for centuries. Thus John Stuart Mill [65] explored concepts that are related towhat the next chapter calls *probabilistic*

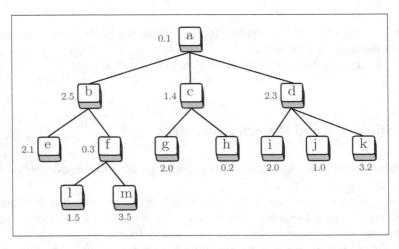

Fig. 1.8 Determine the order in which these search states are visited by heuristic search algorithms. The numbers next to the "boxes" give the values of the evaluation function for the individual search states

representation; and William Whewel [96] advocated *prototypical* representations that are close to the subject of our Chap. 3.

1.7 Solidify Your Knowledge

The exercises are to solidify the acquired knowledge. The suggested thought experiments will help the reader see this chapter's ideas in a different light and provoke independent thinking. Computer assignments will force the readers to pay attention to seemingly insignificant details they might otherwise overlook.

Exercises

1. In the sliding-tiles puzzle, suggest a better evaluation function than the one used in the text.
2. Figure 1.8 shows a search tree where each node represents one search state and is tagged with the value of the evaluation function. In what order will these states be visited by hill-climbing search?
3. Suppose the evaluation function in the "pies" domain calculates the percentage of correctly classified training examples. Let the initial state be the expression describing the second positive example in Table 1.1. Hand-simulate the hill-climbing search that uses generalization and specialization operators.
4. What is the size of the instance space in a domain where examples are described by ten boolean attributes? How large is then the space of classifiers?

Give It Some Thought

1. In the "pies" domain, the size of the space of all classifiers is 2^{108}, provided that each subset of the instance space can be represented by a distinct classifier. How much will the search space shrink if we permit only classifiers in the form of conjunctions of attribute-value pairs?
2. What kind of noise can you think of in the "pies" domain? What can be the source of this noise? What other issues may render training sets of this kind less than perfect?
3. Some classifiers behave as black boxes that do not offer much in the way of explanations. This, for instance, was the case of the "circles" domain. Suggest examples of domains where black-box classifiers are impractical, and suggest domains where this limitation does not matter.
4. Consider the data-related difficulties summarized in Sect. 1.5. Which of them are really serious, and which can perhaps be tolerated?
5. What is the difference between redundant attributes and irrelevant attributes?
6. Take a class that you think is difficult to describe—for instance, the recognition of a complex biological object (oak tree, ostrich, etc.) or the recognition of a music genre (rock, folk, jazz, etc.). Suggest the list of attributes to describe the training examples. Are the values of these attributes easy to obtain? Which of the problems discussed in this chapter do you expect will complicate the learning process?

Computer Assignments

1. Write a program implementing hill climbing and apply it to the sliding-tiles puzzle. Choose appropriate representation for the search states, write a module that decides whether a state is a final state, and implement the search operators. Define two or three alternative evaluation functions and observe how each of them leads to a different sequence of search steps.
2. Write a program that will implement the "growing circles" algorithm from Sect. 1.3. Create a training set of two-dimensional examples such as those in Fig. 1.5. The learning program will use the hill-climbing search. The evaluation function will calculate the percentage of training examples correctly classified by the classifier. Consider the following search operators: (1) increase/decrease the radius of the circle, (2) use a different training example as the circle's center.
3. Write a program that will implement the search for the description of the "pies that Johnny likes." Define your own generalization and specialization operators. The evaluation function will rely on the error rate observed on the training examples.

Chapter 2
Probabilities: Bayesian Classifiers

The earliest attempts to predict an example's class based on the known attribute values go back to well before World War II—prehistory, by the standards of computer science. Of course, nobody used the term "machine learning," in those days, but the goal was essentially the same as the one addressed in this book.

Here is the essence of the solution strategy they used: using the Bayesian probabilistic theory, calculate for each class the probability of the given object belonging to it, and then choose the class with the highest value.

2.1 The Single-Attribute Case

Let us start with something so simple as to be unrealistic: a domain where each example is described with a single attribute. Once we have developed the basic principles, we will generalize them so that they can be used in more practical domains.

Probabilities The basics are easily explained using the toy domain from the previous chapter. The training set consists of twelve pies ($N_{all} = 12$), of which six are positive examples of the given concept ($N_{pos} = 6$) and six are negative ($N_{neg} = 6$). Assuming that the examples represent faithfully the real situation, the probability of Johnny liking a randomly picked pie is therefore 50%:

$$P(\text{pos}) = \frac{N_{pos}}{N_{all}} = \frac{6}{12} = 0.5 \qquad (2.1)$$

Let us now take into consideration one of the attributes, say, filling-size. The training set contains eight examples with thick filling ($N_{thick} = 8$). Out of these, three are labeled as positive ($N_{pos|thick} = 3$). This means that the "*conditional probability*

© Springer International Publishing AG 2017
M. Kubat, *An Introduction to Machine Learning*,
DOI 10.1007/978-3-319-63913-0_2

Fig. 2.1 The prior probabilities, $P(\text{pos}) = \frac{6}{12}$ and $P(\text{thick}) = \frac{8}{12}$; the conditional probabilities, $P(\text{pos}|\text{thick}) = \frac{3}{8}$ and $P(\text{thick}|\text{pos}) = \frac{3}{6}$; and the joint probability, $P(\text{likes},\text{thick}) = \frac{3}{12}$

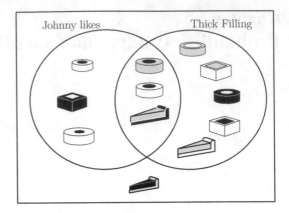

of an example being positive given that `filling-size=thick`" is 37.5%—this is what the relative frequency of positive examples among those with thick filling implies:

$$P(\text{pos}|\text{thick}) = \frac{N_{pos|thick}}{N_{thick}} = \frac{3}{8} = 0.375 \qquad (2.2)$$

Applying Conditional Probability to Classification Importantly, the relative frequency is calculated only for pies with the given attribute value. Among these same eight pies, five represented the negative class, which means that $P(\text{neg}|\text{thick}) = 5/8 = 0.625$. Observing that $P(\text{neg}|\text{thick}) > P(\text{pos}|\text{thick})$, we conclude that the probability of Johnny disliking a pie with thick filling is greater than the probability of the opposite case. It thus makes sense for the classifier to label all examples with `filling-size=thick` as negative instances of the "pie that Johnny likes."

Note that conditional probability, $P(\text{pos}|\text{thick})$, is more trustworthy than the prior probability, $P(\text{pos})$, because of the additional information that goes into its calculation. This is only natural. In a DayCare center where the number of boys is about the same as that of girls, we expect a randomly selected child to be a boy with $P(\text{boy}) = 0.5$. But the moment we hear someone call the child Johnny, we increase this expectation, knowing that it is rare for a girl to have this name. This is why $P(\text{boy}|\text{Johnny}) > P(\text{boy})$.

Joint Probability Conditional probability should not be confused with *joint probability* of two events occurring simultaneously. Be sure to use the right notation: in joint probability, the terms are separated by commas, $P(\text{pos},\text{thick})$; in conditional probability, by a vertical bar, $P(\text{pos}|\text{thick})$. For a randomly picked pie, $P(\text{pos},\text{thick})$ denotes the probability that the example is positive *and* its filling is thick; whereas $P(\text{pos}|\text{thick})$ refers to the occurrence of a positive example among those that have `filling-size=thick`.

A Concrete Example Figure 2.1 illustrates the terms. The rectangle represents all pies. The positive examples are contained in one circle and those with filling-size=thick in the other; the intersection contains three instances that satisfy both conditions; one pie satisfies neither, and is therefore left outside both circles. The conditional probability, $P(\text{pos}|\text{thick}) = 3/8$, is obtained by dividing the size of the intersection (three) by the size of the circle thick (eight). The joint probability, $P(\text{pos}, \text{thick}) = 3/12$, is obtained by dividing the size of the intersection (three) by the size of the entire training set (twelve). The prior probability of $P(\text{pos}) = 6/12$ is obtained by dividing the size of the circle pos (six) with that of the entire training set (twelve).

Obtaining Conditional Probability from Joint Probability The picture convinces us that joint probability can be obtained from prior probability and conditional probability:

$$P(\text{pos}, \text{thick}) = P(\text{pos}|\text{thick}) \cdot P(\text{thick}) = \frac{3}{8} \cdot \frac{8}{12} = \frac{3}{12}$$

$$P(\text{thick}, \text{pos}) = P(\text{thick}|\text{pos}) \cdot P(\text{pos}) = \frac{3}{6} \cdot \frac{6}{12} = \frac{3}{12}$$

Note that joint probability can never exceed the value of the corresponding conditional probability: $P(\text{pos}, \text{thick}) \leq P(\text{pos}|\text{thick})$. This is because conditional probability is multiplied by prior probability, $P(\text{thick})$ or $P(\text{pos})$, which can never be greater than 1.

Another fact to notice is that $P(\text{thick}, \text{pos}) = P(\text{pos}, \text{thick})$ because both represent the same thing: the probability of thick and pos co-occurring. Consequently, the left-hand sides of the previous two formulas have to be equal, which implies the following:

$$P(\text{pos}|\text{thick}) \cdot P(\text{thick}) = P(\text{thick}|\text{pos}) \cdot P(\text{pos})$$

Dividing both sides of this last equation by $P(\text{thick})$, we obtain the famous Bayes formula, the foundation for the rest of this chapter:

$$P(\text{pos}|\text{thick}) = \frac{P(\text{thick}|\text{pos}) \cdot P(\text{pos})}{P(\text{thick})} \tag{2.3}$$

If we derive the analogous formula for the probability that pies with filling-size = thick will belong to the negative class, we obtain the following:

$$P(\text{neg}|\text{thick}) = \frac{P(\text{thick}|\text{neg}) \cdot P(\text{neg})}{P(\text{thick})} \tag{2.4}$$

Comparison of the values calculated by these two formulas will tell us which class, pos of neg, is more probable. Things are simpler than they look: since the denominator, $P(\text{thick})$, is the same for both classes, we can just as well ignore it and simply choose the class for which the numerator is higher.

A Trivial Numeric Example   That this formula leads to correct values is illustrated in Table 2.1 which, for the sake of simplicity, deals with the trivial case where the examples are described by a single boolean attribute. So simple is this single-attribute world, actually, that we might easily have obtained $P(\text{pos}|\text{thick})$ and $P(\text{neg}|\text{thick})$ directly from the training set, without having to resort to the mighty Bayes formula—this makes it easy to verify the correctness of the results.

When the examples are described by two or more attributes, the way of calculating the probabilities is essentially the same, but we need at least one more trick. This will be introduced in the next section.

What Have You Learned?

To make sure you understand the topic, try to answer the following questions. If needed, return to the appropriate place in the text.

- How is the Bayes formula derived from the relation between the conditional and joint probabilities?
- What makes the Bayes formula so useful? What does it enable us to calculate?
- Can the joint probability, $P(x, y)$, have a greater value than the conditional probability, $P(x|y)$? Under what circumstances is $P(x|y) = P(x, y)$?

2.2   Vectors of Discrete Attributes

Let us now proceed to the question how to apply the Bayes formula in domains where the examples are described by vectors of attributes such as $\mathbf{x} = (x_1, x_2, \ldots, x_n)$.

Multiple Classes   Many realistic applications have more than two classes, not just the pos and neg from the "pies" domain. If c_i is the label of the i-th class, and if $\mathbf{x}$ is the vector describing the object we want to classify, the Bayes formula acquires the following form:

$$P(c_i|\mathbf{x}) = \frac{P(\mathbf{x}|c_i)P(c_i)}{P(\mathbf{x})}$$

Table 2.1 Illustrating the principle of Bayesian decision making

Let the training examples be described by a single attribute, filling-size, whose value is either thick or thin. We want the machine to recognize the positive class (pos). Here are the eight available training examples:

	ex1	ex2	ex3	ex4	ex5	ex6	ex7	ex8
Size	thick	thick	thin	thin	thin	thick	thick	thick
Class	pos	pos	pos	pos	neg	neg	neg	neg

The probabilities of the individual attribute values and class labels are obtained by their relative frequencies. For instance, three out of the eight examples are characterized by filling-size=thin; therefore, $P(\text{thin}) = 3/8$.

$P(\text{thin}) = 3/8$
$P(\text{thick}) = 5/8$
$P(\text{pos}) = 4/8$
$P(\text{neg}) = 4/8$

The conditional probability of a concrete attribute value within a given class is, again, determined by relative frequency. Our training set yields the following values:

$P(\text{thin}|\text{pos}) = 2/4$
$P(\text{thick}|\text{pos}) = 2/4$
$P(\text{thin}|\text{neg}) = 1/4$
$P(\text{thick}|\text{neg}) = 3/4$

Using these values, the Bayes formula gives the following conditional probabilities:

$P(\text{pos}|\text{thin}) = 2/3$
$P(\text{pos}|\text{thick}) = 2/5$
$P(\text{neg}|\text{thin}) = 1/3$
$P(\text{neg}|\text{thick}) = 3/5$

(note that $P(\text{pos}|\text{thin}) + P(\text{neg}|\text{thin}) = P(\text{pos}|\text{thick}) + P(\text{neg}|\text{thick}) = 1$)

Based on these results, we conclude that an example with filling-size=thin should be classified as positive because $P(\text{pos}|\text{thin}) > P(\text{neg}|\text{thin})$. Conversely, an example with filling-size = thick should be classified as negative because $P(\text{neg}|\text{thick}) > P(\text{pos}|\text{thick})$.

The denominator being the same for each class, we choose the class that maximizes the numerator, $P(\mathbf{x}|c_i)P(c_i)$. Here, $P(c_i)$ is easy to estimate by the relative frequency of c_i in the training set. As for $P(\mathbf{x}|c_i)$, however, things are not so simple.

A Vector's Probability $P(\mathbf{x}|c_i)$ is the probability that a randomly selected representative of class c_i is described by vector $\mathbf{x}$. Can its value be estimated by relative frequency? Not really. In the "pies" domain, the size of the instance space was 108 different examples, of which the training set contained twelve. These twelve vectors were each represented by one training example, while none of the other vectors (the vast majority!) was represented at all. The relative frequency of $\mathbf{x}$ among the six positive examples was thus either $P(\mathbf{x}|\text{pos}) = 1/6$, when $\mathbf{x}$ was among them, or $P(\mathbf{x}|\text{pos}) = 0$, when it was not. Any $\mathbf{x}$ identical to a training example "inherits" this example's class label; if the vector is *not* found in the training set, we have $P(\mathbf{x}|c_i) = 0$ for any c_i. The numerator in the Bayes formula thus being always $P(\mathbf{x}|c_i)P(c_i) = 0$, we are unable to choose the most probable class. Evidently, we will not get very far calculating the probability of an event that occurs only once or not at all.

This, fortunately, is not the case with the individual attributes. For instance, shape=circle occurs four times among the positive examples and twice among the negative, the corresponding probabilities thus being $P(\text{shape} = \text{circle}|\text{pos}) = 4/6$ and $P(\text{shape} = \text{circle}|\text{neg}) = 2/6$. If an attribute can acquire only two or three values, chances are high that each of these values is represented in the training set more than once, thus offering better grounds for probability estimates.

Mutually Independent Attributes What is needed is a formula that combines probabilities of individual attribute values into the probability of the given attribute vector in the given class: $P(\mathbf{x}|c_i)$. As long as the attributes are independent of each other, this is simple. If $P(x_i|c_j)$ is the probability that the value of the i-th attribute of an example from class c_j is x_i, then the probability, $P(\mathbf{x}|c_j)$, that a random representative of c_j is described by $\mathbf{x} = (x_1, x_2, \ldots, x_n)$, is calculated as follows:

$$P(\mathbf{x}|c_j) = \prod_{i=1}^{n} P(x_i|c_j) \tag{2.5}$$

An object will be labeled with c_j if this class maximizes the following version of the Bayes formula's numerator:

$$P(c_j) \cdot \prod_{i=1}^{n} P(x_i|c_j) \tag{2.6}$$

The Naive Bayes Assumption The reader may complain that the assumption of mutually independent attributes is rarely justified. Indeed, can the interrelation of diverse variables ever be avoided? An object's weight grows with its size, the quality of health care may be traced to an individual's income, an object's color can be derived from its physical properties. In short, domains where no two attributes are in any way related to each other are rare. No wonder that the above-described approach is known under the unflattering name, *Naive Bayes*.

Yet practical experience is not bad at all. True, the violation of the "independence requirement" renders the probability estimates inaccurate. However, this does not necessarily make them point to the wrong classes. Remember? $\mathbf{x}$ is labeled with the class that maximizes $P(\mathbf{x}|c_i) \cdot P(c_i)$. If the product's value is 0.8 for one class and 0.2 for the other, then the classifier's behavior will not change even if the probability estimates miss the accuracy mark by ten or 20%. And so, while requesting that the attributes in principle be independent, we will do reasonably well even if they are not.

When Mutual Independence Cannot Be Assumed   This said, we have to ask how to handle the case where attribute interdependence *cannot* be ignored. A scientist's first instinct may be to suggest more sophisticated ways of estimating $P(\mathbf{x}|c_i)$. These do indeed exist, but their complexity grows with the number of attributes, and they contain terms whose values are hard to determine. The practically minded engineer doubts that the trouble is justified by the benefits it brings.

A more pragmatic approach will therefore seek to reduce the attribute dependence by appropriate data pre-processing. A good way to start is to get rid of redundant attributes, those whose values are known to depend on others. For instance, if the set of attributes contains age, date-of-birth, and current-date, chances are that Naive Bayes will do better if we use only age.

We can also try to replace two or three attributes by an artificially created one that combines them. Thus in the "pies" domain, a baker might have told us that filling-size is not quite independent of crust-size: if one is thick, the other is thin and vice versa. In this event, we may benefit from replacing the two attributes with a new one, say, CF-size, that acquires only two values: thick-crust-and-thin-filling or thin-crust-and-thick-filling.

In the last resort, if we are prejudiced against advanced methods of multivariate probability estimates, and if we want to avoid data pre-processing, there is always the possibility of giving up on Bayesian classifiers altogether, preferring some of the machine-learning paradigms from the later chapters of this book.

A Numeric Example   To get used to the mechanism in which Naive Bayes is used for classification purposes, the reader may want to go through the example in Table 2.2. Here the class of a previously unseen pie is established based on the training set from Table 1.1. The Bayes formula is used, and the attributes are assumed to be mutually independent.

The procedure is summarized by the pseudocode in Table 2.3.

What Have You Learned?

To make sure you understand the topic, try to answer the following questions. If needed, return to the appropriate place in the text.

- Under what circumstances shall we assume that the individual attributes are mutually independent? What benefit does this assumption bring for the estimates of $P(\mathbf{x}|c_i)$?
- Discuss the conflicting aspects of this assumption.

2.3 Probabilities of Rare Events: Exploiting the Expert's Intuition

In the first approximation, probability is almost identified with relative frequency: having observed x thirty times in one hundred trials, we assume that $P(x) = 0.3$. This is how we did it in the previous sections.

Table 2.2 Bayesian classification: examples described by vectors of independent attributes

Suppose we want to apply the Bayesian formula to the training set from Table 1.1 in order to determine the class of the following object:

```
x = [shape=square, crust-size=thick, crust-shade=gray
      filling-size=thin, filling-shade=white]
```

There are two classes, pos and neg. The procedure is to calculate the numerator of the Bayes formula separately for each of them, and then choose the class with the higher value. In the training set, each class has the same number of representatives: $P(\text{pos}) = P(\text{neg}) = 0.5$. The remaining terms, $\prod_{i=1}^{n} P(x_i|\text{pos})$ and $\prod_{i=1}^{n} P(x_i|\text{neg})$, are calculated from the following conditional probabilities:

P(shape=square|pos)	= 1/6	P(shape=square|neg)	= 2/6
P(crust-size=thick|pos)	= 5/6	P(crust-size=thick|neg)	= 5/6
P(crust-shade=gray|pos)	= 1/6	P(crust-shade=gray|neg)	= 2/6
P(filling-size=thin|pos)	= 3/6	P(filling-size=thin|neg)	= 1/6
P(filling-shade=white|pos)	= 1/6	P(filling-shade=white|neg)	= 2/6

Based on these values, we obtain the following probabilities:

$$P(\mathbf{x}|\text{pos}) = \prod_{i=1}^{n} P(x_i|\text{pos}) = \frac{1}{6} \cdot \frac{5}{6} \cdot \frac{1}{6} \cdot \frac{3}{6} \cdot \frac{1}{6} = \frac{15}{6^5}$$

$$P(\mathbf{x}|\text{neg}) = \prod_{i=1}^{n} P(x_i|\text{neg}) = \frac{2}{6} \cdot \frac{5}{6} \cdot \frac{2}{6} \cdot \frac{1}{6} \cdot \frac{2}{6} = \frac{40}{6^5}$$

Since $P(\mathbf{x}|\text{pos}) < P(\mathbf{x}|\text{neg})$, we label $\mathbf{x}$ with the negative class.

Table 2.3 Classification with the Naive-Bayes principle

The example to be classified is described by $\mathbf{x} = (x_1, \ldots, x_n)$.

1. For each x_i, and for each class c_j, calculate the conditional probability, $P(x_i|c_j)$, as the relative frequency of x_i among those training examples that belong to c_j.
2. For each class, c_j, carry out the following two steps:

   i) estimate $P(c_j)$ as the relative frequency of this class in the training set;
   ii) calculate the conditional probability, $P(\mathbf{x}|c_j)$, using the "naive" assumption of mutually independent attributes:

$$P(\mathbf{x}|c_j) = \prod_{i=1}^{n} P(x_i|c_j)$$

3. Choose the class with the highest value of $P(c_j) \cdot \prod_{i=1}^{n} P(x_i|c_j)$.

To be fair, though, such estimates can be trusted only when supported by a great many observations. It is conceivable that a coin flipped four times comes up heads three times, and yet it will be overhasty to interpret this observation as meaning that $P(\text{heads}) = 0.75$; the physics of the experiment suggests that a fair coin should come up heads 50% of the time. Can this *prior expectation* help us improve probability estimates in domains with insufficient numbers of observations?

The answer is, "Yes, we can use the m-estimate."

The Essence of an m-Estimate Let us illustrate the principle using the case of an unfair coin where one side comes up somewhat more frequently than the other. In the absence of any better guidance, the prior expectation of heads is $\pi_{head} = 0.5$. An auxiliary parameter, m, helps the engineer tell the class-predicting program *how confident* he is in this value, how much the prior expectation can be trusted (higher m indicating higher confidence).

Let us denote by N_{all} the number of times the coin was flipped, and by N_{heads} the number of times the coin came up heads. The way to combine these values with the prior expectation and confidence is summarized by the following formula:

$$P_{heads} = \frac{N_{heads} + m\pi_{heads}}{N_{all} + m} \tag{2.7}$$

Note that the formula degenerates to the prior expectation, π_{heads}, if $N_{all} = N_{heads} = 0$. Conversely, it converges to that of relative frequency if N_{all} and N_{heads} are so large as to render the terms $m\pi_{heads}$ and m negligible. Using the values $\pi_{heads} = 0.5$ and $m = 2$, we obtain the following:

$$P_{heads} = \frac{N_{heads} + 2 \times 0.5}{N_{all} + 2} = \frac{N_{heads} + 1}{N_{all} + 2}$$

Illustrating Probability Estimates Table 2.4 shows how the values thus calculated gradually evolve in the course of five trials. The reader can see that the m-estimate is for small numbers of experiments more in line with common sense than relative

Table 2.4 For each successive trial, the second row gives the observed outcome; the third, the relative frequency of *heads*; the last, the m-estimate of the probability, assuming $\pi_{heads} = 0.5$ and $m = 2$

Toss number	1	2	3	4	5
Outcome	Heads	Heads	Tails	Heads	Tails
Relative frequency	1.00	1.00	0.67	0.75	0.60
m-estimate	0.67	0.75	0.60	0.67	0.57

frequency. Thus after two trials, m-estimate suggests a 0.75 chance of heads, whereas anybody espousing relative frequency will have to concede that, based on the two experiments, there is a zero chance that the coin will come up tails. As the number of trials increases, though, the values returned by m-estimate and relative frequency tend to converge.

The Impact of the User's Confidence Let us take a closer look at the effect of m, the user's confidence. A lot is revealed if we compare the two different settings below: $m = 100$ on the left and $m = 1$ on the right (in both cases, $\pi_{heads} = 0.5$).

$$\frac{N_{heads} + 50}{N_{all} + 100} \qquad\qquad \frac{N_{heads} + 0.5}{N_{all} + 1}$$

The version with $m = 100$ allows the prior estimate to be modified only if really substantial evidence is available ($N_{heads} \gg 50, N_{all} \gg 100$). By contrast, the version with $m = 1$ allows the user's opinion to be controverted with just a few experimental trials.

Domains with More Than Two Outcomes Although we have used a two-outcome domain, the formula is applicable also in multi-outcome domains. Rolling a fair die can result in six different outcomes, and we expect that the probability of seeing, say, three points is $\pi_{three} = 1/6$. Using $m = 6$, we obtain the following:

$$P_{three} = \frac{N_{three} + m\pi_{three}}{N_{all} + m} = \frac{N_{three} + 6 \cdot \frac{1}{6}}{N_{all} + 6} = \frac{N_{three} + 1}{N_{all} + 6}$$

Again, if N_{all} is so high that $m = 6$ and $m\pi_{three} = 1$ can be neglected, the formula converges to relative frequency: $P_{three} = \frac{N_{three}}{N_{all}}$. If we do not want this to happen prematurely (perhaps because we have high confidence in the prior estimate, π_{three}), we prevent it by choosing a higher m.

The Limits of m-Estimates We should not forget that the m-estimate is only as good as the parameters it relies on. If we start from an unrealistic prior estimate, the result can be disappointing. Suppose that $\pi_{heads} = 0.9$ and $m = 10$. Equation (2.7) then turns into the following:

$$P_{heads} = \frac{N_{heads} + 9}{N_{all} + 10}$$

When we use this formula to recalculate the values from Table 2.4, we will realize that, after five trials, the probability is estimated as $P_{heads} = \frac{3+9}{5+10} = \frac{12}{15} = 0.8$, surely a less plausible value than the one obtained in the case of $\pi_{heads} = 0.5$ where we got $P_{heads} = 0.57$. The reader is encouraged to verify that the situation will somewhat improve if we reduce m.

Mathematical Soundness Let us make one last comment. A common understanding in mathematics is that the probabilities of all possible events should sum up to 1: if an experiment can have N different outcomes, and if P_i is the probability of the i-th outcome, then $\sum_{i=1}^{N} P_i = 1$. It is easy to verify that Eq. (2.7) satisfies this condition for any value of m. Suppose we are dealing with the coin-tossing domain where there are only two possible outcomes. If the prior estimates sum up to 1 ($\pi_{heads} + \pi_{tails} = 1$), then, given that $N_{heads} + N_{tails} = N_{all}$, we derive the following:

$$P_{heads} + P_{tails} = \frac{N_{heads} + m\pi_{heads}}{N_{all} + m} + \frac{N_{tails} + m\pi_{tails}}{N_{all} + m}$$

$$= \frac{N_{heads} + N_{tails} + m(\pi_{heads} + \pi_{tails})}{N_{all} + m} = 1$$

The interested reader will easily generalize this to any finite number of classes.

Why This May Be Useful In the problem presented in Table 2.5, we want to classify example x using the Bayesian classifier. To be able to do that, we first need to calculate the requisite conditional probabilities. Trying to do so for the positive class, however, we realize that, since the training set is so small, none of the training examples has crust-shade=gray, the value observed in x. If the probabilities are estimated by relative frequency, this concrete conditional probability would be 0. As a result, $P(x|pos) = 0$, regardless of all the other probabilities. This simply does not seem right.

The problem disappears if we use m-estimate instead of relative frequency because the m-estimate is non-zero even if the concrete value has never being observed in the training set.

What Have You Learned?

To make sure you understand the topic, try to answer the following questions. If needed, return to the appropriate place in the text.

- Under what circumstances is relative frequency ill-suited for estimates of discrete probabilities?
- What is the impact of parameter m in Eq. (2.7)? Under what circumstances will you prefer large m, and when will you rather go for small m?
- What is the impact of the prior estimate, π_{heads}, in Eq. (2.7)? How is the credibility of m-estimates affected by unrealistic values of π_{heads}?

Table 2.5 An example of one reason for using m-estimates in Bayesian classification

Let us return to the "pies" domain from Table 1.1. Remove from the table the first example, then use the rest for the calculation of the probabilities.

```
x = [shape=circle, crust-size=thick, crust-shade=gray
        filling-size=thick, filling-shade=dark]
```

Let us first calculate the probabilities of the individual attribute values:

```
P(shape=circle|pos)          = 3/5
P(crust-size=thick|pos)      = 4/5
P(crust-shade=gray|pos)      = 0/5
P(filling-size=thick|pos)    = 2/5
P(filling-shade=dark|pos)    = 3/5
```

Based on these values, we obtain the following probabilities:

$$P(x|pos) = \frac{3 \times 4 \times 0 \times 2 \times 3}{5^5} = 0.$$

We see that the circumstance that none of the five positive examples has crust-shade=gray causes the corresponding conditional probability to equal 0.

The problem is solved if we calculate the probabilities using the m-estimate. In this case, none of the conditional probabilities will be 0.

2.4   How to Handle Continuous Attributes

Up till now, we limited our considerations to attributes that assume *discrete* values, estimating their probabilities either by relative frequency or by the m-estimate. This, however, is not enough. In many applications, we encounter attributes (such as age, price or weight) that acquire values from continuous domains.

Relative frequency is then impractical. While it is easy to establish that the probability of an engineering student being male is $P_{male} = 0.7$, the probability that this student's body weight is 184.5 pounds cannot be specified so readily: the number of different weight values being infinite, the probability of any one of them is infinitesimally small. What to do in this case?

Discretizing Continuous Attributes   One possibility is to *discretize*. The simplest "trick" will split the attribute's original domain in two; for instance, by replacing age with the boolean attribute old that is *true* for age > 60 and *false* otherwise. However, at least part of the available information then gets lost: a person may be old, but we no longer know *how old*; nor do we know whether one old person is older than another old person.

The loss will be mitigated if we divide the original domain into not two, but several intervals, say, $(0, 10], \ldots (90, 100]$.[1] Suppose we get ourselves a separate bin for each of these, and place a little black ball into the i-th bin for each training example whose value of age falls into the i-th interval.

Having done so, we may reach the situation depicted in Fig. 2.2. The upper part shows the bins, side by side, and the bottom part shows a piecewise constant function

Fig. 2.2 A simple discretization method that represents each subinterval by a separate bin. The *bottom chart* plots the histogram over the individual subintervals

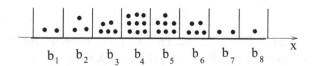

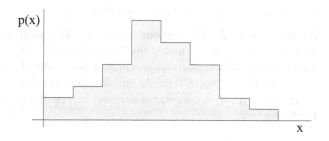

created in the following manner: if N is the size of the training set, and N_i is the number of balls in the i-th bin, then the function's value in the i-th interval is N_i/N—the relative frequency of the balls in the i-th bin. Since the area under the function is $\frac{\Sigma N_i}{N} = 1$, we have a mechanism to estimate the probability *not* of a concrete value of age, but rather of this value falling into the given interval.

Probability Density Function If the step-function thus constructed seems too crude, we may fine-tune it by dividing the original domain into shorter—and thus more numerous—intervals, provided that the number of balls in each bin is sufficient for reliable probability estimates. If the training set is infinitely large, we can, theoretically speaking, keep reducing the lengths of the intervals until they become infinitesimally small. The result of the bin-filling exercise will then no longer be a step-function, but rather a continuous function, $p(x)$, such as the one in Fig. 2.3. Its interpretation follows from the way it has been created: a high value of $p(x)$ indicates that there are many examples with age close to x; conversely, a low value of $p(x)$ tells us that age in the vicinity of x is rare. This is why we call $p(x)$ a *probability density function*, often avoiding this mouthful by preferring the acronym *pdf*.

---

[1] We assume here that 100 is the maximum value observed in the training set. Alternatively, our background knowledge may inform us that the given attribute's value cannot exceed 100.

Fig. 2.3 When using the *pdf*, we identify the probability of $x \in [a, b]$ with the relative size of the area below the corresponding section of the *pdf*

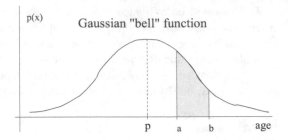

Let us be careful about notation. The discrete probability of x is indicated by an uppercase letter, $P(x)$. The *pdf* at x is denoted by a lowercase letter, $p(x)$. And if we want to point out that the *pdf* has been created exclusively from examples belonging to class c_i, we do so by using a subscript, $p_{c_i}(x)$.

Bayes Formula for Continuous Attributes The good thing about the *pdf* is that it makes it possible to employ the Bayes formula even in the case of continuous attributes. We only replace the conditional probability $P(x|c_i)$ with $p_{c_i}(x)$, and $P(x)$ with $p(x)$. Let us begin with the trivial case where the object to be classified is described by a single continuous attribute, x. The Bayes formula then assumes the following form:

$$P(c_i \mid x) = \frac{p_{c_i}(x) \cdot P(c_i)}{p(x)} \tag{2.8}$$

Here, $P(c_i)$ is estimated by the relative frequency of c_i in the training set, $p(x)$ is the *pdf* created from all training examples, and $p_{c_i}(x)$ is the *pdf* created from those training examples that belong to c_i.

Again, the denominator can be ignored because it has the same value for any class. The classifier simply calculates, separately for each class, the value of the numerator, $p_{c_i}(x) \cdot P(c_i)$, and then labels the object with the class for which the product is maximized.

Naive Bayes Revisited When facing the more realistic case where the examples are described by vectors of attributes, we will avail ourselves of the same "trick" as before: the assumption that all attributes are mutually independent. Suppose we encounter an example described by $\mathbf{x} = (x_1, \ldots, x_n)$. The *pdf* at $\mathbf{x}$ is approximated by the product along the individual attributes:

$$p_{c_j}(\mathbf{x}) = \Pi_{i=1}^{n} p_{c_j}(x_i) \tag{2.9}$$

A statistician will be able to suggest formulas that are theoretically sounder; however, higher sophistication often fails to give satisfaction. For one thing, we may commit the sin of accurate calculations with imprecise numbers. Besides, the more complicated the technique, the higher the danger it will be applied incorrectly.

What Have You Learned?

To make sure you understand the topic, try to answer the following questions. If needed, return to the appropriate place in the text.

- What is the probability density function, *pdf*, and how does it help us in the context of Bayesian classification?
- Explain the discretization mechanism that helped us arrive at an informal definition of a *pdf*.
- How does the Bayes formula change in domains with continuous attributes? How do we estimate the values of the individual terms?

2.5 Gaussian "Bell" Function: A Standard *pdf*

One way to approximate a *pdf* is to employ the discretization technique from the previous section. Alternatively, we can capitalize on standardized models known to be applicable to many realistic situations. Perhaps the most popular among them is the *gaussian function*, named after the great German mathematician.

The Shape and the Formula Describing It The curve in Fig. 2.3 is an example; its shape betrays why many people call it a "bell function." The maximum is reached at the mean, $x = \mu$, and the curve slopes down gracefully with the growing distance of x from μ. It is reasonable to expect that this is a good model of the *pdf* of such variables as the body temperature where the density peaks around $x = 99.7$ degrees Fahrenheit.

Expressed mathematically, the gaussian function is defined by the following formula where e is the base of the natural logarithm:

$$p(x) = k \cdot e^{-\frac{(x-\mu)^2}{2\sigma^2}} \tag{2.10}$$

Parameters Note that the greater the difference between x and μ, the greater the exponent's numerator, and thus the smaller the value of $p(x)$ because the exponent is negative. The reason the numerator is squared, $(x - \mu)^2$, is to make sure that the value slopes down with the same angle on both sides of the mean, μ; the curve is symmetric. How steep the slope is depends on σ^2, a parameter called *variance*. Greater variance means smaller sensitivity to the difference between x and μ, and thus a "flatter" bell curve; conversely, smaller variance defines a narrower bell curve.

The task for the coefficient k is to make the area under the bell function equal to 1 as required by the theory of probability. It would be relatively easy to prove that this happens when k is determined by the following formula:

$$k = \frac{1}{\sqrt{2\pi\sigma^2}} \tag{2.11}$$

Setting the Parameter Values To be able to use this model when approximating $p_{c_i}(x)$ in a concrete application, we only need to estimate the values of its parameters, μ and σ^2. This is easy. Suppose that class c_i has m representatives among the training examples. If x_i is the value of the given attribute in the i-th example, then the mean and variance, respectively, are calculated using the following formulas:

$$\mu = \frac{1}{m} \sum_{i=1}^{m} x_i \qquad (2.12)$$

$$\sigma^2 = \frac{1}{m-1} \sum_{i=1}^{m} (x_i - \mu)^2 \qquad (2.13)$$

In plain English, the gaussian center, μ, is obtained as the arithmetic average of the values observed in the training examples, and the variance is obtained as the average of the squared differences between x_i and μ. Note that, when calculating variance, we divide the sum by $m - 1$, and not by m, as we might expect. The intention is to compensate for the fact that μ itself is only an estimate. The variance should therefore be somewhat higher than what it would be if we divided by m. Of course, this matters only if the training set is small: for large m, the difference between m and $m - 1$ is negligible.

What Have You Learned?

To make sure you understand the topic, try to answer the following questions. If needed, return to the appropriate place in the text.

- Give an example of a continuous variable whose *pdf* can be expected to follow the gaussian distribution.
- What parameters define the bell function? How can we establish their values using the training set?
- How—and why—do we normalize the bell function?

2.6   Approximating PDFs with Sets of Gaussians

While the bell function represents a good mechanism to approximate the *pdf* in many realistic domains, it is not a panacea. Some variables simply do not behave that way. Just consider the distribution of body-weight in a group that mixes grade-school children with their parents. If we create the *pdf* using the discretization method, we will observe two peaks: one for the kids, and the other for the grown-ups. There may be three peaks if it turns out that body-weight of fathers is distributed around a higher mean than that of the mothers. And the number of peaks can be higher still if the families come from diverse ethnic groups.

Combining Gaussian Functions No doubt, a single bell function would misrepresent the situation. But what if we combine two or more of them? If we knew the diverse sources of the examples, we might create a separate gaussian for each source, and then superimpose the bell functions on each other. Would this solve our problem?

The honest answer is, "yes, in this ideal case." In reality, though, prior knowledge about diverse sources is rarely available. A better solution will divide the body-weight values into great many random groups. In the extreme, we may even go as far as to make each example a "group" of its own, and then identify a gaussian center with this example's body-weight, thus obtaining m bell functions (for m examples).

The Formula to Combine Them Suppose we want to approximate the *pdf* of a continuous attribute, x. If we denote by μ_i the value of x in the i-th example, then the *pdf* is approximated by the following sum of m functions:

$$p(x) = k \cdot \Sigma_{i=1}^{m} e^{-\frac{(x-\mu_i)^2}{2\sigma^2}} \qquad (2.14)$$

As before, the normalization constant, k, is here to make sure that the area under the curve is 1. This is achieved when k is calculated as follows:

$$k = \frac{1}{m\sigma\sqrt{2\pi}} \qquad (2.15)$$

From mathematics, we know that if m is sufficiently high, Eq. (2.14) approximates the *pdf* with almost arbitrary accuracy.

Illustrating the Point Figure 2.4 illustrates the approach using a training set consisting of $m = 3$ examples, the values of attribute x being $x_1 = 0.4, x_2 = 0.5$ and $x_3 = 0.7$. The upper three charts show three bell functions, each centered at one of these points, the variance always being $\sigma^2 = 1$. The bottom chart shows the composed *pdf* created by putting together Eqs. (2.14) and (2.15), using the means, $\mu_1 = 0.4, \mu_2 = 0.5$, and $\mu_3 = 0.7$, and $\sigma^2 = 1$:

$$p(x) = \frac{1}{3\sqrt{2\pi}} \cdot [e^{-\frac{(x-0.4)^2}{2}} + e^{-\frac{(x-0.5)^2}{2}} + e^{-\frac{(x-0.7)^2}{2}}]$$

The Impact of Concrete Parameter Values The practical utility of the *pdf* thus obtained (its success when used in the Bayes formula) depends on the choice of σ^2. In Fig. 2.4, we used $\sigma^2 = 1$, but there is no guarantee that this will work in any future application. To be able to adjust it properly, we need to understand how it affects the shape of the composite *pdf*.

Inspecting the gaussian formula, we realize that the choice of a very small value of σ^2 causes great sensitivity to the difference between x and μ_i; the individual bell functions will be "narrow," and the resulting *pdf* will be marked by steep peaks separated by extended "valleys." Conversely, the consequence of a high σ^2 will be an almost flat *pdf*. Seeking a compromise between the two extremes, we will do well if we make σ^2 dependent on the distances between examples.

The simplest solution will use $\sigma^2 = \mu_{max} - \mu_{min}$, where μ_{max} and μ_{min} are the maximum and minimum values of μ_i, respectively. If you think this too crude, you may consider normalizing the difference by the number of examples: $\sigma^2 = (\mu_{max} - \mu_{min})/m$. Large training sets (with high m) will then lead to smaller variations that will narrow the contributing gaussians. Finally, in some domains we might argue that each of the contributing bell functions should have a variance of its own, proportional to the distance from the center of the nearest other bell function. In this case, however, we are no longer allowed to set the value of k by Eq. (2.15).

A Numeric Example The example in Table 2.6 illustrates the whole procedure on a concrete case of a small training set and a vector to be classified. The reader is encouraged to go through all its details to get used to the way the formulas are put together. See also the illustration in Fig. 2.5.

When There Are Too Many Examples For a training set of realistic size, it is impractical to identify each training example with one gaussian centers; nor is it necessary. More often than not, the examples are grouped in *clusters* that can be detected by *cluster analysis* techniques—see Chap. 14. Once the clusters have been found, we identify the gaussian centers with the centroids of the clusters.

What Have You Learned?

To make sure you understand the topic, try to answer the following questions. If needed, return to the appropriate place in the text.

- Under what circumstances is the gaussian function a poor model of the *pdf*?
- Why does the composite *pdf* have to be normalized by k?
- How do we establish the centers and variances of the individual bell functions?

2.7 Summary and Historical Remarks

- Bayesian classifiers calculate the product $P(\mathbf{x}|c_i)P(c_i)$ separately for each class, c_i, and then label the example, $\mathbf{x}$, with the class where this product has the highest value.
- The main problem is how to calculate the probability, $P(\mathbf{x}|c_i)$. Most of the time, the job is simplified by making the assumption that the individual attributes are mutually independent, in which case $P(\mathbf{x}|c_i) = \prod_{j=1}^{n} P(x_j|c_i)$, where n is the number of attributes.

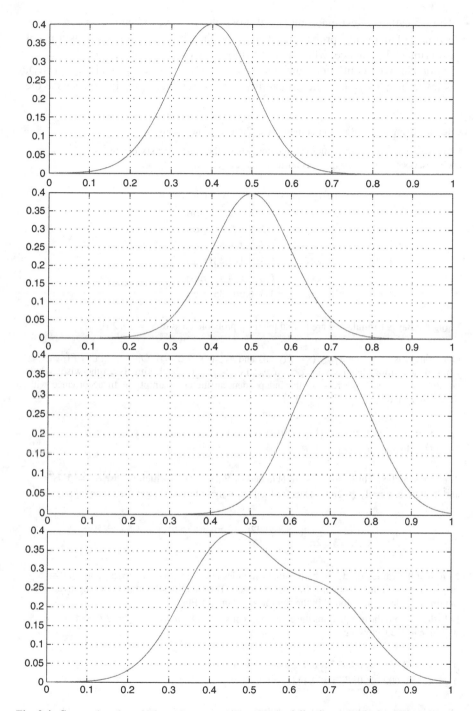

Fig. 2.4 Composing the *pdf* from three examples with the following values of attribute x: $\mu_1 = 0.4$, $\mu_2 = 0.5$, and $\mu_3 = 0.7$. The *upper three charts* show the contributing gaussians; the *bottom chart*, the composition. The variance is $\sigma^2 = 1$

- The so-called *m*-estimate makes it possible to take advantage of a user's estimate of an event's probability. This comes handy in domains with insufficient experimental evidence, where relative frequency cannot be relied on.
- In domains with continuous attributes, the role of the discrete probability, $P(\mathbf{x}|c_i)$, is taken over by $p_{c_i}(\mathbf{x})$, the probability density function, *pdf*, but otherwise the

Table 2.6 Using Naive Bayes in domains with three continuous attributes

Suppose we are given a training set that consists of the following six examples, $ex_1, \ldots, ex_6$, each described by three continuous attributes, $at_1, at_2, \ldots at_3$:

Example	at_1	at_2	at_3	Class
ex_1	3.2	2.1	2.1	pos
ex_2	5.2	6.1	7.5	pos
ex_3	8.5	1.3	0.5	pos
ex_4	2.3	5.4	2.45	neg
ex_5	6.2	3.1	4.4	neg
ex_6	1.3	6.0	3.35	neg

Using the Bayes formula, we are to find the most probable class of $\mathbf{x} = (9, 2.6, 3.3)$.

Our strategy is to evaluate $p_{pos}(\mathbf{x}) \cdot P(pos)$ and $p_{neg}(\mathbf{x}) \cdot P(neg)$. Observing that $P(pos) = P(neg)$, we simply label $\mathbf{x}$ with pos if $p_{pos}(\mathbf{x}) > p_{neg}(\mathbf{x})$ and with neg otherwise. When constructing the *pdf*, we rely on the independent-attributes assumption. In accordance with Sect. 2.2, we have:

$p_{pos}(\mathbf{x}) = p_{pos}(at_1) \cdot p_{pos}(at_2) \cdot p_{pos}(at_3)$ and
$p_{neg}(\mathbf{x}) = p_{neg}(at_1) \cdot p_{neg}(at_2) \cdot p_{neg}(at_3)$

The terms on the right-hand sides are obtained by Eq. (2.14), in which we use $\sigma^2 = 1, m = 3$, and, therefore, $k = 1/\sqrt{(2\pi)^3}$. Thus for the first of these terms, we get the following:

$$p_{pos}(at_1) = \frac{1}{3\sqrt{2\pi}}[e^{-0.5(at_1-3.2)^2} + e^{-0.5(at_1-5.2)^2} + e^{-0.5(at_1-8.5)^2}]$$

Note that the values of at_1 in the positive examples are $\mu_1 = 3.2, \mu_2 = 5.2$, and $\mu_3 = 8.5$, respectively—see the exponents in the expression. The functions for the remaining five terms, obtained similarly, are plotted in the rightmost column of Fig. 2.5.

Substituting into these equations the coordinates of $\mathbf{x}$, namely $at_1 = 9, at_2 = 3.6$, and $at_3 = 3.3$, will give us the following:

$p_{pos}(\mathbf{x}) = 0.0561 \times 0.0835 \times 0.0322 = 0.00015$
$p_{neg}(\mathbf{x}) = 0.0023 \times 0.0575 \times 0.1423 = 0.00001$

Observing that $p_{pos}(\mathbf{x}) > p_{neg}(\mathbf{x})$, we label $\mathbf{x}$ with the class pos.

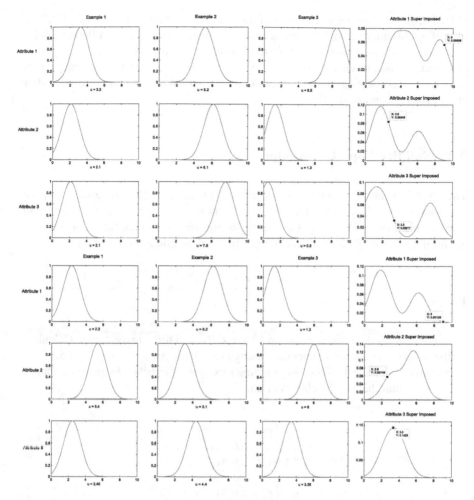

Fig. 2.5 Composing the *pdf*'s separately for the positive and negative class (with $\sigma^2 = 1$). Each row represents one attribute, and each of the *left three columns* represents one example. The *rightmost column* shows the composed *pdf*'s

procedure is the same: the example is labeled with the class that maximizes the product, $p_{c_i}(\mathbf{x})P(c_i)$.

- The concrete shape of the *pdf* is approximated by discretization, by the use of standardized *pdf*s, or by the sum of gaussians.
- The estimates of probabilities are far from perfect, but the results are often satisfactory even when rigorous theoretical assumptions are not satisfied.

Historical Remarks   The first papers to use Bayesian decision theory for classification purposes were Neyman and Pearson [72] and [25], but the paradigm gained momentum only with the advent of the computer, when it was advocated by Chow

[14]. The first to use the assumption of independent attributes was Good [33]. The idea of approximating *pdf*'s by the sum of bell functions comes from Parzen [74].

When provided with perfect information about the probabilities, the Bayesian classifier is guaranteed to provide the best possible classification accuracy. This is why it is sometimes used as a reference to which the performance of other approaches is compared.

2.8   Solidify Your Knowledge

The exercises are to solidify the acquired knowledge. The suggested thought experiments will help the reader see this chapter's ideas in a different light and provoke independent thinking. Computer assignments will force the readers to pay attention to seemingly insignificant details they might otherwise overlook.

Exercises

1. A coin tossed three times came up *heads, tails,* and *tails*, respectively. Calculate the *m*-estimate for these outcomes, using $m = 3$ and $\pi_{heads} = \pi_{tails} = 0.5$.
2. Suppose you have the following training examples, described by three attributes, x_1, x_2, x_3, and labeled by classes c_1 and c_2.

x_1	x_2	x_3	Class
2.1	0.2	3.0	c_1
3.3	1.0	2.9	c_1
2.7	1.2	3.4	c_1
0.5	5.3	0.0	c_2
1.5	4.7	0.5	c_2

Using these data, do the following:

(a) Assuming that the attributes are mutually independent, approximate the following probability density functions: $p_{c_1}(\mathbf{x}), p_{c_2}(\mathbf{x}), p(\mathbf{x})$. Hint: use the idea of superimposed bell functions.
(b) Using the *pdf*'s from the previous step, decide whether $\mathbf{x} = [1.4, 3.3, 3.0]$ should belong to c_1 or c_2.

Give It Some Thought

1. How would you apply the m-estimate in a domain with three possible outcomes, $[A, B, C]$, each with the same prior probability estimate, $\pi_A = \pi_B = \pi_C = 1/3$? What if you trust your expectations of A but are not so sure about B and C? Is there a way to reflect this circumstance in the value of the parameter m?
2. Suggest the circumstances under which the accuracy of probability estimates will benefit from the assumption that attributes are mutually independent. Explain the advantages and disadvantages.
3. How would you calculate the probabilities of the output classes in a domain where some attributes are boolean, others discrete, and yet others continuous? Discuss the possibilities of combining different approaches.

Computer Assignments

1. Machine learning researchers often test their algorithms using publicly available benchmark domains. A large repository of such domains can be found at the following address: www.ics.uci.edu/~mlearn/MLRepository.html. Take a look at these data and see how they differ in the numbers of attributes, types of attributes, sizes and so on.
2. Write a computer program that will use the Bayes formula to calculate the class probabilities in a domain where all attributes are discrete. Apply this program to our "pies" domain.
3. For the case of continuous attributes, write a computer program that accepts the training examples in the form of a table such as the one in Exercise 3 above. Based on these, the program approximates the pdfs, and then uses them to determine the class labels of future examples.
4. Apply this program to a few benchmark domains from the UCI repository (choose from those where all attributes are continuous) and observe that the program succeeds in some domains better than in others.

Chapter 3
Similarities: Nearest-Neighbor Classifiers

Two plants that look very much alike probably represent the same species; likewise, it is quite common that patients complaining of similar symptoms suffer from the same disease. In short, similar objects often belong to the same class—an observation that forms the basis of a popular approach to classification: when asked to determine the class of object **x**, find the training example most similar to it. Then label **x** with this example's class.

The chapter explains how to evaluate example-to-example similarities, presents concrete mechanisms that use these similarities for classification purposes, compares the performance of this approach with that of the Bayesian classifier, and introduces methods to overcome some inherent weaknesses.

3.1 The k-Nearest-Neighbor Rule

How do we establish that an object is more similar to **x** than to **y**? Some may doubt that this is at all possible. Is a giraffe more similar to a horse than to a zebra? Questions of this kind raise suspicion. Too many arbitrary and subjective factors are involved in answering them.

Similarity of Attribute Vectors The machine-learning task, as formulated in this book, reduces these objections to a minimum. Rather than real objects, the classifier will compare their attribute-based descriptions—hardly an insurmountable problem. Thus in the toy domain from Chap. 1, the similarity of two pies can be established by counting the attributes in which they differ: the fewer the differences, the greater the similarity. The first row in Table 3.1 gives the attribute values of object **x**. For each of the twelve training examples that follow, the rightmost column specifies the number of differences between the given example and **x**. The smallest value being found in the case of ex_5, we conclude that this is the training example most similar to **x**, and this suggests that we should label **x** with pos, the class of ex_5.

© Springer International Publishing AG 2017
M. Kubat, *An Introduction to Machine Learning*,
DOI 10.1007/978-3-319-63913-0_3

Table 3.1 Counting the numbers of differences between pairs of discrete-attribute vectors

| Example | Shape | Crust | | Filling | | Class | # differences |
		Size	Shade	Size	Shade		
x	Square	Thick	Gray	Thin	White	?	–
ex_1	Circle	Thick	Gray	Thick	Dark	pos	3
ex_2	Circle	Thick	White	Thick	Dark	pos	4
ex_3	Triangle	Thick	Dark	Thick	Gray	pos	4
ex_4	Circle	Thin	White	Thin	Dark	pos	4
ex_5	Square	Thick	Dark	Thin	White	pos	1
ex_6	Circle	Thick	White	Thin	Dark	pos	3
ex_7	Circle	Thick	Gray	Thick	White	neg	2
ex_8	Square	Thick	White	Thick	Gray	neg	3
ex_9	Triangle	Thin	Gray	Thin	Dark	neg	3
ex_{10}	Circle	Thick	Dark	Thick	White	neg	3
ex_{11}	Square	Thick	White	Thick	Dark	neg	3
ex_{12}	Triangle	Thick	White	Thick	Gray	neg	4

Of the 12 training examples, ex_5 is the one most similar to x

Table 3.2 The simplest version of the k-NN classifier

Suppose we have a mechanism to evaluate the similarly between attribute vectors. Let x denote the object whose class we want to determine.

1. Among the training examples, identify the k nearest neighbors of x (examples most similar to x).
2. Let c_i be the class most frequently found among these k nearest neighbors.
3. Label x with c_i.

Dealing with continuous attributes is just as simple. The fact that each example can be represented by a point in an n-dimensional space makes it possible to calculate the geometric distance between any pair of examples, for instance, by the Euclidean distance (Sect. 3.2 will have more to say about how to measure distance between vectors). And again, the closer to each other the examples are in the instance space, the greater their mutual similarity. This, by the way, is how the *nearest-neighbor classifier* got its name: the training example with the smallest distance from x in the instance space is, geometrically speaking, x's nearest neighbor.

From a Single Neighbor to k Neighbors In noisy domains, the testimony of the nearest neighbor cannot be trusted. What if this single specimen's class label is incorrect? A more robust approach identifies not one, but several nearest neighbors, and then lets them vote. This is the essence of the so-called k-NN classifier, where k is the number of the voting neighbors—usually a user-specified parameter. The pseudocode in Table 3.2 summarizes the algorithm.

Note that, in a two-class domain, k should be an odd number so as to prevent ties. For instance, a 4-NN classifier might face a situation where the number of positive neighbors is the same as the number of negative neighbors. This will not happen to a 5-NN classifier.

As for domains that have more than two classes, using an odd number of nearest neighbors does not prevent ties. For instance, the 7-NN classifier can realize that three neighbors belong to class C_1, three neighbors belong to class C_2, and one neighbor belongs to class C_3. The engineer designing the classifier then needs to define a mechanism to choose between C_1 and C_2.

An Illustration Certain "behavioral aspects" of this paradigm can be made obvious with the help of a fictitious domain where the examples are described by two numeric attributes, a situation easy to visualize. Figure 3.1 shows several positive and negative training examples, and also some objects (the big black dots) whose classes the k-NN classifier is to determine. The reader can see that objects **1** and **2** are surrounded by examples from the same class, and their classification is therefore straightforward. On the other hand, object **3** is located in the "no man's land" between the positive and negative regions, so that even a small amount of attribute noise can send it to either side. The classification of such *borderline examples* is unreliable.

In the right-hand part of the picture, object **4** finds itself deep in the positive region, but class noise has mislabeled its nearest neighbor in the training set as negative. Whereas the 1-NN classifier will go wrong, here, the 3-NN classifier will give the correct answer because the other two neighbors, which are positive, will outvote the single negative neighbor.

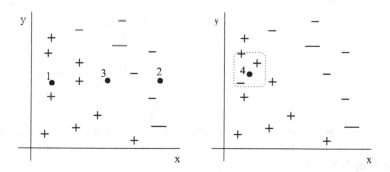

Fig. 3.1 Object **3**, finding itself in the borderline region, is hard to classify. In the noisy domain shown in the *right-hand part*, the 1-NN classifier will misclassify object **4**, but the mistake is corrected if the 3-NN classifier is used

What Have You Learned?

To make sure you understand this topic, try to answer the following questions. If you have problems, return to the corresponding place in the preceding text.

- How can we measure example-to-example similarity in domains where all attributes are discrete, and how in domains where they are all continuous?
- Under what circumstances will the k-NN classifier (with $k > 1$) outperform the 1-NN classifier and why?
- Explain why, in 2-class domains, the k in the k-NN classifier should be an odd number.
- How does attribute noise affect the classification of a "borderline example"? What is the impact of class-label noise?

3.2   Measuring Similarity

As indicated, a natural way to find the nearest neighbors of object $\mathbf{x}$ is to compare the geometrical distances of the individual training examples from $\mathbf{x}$. Figure 3.1 illustrated this principleusing a domain so simple that the distances could be

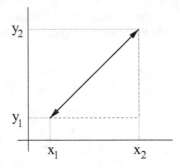

Fig. 3.2 The Euclidean distance between two points in a two-dimensional space is equal to the length of the triangle's hypotenuse

measured by a ruler. Yet the ruler will hardly be our tool of choice if the examples are described by more than two attributes. What we need is an expression to calculate the similarity based on attribute values.

Euclidean Distance   In a plane, the geometric distance between two points, $\mathbf{x} = (x_1, x_2)$ and $\mathbf{y} = (y_1, y_2)$, is obtained with the help of the pythagorean theorem as indicated in Fig. 3.2: $d(\mathbf{x}, \mathbf{y}) = \sqrt{(x_1 - y_1)^2 + (x_2 - y_2)^2}$. This formula is easy to generalize to a domain with n continuous attributes where the *Euclidean distance* between $\mathbf{x} = (x_1, \ldots, x_n)$ and $\mathbf{y} = (y_1, \ldots, y_n)$ is defined as follows:

$$d_E(\mathbf{x}, \mathbf{y}) = \sqrt{\Sigma_{i=1}^n (x_i - y_i)^2} \qquad (3.1)$$

Table 3.3 Using the nearest-neighbor principle in a 3-dimensional Euclidean space

Using the following training set of four examples described by three numeric attributes, determine the class of object $\mathbf{x} = [2, 4, 2]$.

Distance between ex_i and $[2, 4, 2]$		
ex_1	$\{[1, 3, 1], \text{pos}\}$	$\sqrt{(2-1)^2 + (4-3)^2 + (2-1)^2} = \sqrt{3}$
ex_2	$\{[3, 5, 2], \text{pos}\}$	$\sqrt{(2-3)^2 + (4-5)^2 + (2-2)^2} = \sqrt{2}$
ex_3	$\{[3, 2, 2], \text{neg}\}$	$\sqrt{(2-3)^2 + (4-2)^2 + (2-2)^2} = \sqrt{5}$
ex_4	$\{[5, 2, 3], \text{neg}\}$	$\sqrt{(2-5)^2 + (4-2)^2 + (2-3)^2} = \sqrt{4}$

Calculating the Euclidean distances between $\mathbf{x}$ and the training examples, we realize that $\mathbf{x}$'s nearest neighbor is ex_2. Its label being pos, the 1-NN classifier returns the positive label.

The same result is obtained by the 3-NN classifier because two of $\mathbf{x}$'s three nearest neighbors (ex_1 and ex_2) are positive, and only one (ex_4), is negative.

The way this metric is used in the context of k-NN classifiers is illustrated in Table 3.3 where the training set consists of four examples described by three numeric attributes.

A More General Formulation The reader can see that the term under the square-root symbol is a sum of the squared distances along corresponding attributes.[1] This observation is mathematically expressed as follows:

$$d_M(\mathbf{x}, \mathbf{y}) = \sqrt{\Sigma_{i=1}^n d(x_i, y_i)} \tag{3.2}$$

This is how the distance is usually calculated in the case of vectors in which discrete and continuous attributes are mixed (in the symbol, $d_M(\mathbf{x}, \mathbf{y})$, this is indicated by the subscript, M). For instance, we can use $d(x_i, y_i) = (x_i - y_i)^2$ for continuous attributes, whereas for discrete attributes, we put $d(x_i, y_i) = 0$ if $x_i = y_i$ and $d(x_i, y_i) = 1$ if $x_i \neq y_i$.

Note that if all attributes are continuous, the formula is identical to Euclidean distance; and if the attributes are all discrete, the formula simply specifies the number of attributes in which the two vectors differ. In purely Boolean domains, where for any attribute only the values *true* or *false* are permitted (let us abbreviate these values as t and f, respectively), this latter case is called *Hamming distance*, d_H. For instance, the Hamming distance between the vectors $\mathbf{x} = (t, t, f, f)$ and $\mathbf{y} = (t, f, t, f)$ is $d_H(\mathbf{x}, \mathbf{y}) = 2$. In general, however, Eq. (3.2) is meant to be employed in domains where the examples are described by a mixture of discrete and continuous attributes.

---

[1]One benefit of these differences being squared, and thus guaranteed to be positive, is that this prevents negative differences, $x_i - y_i < 0$, to be subtracted from positive differences, $x_i - y_i > 0$.

Attribute-to-Attribute Distances Can Be Misleading We must be careful not to apply Formula (3.2) mechanically, ignoring the specific aspects of the given domain. Let us briefly discuss two circumstances that make it is easy to go wrong.

Suppose our examples are described by three attributes, size, price, and season. Of these, the first two are obviously continuous, and the last, discrete. If $\mathbf{x} = (2, 1.5, \text{summer})$ and $\mathbf{y} = (1, 0.5, \text{winter})$, then Eq. (3.2) gives the following distance between the two:

$$d_M(\mathbf{x}, \mathbf{y}) = \sqrt{(2-1)^2 + (1.5 - 0.5)^2 + 1} = \sqrt{3}$$

Let us first consider the third attribute: summer being different from winter, our earlier considerations lead us to establish that $d(\text{summer}, \text{winter}) = 1$. In reality, though, the difference between summer and winter is sometimes deemed greater than the difference between, say, summer and fall which are "neighboring seasons." Another line of reasoning, however, may convince us that spring and fall are more similar to each other than summer and winter— at least as far as weather is concerned. We can see that the two values, 0 and 1, will clearly not suffice, here. Intermediate values should perhaps be considered, the concrete choice depending on the specific needs of the given application. The engineer who does not pay attention to factors of this kind may fail to achieve good results.

Mixing continuous and discrete attributes can be risky in another way. A thoughtless application of Eq. (3.2) can result in a situation where the difference between two sizes (e.g., $\text{size}_1 = 1$ and $\text{size}_1 = 12$, which means that $d(\text{size}_1, \text{size}_2) = 11^2 = 121$) can totally dominate the difference between two seasons (which, in the baseline version, could not exceed 1). This observation is closely related to the problem of *scaling*—discussed in the next section.

Distances in General The reader is beginning to understand that the issue of similarity is far from trivial. Apart from Eq. (3.2), quite a few other formulas have been suggested, some of them fairly sophisticated.[2] While it is good to know they exist, we will not examine them here because we do not want to digress too far from our main topic. Suffice it so say that any distance metric has to satisfy the following requirements:

1. the distance must never be negative;
2. the distance between two identical vectors, $\mathbf{x}$ and $\mathbf{y}$, is zero;
3. the distance from $\mathbf{x}$ to $\mathbf{y}$ is the same as the distance from $\mathbf{y}$ to $\mathbf{x}$;
4. the metric must satisfy the triangular inequality: $d(\mathbf{x}, \mathbf{y}) + d(\mathbf{y}, \mathbf{z}) \geq d(\mathbf{x}, \mathbf{z})$.

---

[2]Among these, perhaps the best-known are the polar distance, the Minkowski metric, and the Mahalanobis distance.

What Have You Learned?

To make sure you understand this topic, try to answer the following questions. If you have problems, return to the corresponding place in the preceding text.

- What is Euclidean distance, and what is Hamming distance? In what domains can they be used? How is distance related to similarity?
- Write down the distance formula for domains where examples are described by a mixture of continuous and discrete attributes. Discuss the difficulties that complicate its straightforward application in practice.
- What fundamental properties have to be satisfied by any distance metric?

3.3   Irrelevant Attributes and Scaling Problems

By now, the reader understands the principles of the k-NN classifier well enough to be able to write a computer program implementing the tool. This is not enough, though. If applied mechanically, the software may disappoint. It is necessary to understand why.

The rock-bottom of the nearest-neighbor paradigm is that, "objects are similar if the geometric distance between the vectors describing them is small." This said, we must be careful not to let this principle degenerate into a dogma. In certain situations, the geometric distance can be misleading. Two of them are quite common.

Irrelevant Attributes It would be a mistake to think that all attributes are created equal. They are not. In the context of machine learning, some are *irrelevant* in the sense that their values have nothing to do with the given example's class. But they do affect the geometric distance between vectors.

A simple illustration will clarify the point. In the training set from Fig. 3.3, the examples are characterized by two numeric attributes: body-temperature (the horizontal axis) and shoe-size (the vertical axis). The black dot stands for the object that the k-NN classifier is expected to label as healthy (pos) or sick (neg).

As expected, all positive examples find themselves in the shaded area delimited by two critical points along the "horizontal" attribute: temperatures exceeding the maximum indicate fever; those below the minimum, hypothermia. As for the "vertical" attribute, though, we see that the positive and negative examples alike are distributed along the entire range, show-size being unable to betray anything about a person's health. The object we want to classify is in the highlighted region, and common sense requires that it should be labeled as positive—despite the fact that its nearest neighbor happens to be negative.

The Lesson If only the first attribute is used, the Euclidean distance between the two examples is $d_E(x, y) = \sqrt{(x_1 - y_1)^2} = |x_1 - y_1|$. If both attributes are used, the Euclidean distance will be $d_E(\mathbf{x}, \mathbf{y}) = \sqrt{(x_1 - y_1)^2 + (x_2 - y_2)^2}$. If the second

Fig. 3.3 The "vertical" attribute is irrelevant for classification, and yet it affects the geometric distances between examples. Object **1** is positive, even though its nearest neighbor in the 2-dimensional space is negative

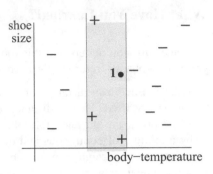

attribute is irrelevant, then the term $(x_2 - y_2)^2$ is superfluous—and yet it affects, adversely, k-NN's notion of similarity! This is what occurred in Fig. 3.3, and this is why object **1** was misclassified.

How much damage is caused by irrelevant attributes depends on how many of them are used to describe the examples. In a domain with hundreds of attributes, of which only one is irrelevant, there is no need to panic: one lonely culprit is unlikely to distort the value of $d_E(\mathbf{x}, \mathbf{y})$ in any meaningful way. But things will change as the percentage of irrelevant attributes grows. If the vast majority of the attributes have nothing to do with the class we want to recognize, then the geometric distance will become almost meaningless, and the classifier's performance will be dismal.

The Scales of Attribute Values Suppose we want to evaluate the similarity of two examples, $\mathbf{x} = (t, 0.2, 254)$ and $\mathbf{y} = (f, 0.1, 194)$, described by three attributes, of which the first is boolean, the second is continuous with values from interval $[0; 1]$, and the third is continuous with values from interval $[0; 1000]$. Using Eq. (3.2), the reader will find it easy to calculate the distance between $\mathbf{x}$ and $\mathbf{y}$, obtaining the following:

$$d_M(\mathbf{x}, \mathbf{y}) = \sqrt{(1 - 0)^2 + (0.2 - 0.1)^2 + (254 - 194)^2}$$

Inspecting this expression, we notice that the third attribute completely dominates, reducing the other two to virtual insignificance. No matter how we modify their values within their ranges, the distance, $d_M(\mathbf{x}, \mathbf{y})$, will hardly change. Fortunately, the situation is easy to rectify. If we divide, in the training set, all values of the third attribute by 1000, thus "squeezing" its range to $[0; 1]$, the impacts of the attributes will become more balanced. We can see that the scales of the attribute values can radically affect the k-NN classifier's behavior.

Another Aspect of Attribute Scaling There is more to it. Consider the following two training examples, ex_1 and ex_2, and the object $\mathbf{x}$ whose class we want to determine:

$ex_1 = [(10, 10), \texttt{pos}]$
$ex_2 = [(20, 0), \texttt{neg}]$
$\mathbf{x} = (32, 20)$

The distances are $d_M(\mathbf{x}, \text{ex}_1) = \sqrt{584}$ and $d_M(\mathbf{x}, \text{ex}_2) = \sqrt{544}$. The latter being smaller, the 1-NN classifier will label $\mathbf{x}$ as neg. Suppose, however, that the second attribute expresses temperature, and does so in centigrades. If we decide to use Fahrenheits instead, the three vectors will change as follows:

$$\text{ex}_1 = [(10, 50), \text{pos})]$$
$$\text{ex}_2 = [(20, 32), \text{neg})]$$
$$\mathbf{x} = (32, 68)$$

Recalculating the distances, we obtain $d_M(\mathbf{x}, \text{ex}_1) = \sqrt{808}$ and $d_M(\mathbf{x}, \text{ex}_2) = \sqrt{1440}$. This time, it is the first distance that is smaller, and 1-NN will therefore classify $\mathbf{x}$ as positive. This seems a bit silly. The examples are still the same, except that we chose different units for temperature; and yet the classifier's verdict has changed.

Normalizing Attribute Scales One way out of this trouble is to *normalize* the attributes: to re-scale them in a way that makes all values fall into the same interval, $[0; 1]$. From the several mechanisms that have been used to this end, perhaps the simplest is the one that first identifies, for the given attribute, its maximum (*MAX*) and minimum (*MIN*), and then replaces each value, x, of this attribute using the following formula:

$$x = \frac{x - MIN}{MAX - MIN} \tag{3.3}$$

A simple illustration will show us how this works. Suppose that, in the training set consisting of five examples, a given attribute acquires the following values, respectively:

$$[7, 4, 25, -5, 10]$$

We see that $MIN = -5$ and $MAX = 25$. Subtracting MIN from each of the values, we obtain the following:

$$[12, 9, 30, 0, 15]$$

The reader can see that the "new minimum" is 0, and the "new maximum" is $MAX - MIN = 25 - (-5) = 30$. Dividing the obtained values by $MAX - MIN$, we obtain a situation where all the values fall into $[0; 1]$:

$$[0.4, 0.3, 1, 0, 0.5]$$

One Potential Weakness of Normalization Normalization reduces error rate in many practical applications, especially if the scales of the original attributes vary significantly. The downside is that the description of the examples thus becomes distorted. Moreover, the pragmatic decision to make all values fall between 0 and 1 may not be justified. For instance, if the difference between summer and fall is 1, it will always be bigger than, say, the difference between two normalized body temperatures. Whether this matters or not is up to the engineer's common sense—assisted by his or her experience and perhaps a little experimentation.

What Have You Learned?

To make sure you understand this topic, try to answer the following questions. If you have problems, return to the corresponding place in the preceding text.

- Why do irrelevant attributes impair the k-NN classifier's performance? How does the performance depend on the number of irrelevant attributes?
- Explain the basic problems pertaining to attribute scaling. Describe a simple approach to normalization.

3.4   Performance Considerations

Having spent all this time exploring various aspects of the k-NN classifier, the reader is bound to ask: should I really care? Sure enough, the technique is easy to implement in a computer program, and its function is easy to grasp. But is there a reason to believe that its classification performance is good enough?

The 1-NN Classifier Versus Ideal Bayes Let us give the question some thought. The ultimate yardstick by which to measure k+-NN's potential is the Bayesian approach. We will recall that if the probabilities and *pdf*'s employed by the Bayesian formula are known with absolute accuracy, then this classifier—let us call it *Ideal Bayes*—exhibits the lowest error rate theoretically achievable on the given (noisy) data. It would be reassuring to know that the k-NN paradigm does not trail too far behind.

The question has been subjected to rigorous mathematical analysis, and here are the results. Figure 3.4 shows what the comparison will look like under certain ideal circumstances such as an infinitely large training set which fills the instance space with infinite density. The solid curve represents the two-class case where each example is either positive or negative. We can see that if the error rate of the *Ideal Bayes* is 5%, the error rate of the 1-NN classifier (vertical axis) is 10%. With the growing amount of noise, the difference between the two approaches decreases, only to disappear when the *Ideal Bayes* suffers 50% error rate—in which event, of course, the labels of the training examples are virtually random, and any attempt at automated classification is doomed anyway. The situation is not any better in multi-class domains, represented in the graph by the dotted curve. Again, the *Ideal Bayes* outperforms the 1-NN classifier by a comfortable margin.

Increasing the Number of Neighbors From the perspective of the 1-NN classifier, the results from Fig. 3.4 are rather discouraging. On the other hand, we know that the classifier's performance might improve when we use the more general k-NN (for $k > 1$), where some of the noisy nearest neighbors get outvoted by better-behaved ones. Does mathematics lend some support to this expectation?

Yes it does. Under the above-mentioned ideal circumstances, the error rate has been proven to decrease with the growing value of k, until it converges to that of *Ideal Bayes* for $k \to \infty$. At least in theory, then, the performance of the nearest-neighbor classifier is able to reach the absolute maximum.

Practical Limitations of Theories The engineer's world is indifferent to theoretical requirements. In a realistic application, the training examples will but sparsely populate the instance space, and increasing the number of voting neighbors can be counterproductive. More often than not, the error rate *does* improve with the growing k, but only up to a certain point from which it starts growing again—as illustrated in Fig. 3.5 where the horizontal axis represents the values of k, and the vertical axis represents the error rate that is measured on an independent testing set.

The explanation is simple: the more distant "nearest neighbors" may find themselves in regions (in the instance space) that are already occupied by other classes; as such, they only mislead the classifier. Consider the extreme: if the training

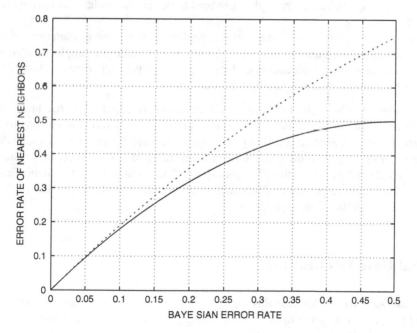

Fig. 3.4 The theoretical error rate of the 1-NN rule compared to that of the *Ideal Bayes*. Two classes: the *solid curve*; many classes: the *dotted curve*

Fig. 3.5 With the growing
number of voting neighbors
(k), the error rate of the k-NN
classifier decreases until it
reaches a level from which it
starts growing again

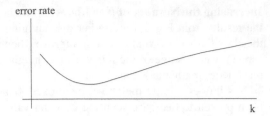

set contains 25 training examples, then the 25-NN classifier is useless because it simply labels any object with the class that has the highest number of representatives in the training data.[3]

The Curse of Dimensionality Obviously, when classifying object **x**, some of its nearest neighbors may *not* be similar enough to **x** to deserve to participate in the vote. This often happens in domains marked by many attributes. Suppose that the values of each attribute are confined to the unit-length interval, $[0; 1]$. Using the pythagorean theorem, it would be easy to show that the maximum Euclidean distance in the n-dimensional space defined by these attributes is $d_{MAX} = \sqrt{n}$. For $n = 10^4$ (quite reasonable in such domains as, say, text categorization), this means $d_{MAX} = 100$. In view of the fact that no attribute value can exceed 1, this is perhaps surprising. No wonder that examples then tend to be sparse—unless the training set is really very large.

The term sometimes mentioned, in this context, is the *curse of dimensionality*: as we increase the number of attributes, the number of training examples needed to fill the instance space with adequate density grows very fast, perhaps so fast as to render the nearest-neighbor paradigm impractical.

To Conclude Although the ideal k-NN classifier is capable of reaching the performance of the *Ideal Bayes*, the engineer has to be aware of the practical limitations of both approaches. Being able to use the *Ideal Bayes* is unrealistic in the absence of the perfect knowledge of the probabilities and *pdf*'s. On the other hand, the k-NN classifier is prone to suffer from sparse data, from irrelevant attributes, and from scaling-related problems. The concrete choice has to be based on the specific requirements of the given application.

What Have You Learned?

To make sure you understand this topic, try to answer the following questions. If you have problems, return to the corresponding place in the preceding text.

---

[3]The optimal value of k (the one with the minimum error rate) is usually established experimentally.

- How does the performance of the k-NN classifier compare to that of the *Ideal Bayes*? Summarize this separately for $k = 1$ and $k > 1$. What theoretical assumptions do these two paradigms rely on?
- How will the performance of a k-NN classifier depend on the growing values of k in theory and in a realistic application?
- What is understood by the *curse of dimensionality*?

3.5   Weighted Nearest Neighbors

So far, the voting mechanism was assumed to be democratic, each nearest neighbor having the same vote. This seems to be a fair enough arrangement, but from the perspective of classification performance, we can do better.

In Fig. 3.6, the task is to determine the class of object **x**, represented by a black dot. Since three of the nearest neighbors are negative, and only two are positive, the 5-NN classifier will label **x** as negative. And yet, something seems to be wrong, here: the three negative neighbors are quite distant from **x**; as such, they may not deserve to have the same say as the two positive examples in the object's immediate vicinity.

Weighted Nearest Neighbors Situations of this kind motivate the introduction of *weighted voting*, in which the weight of each of the nearest neighbors is made proportional to its distance from **x**: the closer the neighbor, the greater its impact.

Let us denote the weights as $w_1, \ldots w_k$. The *weighted k-NN* classifier sums up the weights of those neighbors that recommend the positive class (let the result be denoted by Σ^+) and then sums up the weights of those neighbors that support the negative class (Σ^-). The final verdict is based on which is higher: if $\Sigma^+ > \Sigma^-$, then **x** is deemed positive, otherwise it is deemed negative. Generalization to the case with $n > 2$ classes is straightforward.

For the sake of illustration, suppose the positive label is found in neighbors with weights 0.6 and 0.7, respectively, and the negative label is found in neighbors with weights $0.1, 0.2$, and 0.3. The weighted k-NN classifier will choose the positive

Fig. 3.6 The testimony of the two "positive" neighbors should outweigh that of the three more distant "negative" neighbors

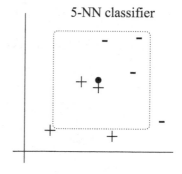

5-NN classifier

class because the combined weight of the positive neighbors, $\Sigma^+ = 0.6+0.7 = 1.3$, is greater than that of the negative neighbors, $\Sigma^- = 0.1 + 0.2 + 0.3 = 0.6$. Just as in Fig. 3.6, the more frequent negative label is outvoted by the positive neighbors because the latter are closer to **x**.

A Concrete Formula Let us introduce a simple formula to calculate the weights. Suppose the k neighbors are ordered according to their distances, $d_1, \ldots, d_k$, from **x** so that d_1 is the smallest distance and d_k is the greatest distance. The weight of the i-th closest neighbor is calculated as follows:

$$
w_i = \begin{cases} \frac{d_k-d_i}{d_k-d_1}, & d_k \neq d_1 \\ 1 & d_k = d_1 \end{cases}
\tag{3.4}
$$

Obviously, the weights thus obtained will range from 0 for the most distant neighbor to 1 for the closest one. This means that the approach actually considers only $k-1$ neighbors (because $w_k = 0$). Of course, this makes sense only for $k > 3$. If we used $k = 3$, then only two neighbors would really participate, and the weighted 3-NN classifier would degenerate into the 1-NN classifier.

Table 3.4 illustrates the procedure on a simple toy domain.

Another important thing to observe is that if all nearest neighbors have the same distance from **x**, then they all get the same weight, $w_i = 1$, on account of the bottom part of the formula in Eq. (3.4). The reader will easily verify that $d_k = d_1$ if and only if all the k nearest neighbors have the same distance from **x**.

Table 3.4 Illustration of the weighted nearest neighbor rule

The task is to use the weighted 5-NN classifier to determine the class of object **x**. Let the distances between **x** and the five nearest neighbors be $d_1 = 1, d_2 = 3, d_3 = 4, d_4 = 5, d_5 = 8$. Since the minimum is $d_1 = 1$ and the maximum is $d_5 = 8$, the individual weights are calculated as follows:

$$
w_i = \frac{d_5 - d_i}{d_5 - d_1} = \frac{8 - d_i}{8 - 1} = \frac{8 - d_i}{7}
$$

This gives the following values:

$$
w_1 = \frac{8-1}{7} = 1, w_2 = \frac{8-3}{7} = \frac{5}{7}, w_3 = \frac{8-4}{7} = \frac{4}{7}, w_4 = \frac{8-5}{7} = \frac{3}{7}, w_5 = \frac{8-8}{7} = 0.
$$

If the two nearest neighbors are positive and the remaining three are negative, then **x** is classified as positive because $\Sigma^+ = 1 + \frac{5}{7} > \Sigma^- = \frac{4}{7} + \frac{3}{7} + 0$.

What Have You Learned?

To make sure you understand this topic, try to answer the following questions. If you have problems, return to the corresponding place in the preceding text.

- Why did we argue that each of the voting nearest neighbors should sometimes have a different weight?
- Discuss the behavior of the formula recommended in the text for the calculation of the individual weights.

3.6 Removing Dangerous Examples

As far as the classification procedure is concerned, the value of each individual training example can be different. Some are typical of the classes they represent, others less so, and yet others are outright misleading. This is why it is often a good thing to pre-process the training set before using it: to remove examples suspected of being less than useful.

The concrete method to pre-process the training set is guided by two essential observations that are illustrated in Fig. 3.7. First, an example labeled with one class

Fig. 3.7 Illustration of two potentially harmful types of examples: those surrounded by examples of a different class, and those in the "borderline region."

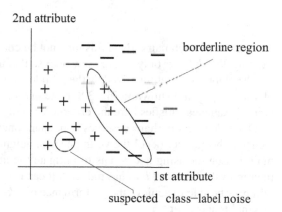

but surrounded by examples of another class is likely to be the result of class-label noise. Second, examples from the borderline region between two classes are unreliable because even small amount of noise in their attribute values can shift their locations in the wrong directions, thus affecting classification. In both cases, the examples better be removed.

The Technique of Tomek Links Before the culprit can be removed, however, it has to be detected. One way to do so is by the technique of *Tomek Links*, named so after the mathematician who first used them a few decades ago.[4] A pair of examples, **x** and **y**, are said to form a *Tomek Link* if the following three requirements are satisfied at the same time:

1. **x** is the nearest neighbor of **y**
2. **y** is the nearest neighbor of **x**
3. the class of **x** is not the same as the class of **y**.

These conditions being characteristic of borderline examples, and also of examples surrounded by examples of anotherclass, it makes sense to remove from the

Fig. 3.8 *Dotted lines* connect all Tomek links. Each participant in a Tomek Link is its partner's nearest neighbor, and each of the two examples has a different class

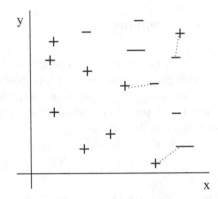

training set all such pairs. Even this may not be enough. Sometimes, the removal of existing *Tomek Links* only creates new ones so that the procedure has to be repeated.

The algorithm is summarized by the pseudocode in Table 3.5, and a few instances of *Tomek Links* are shown in Fig. 3.8. Note that there are only these three; no other pair of examples satisfies here the criteria for being called a *Tomek Link*.

One side-effect perhaps deserves to be mentioned: once the training set has been cleaned, the number (k) of the voting nearest neighbors can be reduced because the main reason for using a $k > 1$ is to mitigate the negative impact of noise—which the removal of *Tomek Links* has reduced. It can even happen that the 1-NN classifier will now be able to achieve the performance of a k-NN classifier that uses the entire original training set.

A Limitation Nothing is perfect. The technique of *Tomek Links* does not identify all misleading examples; and, conversely, some of the removed examples might have been "innocent," and thus deserved to be retained. Still, experience has shown that the removal of *Tomek Links* usually does improve the overall quality of the data.

---

[4]It is fair to mention that he used them for somewhat different purposes.

Table 3.5 The algorithm to identify (and remove) *Tomek Links*

Input: the training set of N examples

1. Let $i = 1$ and let T be an empty set.
2. Let $\mathbf{x}$ be the i-th training example and let $\mathbf{y}$ be the nearest neighbor of $\mathbf{x}$.
3. If $\mathbf{x}$ and $\mathbf{y}$ belong to the same class, go to 5.
4. If $\mathbf{x}$ is the nearest neighbor of $\mathbf{y}$, let $T = T \cup \{\mathbf{x}, \mathbf{y}\}$.
5. Let $i = i + 1$. If $i \leq N$, goto 2.
6. Remove from the training set all examples that are now in T.

The engineer only has to be careful in two specific situations. (1) When the training set is very small, and (2) when one of the classes significantly outnumbers the other. The latter case will be discussed in Sect. 10.2.

What Have You Learned?

To make sure you understand this topic, try to answer the following questions. If you have problems, return to the corresponding place in the preceding text.

- What motivates the attempts to "clean" the training set?
- What are *Tomek Links*, are how do we identify them in the training set? Why does the procedure sometimes have to be repeated?
- How does the removal of *Tomek Links* affect the k-NN classifier?

3.7 Removing Redundant Examples

Some training examples do not negatively affect classification performance, and yet we want to get rid of them. Thing is, they do not reduce the error rate, either; they only add to computational costs.

Redundant Examples To find the nearest neighbor in a domain with 10^6 training examples described by 10^4 attributes (and such domains are not rare), the program relying on the Euclidean distance has to carry out $10^6 \times 10^4 = 10^{10}$ arithmetic operations. When the task is to classify thousands of objects at a time (which, again, is not impossible), the number of arithmetic operations is $10^{10} \times 10^3 = 10^{13}$. This is a lot.

Fortunately, training sets are often redundant in the sense that the k-NN classifier's behavior will not change even if some of the training examples are deleted. Sometimes, a great majority of the examples can be removed with impunity because

they add to classification costs without affecting classification performance. This is
the case of the domain shown in the upper-left corner of Fig. 3.9.

Consistent Subset In an attempt to reduce redundancy, we want to replace the
training set, T, with its *consistent subset*, S. In our context, S is said to be a consistent
subset of T if replacing T with S does not affect what class labels are returned by
the k-NN classifier. Such definition, however, is not very practical because we have
no idea of how the k-NN classifier will behave—using either T or S—on future
examples. Let us therefore modify the criterion: S will be regarded as a consistent
subset of T if any **ex** $\in T$ receives the same label no matter whether the k-NN
classifier has employed $T - \{$**ex**$\}$ or $S - \{$**ex**$\}$.

It is in the nature of things that a realistic training set will have many consistent
subsets to choose from. Of course, the smaller the subset, the better. But a
perfectionist who insists on always finding the smallest one should be warned: this
ideal can often be achieved only at the price of enormous computational costs.
The practically minded engineer doubts that these costs are justified, and will be
content with a computationally efficient algorithm that downsizes the original set to
a "reasonable size," unscientific though such formulation may appear to be.

Creating a Consistent Subset One such pragmatic technique is presented in
Table 3.6. The algorithm starts by choosing one random example from each class.

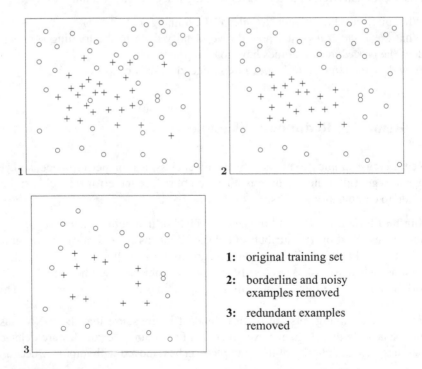

1: original training set

2: borderline and noisy
   examples removed

3: redundant examples
   removed

Fig. 3.9 An illustration of what happens when the borderline, noisy, and redundant examples are
removed from the training set

Table 3.6 Algorithm to create a consistent training subset by the removal of (some) redundant examples

1. Let S contain one positive and one negative example from the training set, T.
2. Using examples from S, re-classify the examples in T with the 1-NN classifier. Let M be the set of those examples that have in this manner received the wrong class.
3. Copy to S all examples from M.
4. If the contents of S have not changed, in the previous step, then *stop*; otherwise go to step 1.

This initial subset, S, is then used by the 1-NN classifier to suggest the labels of all training examples. At this stage, it is more than likely that some training examples will thus be misclassified. These misclassified examples are added to S, and the whole procedure is repeated using this larger set. This is then repeated again and again until, at a certain moment, S becomes representative enough to allow the 1-NN classifier to label all training examples correctly. This is when the search stops.

What Have You Learned?

To make sure you understand this topic, try to answer the following questions. If you have problems, return to the corresponding place in the preceding text.

- What is the benefit of removing redundant examples from the training set?
- What do we mean by the term, "consistent subset of the training set"? Why is it not necessary always to look for the smallest consistent subset?
- Explain the principle of the simple algorithm that creates a reasonably sized consistent subset.

3.8 Summary and Historical Remarks

- When classifying object $\mathbf{x}$, the k-NN classifier finds, in the training set, k examples most similar to $\mathbf{x}$, and then chooses the class label most common among these examples.
- Classification behavior of the k-NN classifier depends to a great extent on how similarities between attribute vectors are calculated. Perhaps the simplest way to establish the similarity between vectors $\mathbf{x}$ and $\mathbf{y}$ is to use their geometric distance obtained by the following formula:

$$d_M(\mathbf{x}, \mathbf{y}) = \sqrt{\Sigma_{i=1}^{n} d(x_i, y_i)} \tag{3.5}$$

Essentially, we use $d(x_i, y_i) = (x_i - y_i)^2$ for continuous attributes, whereas for discrete attributes, we put $d(x_i, y_i) = 0$ if $x_i = y_i$ and $d(x_i, y_i) = 1$ if $x_i \neq y_i$.

- The performance of the k-NN classifier is poor if many of the attributes describing the examples are irrelevant. Another issue is the scaling of the attribute values. The latter problem can be mitigated by normalizing the attribute values to unit intervals.
- Some examples are harmful in the sense that their presence in the training set increases error rate. Others are redundant in that they only add to the computation costs without improving classification performance. Harmful and redundant examples should be removed.
- In some applications, each of the nearest neighbors can have the same vote. In others, the votes are weighted based on the distance of the examples from the classified object.

Historical Remarks The idea of the nearest-neighbor classifier was originally proposed by Fix and Hodges [27], but the first systematic analysis was offered by Cover and Hart [20] and Cover [19]. Exhaustive treatment was then provided by the book by Dasarathy [21]. The weighted k-NN classifier described here was proposed by Dudani [23]. The oldest technique to find a consistent subset of the training set was described by Hart [35]—the one introduced in this chapter is its minor modification. The notion of *Tomek Links* is due to Tomek [89].

3.9   Solidify Your Knowledge

The exercises are to solidify the acquired knowledge. The suggested thought experiments will help the reader see this chapter's ideas in a different light and provoke independent thinking. Computer assignments will force the readers to pay attention to seemingly insignificant details they might otherwise overlook.

Exercises

1. Determine the class of $\mathbf{y} = [1, 22]$ using the 1-NN classifier and the 3-NN classifier, both using the training examples from Table 3.7. Explain why the two classifiers differ in their classification behavior.
2. Use the examples from Table 3.7 to classify object $\mathbf{y} = [3, 3]$ with the 5-NN classifier. Note that two nearest neighbors are positive and three nearest neighbors are negative. Will weighted 5-NN classifier change anything? To see what is going on, plot the locations of the examples in a graph.

Table 3.7 A simple set of training examples for the exercises

x_1	1	1	1	2	3	3	3	4	5
x_2	1	2	4	3	0	2	5	4	3
class	+	−	−	+	+	+	−	−	−

3. Again, use the training examples from Table 3.7. (a) Are there any Tomek links? (b) can you find a consistent subset of the training set by the removal of at least one redundant training example?

Give It Some Thought

1. Discuss the possibility of applying the k-NN classifier to the "pies" domain. Give some thought to how many nearest neighbors to use, and what distance metric to employ, whether to make the nearest neighbors' vote depend on distance, and so on.

2. Suggest other variations on the nearest-neighbor principle. Use the following hints:

   (a) Introduce alternative distance metrics. Do not forget that they have to satisfy the axioms mentioned at the end of Sect. 3.2.

   (b) Modify the voting scheme by assuming that some examples have been created by a knowledgeable "teacher," whereas others have been obtained from a database without any consideration given to their representativeness. The teacher's examples should obviously carry more weight.

3. Design an algorithm that uses hill-climbing search to remove *redundant examples*. Hint: the initial state will contain the entire training set, the search operator will remove a single training example at a time (this removal must not affect behavior).

4. Describe an algorithm that uses hill-climbing search to remove *irrelevant attributes*. Hint: withhold some training examples on which you will test 1-NN's classifier's performance for different subsets of attributes.

Computer Assignments

1. Write a program whose input is the training set, a user-specified value of k, and an object, **x**. The output is the class label of **x**.

2. Apply the program implemented in the previous assignment to some of the benchmark domains from the UCI repository.[5] Always take 40% of the examples out and reclassify them with the 1-NN classifier that uses the remaining 60%.

3. Create a synthetic toy domain consisting of 1000 examples described by a pair of attributes, each from the interval [0,1]. In the square defined by these attribute values, $[0, 1] \times [0, 1]$, define a geometric figure of your own choice and label all examples inside it as positive and all other examples as negative. From this initial

---

[5] www.ics.uci.edu/~mlearn/MLRepository.html.

noise-free data set, create 5 files, each obtained by changing p percent of the class labels, using $p \in \{5, 10, 15, 20, 25\}$ (thus obtaining different levels of class-label noise).

Divide each data file into two parts, the first to be reclassified by the k-NN classifier that uses the second part. Observe how different values of k result in different behaviors under different levels of class-label noise.

4. Implement the Tomek-link method for the removal of harmful examples. Repeat the experiments above for the case where the k-NN classifier uses only examples that survived this removal. Compare the results, observing how the removal affected the classification behavior of the k-NN classifier for different values of k and for different levels of noise.

Chapter 4
Inter-Class Boundaries: Linear and Polynomial Classifiers

When representing the training examples with points in an n-dimensional instance space, we may realize that positive examples tend to be clustered in regions different from those occupied by negative examples. This observation motivates yet another approach to classification. Instead of the probabilities and similarities employed by the earlier paradigms, we can try to identify the *decision surface* that separates the two classes. A very simple possibility is to use to this end a linear function. More flexible are high-order polynomials which are capable of defining very complicated inter-class boundaries. These, however, have to be handled with care.

The chapter introduces two simple mechanisms for induction of *linear classifiers* from examples described by boolean attributes, and then discusses how to use them in more general domains such as those with numeric attributes and more than just two classes. The whole idea is then extended to *polynomial classifiers*.

4.1   The Essence

To begin with, let us constrain ourselves to boolean domains where each attribute is either *true* or *false*. To make it possible for these attributes to participate in algebraic functions, we will represent them by integers: *true* by 1, and *false* by 0.

The Linear Classifier   In Fig. 4.1, one example is labeled as positive and the remaining three as negative. In this particular case, the two classes can be separated by the linear function defined by the following equation:

$$-1.2 + 0.5x_1 + x_2 = 0 \tag{4.1}$$

In the expression on the left-hand side, x_1 and x_2 represent attributes. If we substitute for x_1 and x_2 the concrete values of a given example, the expression, $-1.2 + 0.5x_1 + x_2$, will become either positive or negative. This sign then determines

© Springer International Publishing AG 2017
M. Kubat, *An Introduction to Machine Learning*,
DOI 10.1007/978-3-319-63913-0_4

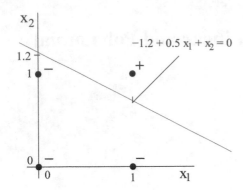

x_1	x_2	$-1.2+0.5x_1 + x_2$	Class
1	1	0.3	pos
1	0	−0.7	neg
0	1	−0.2	neg
0	0	−1.2	neg

Fig. 4.1 A linear classifier in a domain with two classes and two boolean attributes (using 1 instead of *true* and 0 instead of *false*)

the example's class. The table on the right shows how the four examples from the left have thus been classified.

Equation (4.1) is not the only one capable of doing the job. Other expressions, say, $-1.5 + x_1 + x_2$, will label the four examples in exactly the same way. As a matter of fact, the same can be accomplished by infinitely many different classifiers of the following generic form:

$$w_0 + w_1x_1 + w_2x_2 = 0$$

This is easy to generalize to domains with n attributes:

$$w_0 + w_1x_1 + \ldots + w_nx_n = 0 \tag{4.2}$$

If $n = 2$, Eq. (4.2) defines a line; if $n = 3$, a plane; and if $n > 3$, a hyperplane. By the way, if we artificially introduce a "zeroth" attribute, x_0, whose value is always fixed at $x_0 = 1$, the equation can be re-written in the following, more compact, form:

$$\sum_{i=0}^{n} w_ix_i = 0 \tag{4.3}$$

Some Practical Issues When writing a computer program implementing the classifier, the engineer must not forget to decide how to label the rare example that finds itself exactly on the hyperplane—which happens when the expression equals 0. Common practice either chooses the class randomly or gives preference to the one that has more representatives in the training set.

Also, we must not forget that no linear classifier can ever separate the positive and the negative examples if the two classes are *not* linearly separable. Thus if we change in Fig. 4.1 the class label of $\mathbf{x} = (x_1, x_2) = (0, 0)$ from minus to plus, no straight line will ever succeed. Let us defer further discussion of this issue till Sect. 4.5. For the time being, we will consider only domains where the classes *are* linearly separable.

The Parameters The classifier's behavior is determined by the coefficients, w_i. These are usually called *weights*. The task for machine learning is to find their appropriate values.

Not all the weights play the same role. Geometrically speaking, the coefficients $w_1, \ldots w_n$ define the angle of the hyperplane with respect to the system of coordinates; and the last, w_0, called *bias*, determines the *offset*, the hyperplane's distance from the origin of the system of coordinates.

The Bias and the Threshold In the case depicted in Fig. 4.1, the bias was $w_0 = -1.2$. A higher value would "shift" the classifier further from the origin, $[0, 0]$, whereas $w_0 = 0$ would make the classifier intersect the origin. Our intuitive grasp of the role played by bias in the classifier's behavior will improve if we re-write Eq. (4.2) as follows:

$$w_1 x_1 + \ldots w_n x_n = \theta \qquad (4.4)$$

The term on the right-hand side, $\theta = -w_0$, is the *threshold* that the weighted sum has to exceed if the example is to be deemed positive. Note that the threshold equals the negatively taken bias.

Simple Logical Functions Let us simplify our considerations by the (somewhat extreme) requirement that all attributes should have the same weight, $w_i = 1$. Even under this constraint, careful choice of the threshold will implement some useful functions. For instance, the reader will easily verify that the following classifier returns the positive class if and only if every single attribute has $x_i = 1$, a situation known as logical AND.

$$x_1 + \ldots + x_n = n - 0.5 \qquad (4.5)$$

By contrast, the next classifier returns the positive class if *at least one* attribute is $x_i = 1$, a situation known as logical OR.

$$x_1 + \ldots + x_n = 0.5 \qquad (4.6)$$

Finally, the classifier below returns the positive class if *at least k* attributes (out of the total of n attributes) are $x_i = 1$. This represents the so-called *k-of-n* function, of which AND and OR are special cases: AND is *n-of-n*, whereas OR is *1-of-n*.

$$x_1 + \ldots + x_n = k - 0.5 \qquad (4.7)$$

Weights Now that we understand the impact of bias, let us abandon the restriction that all weights be equal, and take a look at the consequences of their concrete values. Consider the linear classifier defined by the following function:

$$2 + 3x_1 - 2x_2 + 0.1x_4 - 0.5x_6 = 0 \tag{4.8}$$

The first thing to notice in the expression on the left side is the absence of attributes x_3 and x_5. These are rendered *irrelevant* by their zero weights, $w_3 = w_5 = 0$. As for the other attributes, their impacts depend on their weights' absolute values as well as on the signs: if $w_i > 0$, then $x_i = 1$ increases the chances of the above expression being positive; and if $w_i < 0$, then $x_i = 1$ increases the chances of the expression being negative. Note that, in the case of the classifier defined by Eq. (4.8), x_1 supports the positive class more strongly than x_4 because $w_1 > w_4$. Likewise, the influence of x_2 is stronger than that of x_6—only in the opposite direction: reducing the value of the overall sum, this weight makes it more likely that an example with $x_2 = 1$ will be deemed negative. Finally, the very small value of w_4 renders attribute x_4 almost irrelevant.

As another example, consider the classifier defined by the following function:

$$2x_1 + x_2 + x_3 = 1.5 \tag{4.9}$$

Here the threshold 1.5 is exceeded either by the sole presence of $x_1 = 1$ (because then $2x_1 = 2 \cdot 1 > 1.5$) or by the combined contributions of $x_2 = 1$ and $x_3 = 1$. This means that x_1 will "outvote" the other two attributes even when x_2 and x_3 join their forces in the support of the negative class.

Low Computational Costs Note the relatively low computational costs of this approach. Whereas the 1-NN classifier had to evaluate many geometric distances, and then search for the smallest among them, the linear classifier only has to determine the sign of a relatively simple expression.

What Have You Learned?

To make sure you understand this topic, try to answer the following questions. If you have problems, return to the corresponding place in the preceding text.

- Write the general expression defining the linear classifier in a domain with four boolean attributes. Why do we prefer to represent the *true* and *false* values by 1 and 0, respectively? How does the classifier determine an example's class?
- How can a linear classifier implement functions AND, OR, and *k-of-n*?
- Explain and discuss the behavior of the linear classifier defined by the expression, $-2.5 + x_2 + 2x_3 = 0$. What do the weights tell us about the role of the individual attributes?
- Compare the computational costs of the linear classifier with those of the nearest-neighbor classifier.

Fig. 4.2 The linear classifier outputs $h(\mathbf{x}) = 1$ when $\sum_{i=0}^{n} w_i x_i > 0$ and $h(\mathbf{x}) = 0$ when $\sum_{i=0}^{n} w_i x_i \leq 0$, signaling the example is positive or negative, respectively

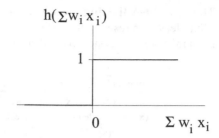

4.2 The Additive Rule: Perceptron Learning

Having developed some understanding of how the linear classifier works, we are ready to take a closer look at how to induce the tool from training data.

The Learning Task We will assume that each training example, $\mathbf{x}$, is described by n binary attributes whose values are either $x_i = 1$ or $x_i = 0$. A positive example is labeled with $c(\mathbf{x}) = 1$, and a negative with $c(\mathbf{x}) = 0$. To make sure we never confuse an example's real class with the one returned by the classifier, we will denote the latter by $h(\mathbf{x})$ where h indicates that this is the classifier's *hypothesis*. If $\sum_{i=0}^{n} w_i x_i > 0$, the classifier "hypothesizes" that the example is positive, and therefore returns $h(\mathbf{x}) = 1$. Conversely, if $\sum_{i=0}^{n} w_i x_i \leq 0$, the classifier returns $h(\mathbf{x}) = 0$. Figure 4.2 serves as a reminder that the classifier labels $\mathbf{x}$ as positive only if the cumulative evidence in favor of this class exceeds 0.

Finally, we will suppose that examples with $c(\mathbf{x}) = 1$ are linearly separable from those with $c(\mathbf{x}) = 0$. This means that there exists a linear classifier that will label correctly all training examples, $h(\mathbf{x}) = c(\mathbf{x})$. The task for machine learning is to find weights, w_i, that will make this happen.

Learning from Mistakes Here is the essence of the most common approach to the induction of linear classifiers. Suppose we already have an interim (even if imperfect) version of the classifier. When presented with a training example, $\mathbf{x}$, the classifier returns its label, $h(\mathbf{x})$. If this differs from the true class, $h(\mathbf{x}) \neq c(\mathbf{x})$, it is because the weights are less than perfect; they thus have to be modified in a way likely to correct this error.

Here is how to go about the weight modification. Let the true class be $c(\mathbf{x}) = 1$. In this event, $h(\mathbf{x}) = 0$ will only happen if $\sum_{i=0}^{n} w_i x_i < 0$, an indication that the weights are too small. If we *increase* the weights, then the sum, $\sum_{i=0}^{n} w_i x_i$, may exceed zero, making the returned label positive, and thus correct. Note that it is enough to increase only the weights of attributes with $x_i = 1$; when $x_i = 0$, then the value of w_i does not matter because anything multiplied by zero is still zero: $0 \cdot w_i = 0$.

Similarly, if $c(\mathbf{x}) = 0$ and $h(\mathbf{x}) = 1$, then the weights of all attributes such that $x_i = 1$ should be *decreased* so as to give the sum the chance to drop below zero, $\sum_{i=0}^{n} w_i x_i < 0$.

The Weight-Adjusting Formula In summary, the presentation of a training example, **x**, can result in three different situations. The technique based on "learning from mistakes" responds to them according to the following table:

Situation	Action
$c(\mathbf{x}) = 1$ while $h(\mathbf{x}) = 0$	Increase w_i for each attribute with $x_i = 1$
$c(\mathbf{x}) = 0$ while $h(\mathbf{x}) = 1$	Decrease w_i for each attribute with $x_i = 1$
$c(\mathbf{x}) = h(\mathbf{x})$	Do nothing

Interestingly, all these actions are carried out by a single formula:

$$w_i = w_i + \eta \cdot [c(\mathbf{x}) - h(\mathbf{x})] \cdot x_i \tag{4.10}$$

Let us take a look at the basic aspects of this weight-adjusting formula.

1. *Correct action.* If $c(\mathbf{x}) = h(\mathbf{x})$, the term in the brackets is $[c(\mathbf{x}) - h(\mathbf{x})] = 0$, which leaves w_i unchanged. If $c(\mathbf{x}) = 1$ and $h(\mathbf{x}) = 0$, the term in the brackets is 1, and the weights are thus increased. And if $c(\mathbf{x}) = 0$ and $h(\mathbf{x}) = 1$, the term in the brackets is negative, and the weights are reduced.
2. *Affecting only relevant weights.* If $x_i = 0$, the term to be added to the i-th weight, $\eta \cdot [c(\mathbf{x}) - h(\mathbf{x})] \cdot 0$, is zero. This means that the formula will affect w_i only when $x_i = 1$.
3. *Amount of change.* This is controlled by the *learning rate*, η, whose user-set value is chosen from the interval $\eta \in (0, 1]$.

Note that the modification of the weights is *additive* in the sense that a term is added to the previous value of the weight. In Sect. 4.3, we will discuss the other possibility: a *multiplicative* formula.

The Perceptron Learning Algorithm Equation (4.10) forms the core of the *Perceptron Learning Algorithm.*[1] The procedure is summarized by the pseudocode in Table 4.1. The principle is simple. Once the weights have been initialized to small random values, the training examples are presented, one at a time. After each example presentation, every weight of the classifier is subjected to Eq. (4.10). The last training example signals that one *training epoch* has been completed. Unless the classifier now labels correctly the entire training set, the learner returns to the first example, thus beginning the second epoch, then the third, and so on. Typically, several epochs are needed to reach the goal.

A Numeric Example Table 4.2 illustrates the procedure on a toy domain where the three training examples, $\mathbf{e}_1$, $\mathbf{e}_2$, and $\mathbf{e}_3$, are described by two binary attributes. After the presentation of $\mathbf{e}_1$, the weights (originally random) are reduced on account

---

[1]Its author, M. Rosenblatt, originally employed this learning technique in a device he called a *Perceptron*.

Table 4.1 The *perceptron learning* algorithm

Assumption: the two classes, $c(\mathbf{x}) = 1$ and $c(\mathbf{x}) = 0$, are linearly separable.

1. Initialize all weights, w_i, to small random numbers.
   Choose an appropriate learning rate, $\eta \in (0, 1]$.
2. For each training example, $\mathbf{x} = (x_1, \ldots, x_n)$, whose class is $c(\mathbf{x})$:

    (i)   Let $h(\mathbf{x}) = 1$ if $\sum_{i=0}^{n} w_i x_i > 0$, and $h(\mathbf{x}) = 0$ otherwise.
    (ii) Update each weight using the formula, $w_i = w_i + \eta[c(\mathbf{x}) - h(\mathbf{x})] \cdot x_i$

3. If $c(\mathbf{x}) = h(\mathbf{x})$ for all training examples, stop; otherwise, return to step 2.

Table 4.2 Illustration of *perceptron learning*

Let the learning rate be $\eta = 0.5$, and let the (randomly generated) initial weights be $w_0 = 0.1, w_1 = 0.3$, and $w_3 = 0.4$. Set $x_0 = 1$.

Task: Using the following training set, the perceptron learning algorithm is to learn how to separate the negative examples, e_1 and e_3, from the positive example, e_2.

Example	x_1	x_2	$c(\mathbf{x})$
e_1	1	0	0
e_2	1	1	1
e_3	0	0	0

The linear classifier's hypothesis about $\mathbf{x}$'s class is $h(\mathbf{x}) = 1$ if $\sum_{i=0}^{n} w_i x_i > 0$ and $h(\mathbf{x}) = 0$ otherwise. After each example presentation, all weights are subjected to the same formula: $w_i = w_i + 0.5 \cdot [c(\mathbf{x}) - h(\mathbf{x})] \cdot x_i$.

The table below shows, step by step, what happens to the weights in the course of learning.

	x_1	x_2	w_0	w_1	w_2	$h(\mathbf{x})$	$c(\mathbf{x})$	$c(\mathbf{x}) - h(\mathbf{x})$
Random classifier			0.1	0.3	0.4			
Example e_1	1	0				1	0	−1
New classifier			−0.4	−0.2	0.4			
Example e_2	1	1				0	1	1
New classifier			0.1	0.3	0.9			
Example e_3	0	0				1	0	−1
Final classifier			−0.4	0.3	0.9			

The final version of the classifier, $-0.4 + 0.3x_1 + 0.9x_2 = 0$, no longer misclassifies any training example. The training has thus been completed in a single epoch.

of $h(\mathbf{e_1}) = 1$ and $c(\mathbf{e_1}) = 0$; however, this happens only to w_0 and w_1 because $x_2 = 0$. In response to $\mathbf{e_2}$, all weights of the classifier's new version are increased because $h(\mathbf{e_2}) = 0$ and $c(\mathbf{e_2}) = 1$, and all attributes have $x_i = 1$. And after $\mathbf{e_3}$, the fact that $h(\mathbf{e_3}) = 1$ and $c(\mathbf{e_1}) = 0$ results in the reduction of w_0, but not of the other weights because $x_1 = x_2 = 0$. From now on, the classifier correctly labels all training examples, and the process can thus be terminated.

Initial Weights and the Number of Attributes The training converged to the separating line in a single epoch, but this was only thanks to a few lucky choices that may have played quite a critical role. Let us discuss them briefly.

First of them is the set of (random) *initial weights*. Different initialization may result in a different number of epochs. Most of the time, the classifier's initial version will be almost useless, and a lot of training is needed before the process converges to something useful. Sometimes, however, the first version may be fairly good, and a single epoch will do the trick. And at the extreme case, there exists the possibility, however remote, that the random-number generator will create a classifier that labels all training examples without a single error, and no training is thus needed.

Another factor is the *length of the attribute vector*. As a rule of thumb, the number of the necessary epochs tends to grow linearly in the number of attributes (assuming the same learning rate, η, is used). For instance, the number of epochs needed in a domain with $3 \times n$ attributes is likely to be about three times the number of epochs that would be needed in a domain with n attributes.

Learning Rate A critical role is played by the *learning rate*, η. Returning to the example from Table 4.2, the reader will note the rapid weight changes. Thus w_0 "jumped" from 0.1 to -0.4 after $\mathbf{e_1}$, then back to 0.1 after $\mathbf{e_2}$, only to return to -0.4 after $\mathbf{e_3}$. Similar changes were experienced by w_1 and w_2. Figure 4.3 visualizes the phenomenon. The reader will easily verify that the four lines represent the four successive versions of the classifier. Note how dramatic, for instance, is the change from classifier 1 to classifier 2, and then from classifier 2 to classifier 3.

Fig. 4.3 The four classifiers from Table 4.2. The classifier defined by the initial weights is denoted by 1; numbers 2 and 3 represent the two intermediate stages; and 4, the final solution. The *arrows* indicate the half-space of positive examples

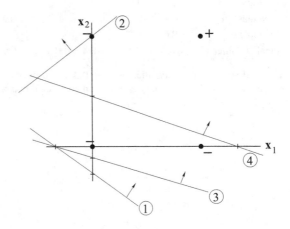

This remarkable sensitivity is explained by the high learning rate, $\eta = 0.5$. A smaller value, such as $\eta = 0.1$, would moderate the changes, thus "smoothing out" the learning. But if we overdo it by choosing, say, $\eta = 0.001$, the training process will become way too slow, and a great many epochs will have to pass before all training examples are correctly classified.

If the Solution Exists, It Will Be Found Whatever the initial weights, whatever the number of attributes, and whatever the learning rate, one thing is always guaranteed. If the positive and negative classes are linearly separable, the perceptron learning algorithm is guaranteed to find one version of the class-separating hyperplane in a finite number of steps.

What Have You Learned?

To make sure you understand this topic, try to answer the following questions. If you have problems, return to the corresponding place in the preceding text.

- Under what circumstances is *perceptron learning* guaranteed to find a classifier that perfectly labels all training examples?
- When does the algorithm reduce the classifier's weights, when does it increase them, and when does it leave them unchanged? Why does it modify w_i only if the corresponding attribute's value is $x_i = 1$?
- What circumstances influence the number of epochs needed by the algorithm to converge to a solution?

4.3 The Multiplicative Rule: WINNOW

Perceptron learning responds to the classifier's error by applying the *additive rule* that added to the weights a positive or negative term. An obvious alternative is the *multiplicative rule* where the weights are multiplied instead of being added to. Such approach has been adopted by WINNOW, an algorithm summarized by the pseudocode in Table 4.3.

The Principle and the Formula The general scenario is the same as in the previous section. A training example, $\mathbf{x}$, is presented, and the classifier returns its hypothesis about the example's label, $h(\mathbf{x})$. The learner compares this hypothesis with the known label, $c(\mathbf{x})$. If the two differ, $c(\mathbf{x}) \neq h(\mathbf{x})$, the weights of the attributes with $x_i = 1$ are modified in the following manner (where $\alpha > 1$ is a user-set parameter):

Situation	Action
$c(\mathbf{x}) = 1$ while $h(\mathbf{x}) = 0$	$w_i = \alpha w_i$
$c(\mathbf{x}) = 0$ while $h(\mathbf{x}) = 1$	$w_i = w_i/\alpha$
$c(\mathbf{x}) = h(\mathbf{x})$	Do nothing

Table 4.3 The WINNOW algorithm

Assumption: the two classes, $c(\mathbf{x}) = 1$ and $c(\mathbf{x}) = 0$, are linearly separable.

1. Initialize the classifier's weights to $w_i = 1$.
2. Set the threshold to $\theta = n - 0.1$ (n being the number of attributes) and choose an appropriate value for parameter $\alpha > 1$ (usually $\alpha = 2$).
3. Present a training example, $\mathbf{x}$, whose class is $c(\mathbf{x})$. The classifier returns $h(\mathbf{x})$.
4. If $c(\mathbf{x}) \neq h(\mathbf{x})$, update the weights of each attribute whose value is $x_i = 1$:

   if $c(\mathbf{x}) = 1$ and $h(\mathbf{x}) = 0$, then $w_i = \alpha w_i$
   if $c(\mathbf{x}) = 0$ and $h(\mathbf{x}) = 1$, then $w_i = w_i/\alpha$

5. If $c(\mathbf{x}) = h(\mathbf{x})$ for all training examples, stop; otherwise, return to step 3.

The reader is encouraged to verify that all these three actions can be carried out by the same formula:

$$w_i = w_i \cdot \alpha^{c(\mathbf{x})-h(\mathbf{x})} \tag{4.11}$$

A Numeric Example Table 4.4 illustrates the principle using a simple toy domain. The training set consists of all possible examples that can be described by three binary attributes. Those with $x_2 = x_3 = 1$ are labeled as positive and all others as negative, regardless of the value of the (irrelevant) attribute x_1.

In *perceptron learning*, the weights were initialized to small random values. In the case of WINNOW, however, they are all initially set to 1. As for the threshold, $\theta = n - 0.1$ is used, slightly less than the number of attributes. In the toy domain from Table 4.4, this means $\theta = 3 - 0.1 = 2.9$ because WINNOW of course has no a priori knowledge of one of the attributes being irrelevant.

When the first four examples are presented, the classifier's initial version labels them all correctly. The first mistake is made in the case of $\mathbf{e}_5$: for this positive example, the classifier incorrectly returns the negative label. The learner therefore increases the weights of attributes with $x_i = 1$ (that is, w_2 and w_3). This new classifier then classifies correctly all the remaining examples, $\mathbf{e}_6$ through $\mathbf{e}_8$. In the second epoch, the classifier errs on $\mathbf{e}_2$, causing a false positive. In response to this error, the algorithm reduces weights w_1 and w_2 (but not w_3 because $x_3 = 0$). After this last weight modification, the classifier labels correctly the entire training set.[2]

---

[2]Similarly as in the case of *perceptron learning*, we could have considered the 0-th attribute, $x_0 = 1$, whose initial weight is $w_0 = 1$.

Table 4.4 Illustration of the WINNOW's behavior

Task. Using the training examples from the table on the left (below), induce the linear classifier. Let $\alpha = 2$, and let $\theta = 2.9$.

Note that the training is here accomplished in two learning steps: presentation of e_5 (false negative), and of e_2 (false positive). After these two weight modifications, the resulting classifier correctly classifies all training examples.

	x_1	x_2	x_3	$c(\mathbf{x})$
e_1	1	1	1	1
e_2	1	1	0	0
e_3	1	0	1	0
e_4	1	0	0	0
e_5	0	1	1	1
e_6	0	1	0	0
e_7	0	0	1	0
e_8	0	0	0	0

	x_1	x_2	x_3	w_1	w_2	w_3	$h(\mathbf{x})$	$c(\mathbf{x})$
Init. class.				1	1	1		
Example e_5	0	1	1				0	1
New weights				1	2	2		
Example e_2	1	1	0				1	0
New weights				0.5	1	2		

Note that the weight of the irrelevant attribute, x_1, is now smaller than the weights of the relevant attributes. Indeed, the ability to penalize irrelevant attributes by significantly reducing their weights, thus "winnowing them out," is one of the main advantages of this technique.

The "Alpha" Parameter Parameter α controls the learner's sensitivity to errors in a manner reminiscent of the learning rate in perceptron learning. The main difference is the requirement that $\alpha > 1$. This guarantees an increase in weight w_i in the case of a false negative, and a decrease in w_i in the case of a false positive. The parameter's concrete value is not completely arbitrary. If it exceeds 1 by just a little (say, if $\alpha = 1.1$), then the weight-updates will be very small, resulting in slow convergence. Increasing α's value accelerates convergence, but risks overshooting the solution. The ideal value is best established experimentally; good results are often achieved with $\alpha = 2$.

No Negative Weights? Let us point out one fundamental difference between WINNOW and perceptron learning. Since the (originally positive) weights are always multiplied by α or $1/\alpha$, none of them can ever drop to zero, let alone become negative. This means that, unless appropriate measures have been taken, a whole class of linear classifiers will thus be eliminated: those with negative or zero coefficients.

The shortcoming is removed if we represent each of the original attributes by a pair of "new" attributes: one copying the original attribute's value, the other having the opposite value. In a domain that originally had n attributes, the total number of attributes will then be $2n$, the value of the $(n + i)$th attribute, x_{n+i}, being the opposite of x_i. For instance, suppose that an example is described by the following three attribute values:

$$x_1 = 1, x_2 = 0, x_3 = 1$$

In the new representation, the same example will be described by six attributes:

$$x_1 = 1, x_2 = 0, x_3 = 1, x_4 = 0, x_5 = 1, x_6 = 0,$$

For these, WINNOW will have to find six weights, $w_1, \ldots, w_6$, or perhaps seven, if w_0 is used.

Comparing It with Perceptron In comparison with *perceptron learning*, WIN-NOW appears to converge faster in domains with irrelevant attributes whose weights are quickly reduced to small values. However, neither WINNOW nor *perceptron learning* is able to recognize (and eliminate) *redundant attributes*. In the event of two attributes always having the same value, $x_i = x_j$, the learning process will converge to the same weight for both, making them look equally important even though it is clear that only one of them is strictly needed.

What Have You Learned?

To make sure you understand this topic, try to answer the following questions. If you have problems, return to the corresponding place in the preceding text.

- What formula is used by the weight-updating mechanism in WINNOW? Why is the formula called *multiplicative*?
- What is the shortcoming of multiplying or dividing the weights by $\alpha > 1$? How is the situation remedied?
- Summarize the differences between WINNOW and *perceptron learning*.

4.4 Domains with More Than Two Classes

Having only two sides, a hyperplane may separate the positive examples from the negative—and that's it; when it comes to *multi-class domains*, the tool seems helpless. Or is it?

Groups of Binary Classifiers What is beyond the powers of an individual can be solved by a team. One approach that is sometimes employed, in this context, is illustrated in Fig. 4.4. The "team" consists of four binary classifiers, each specializing on one of the four classes, C_1 through C_4. Ideally, the presentation of an example from C_i results in the i-th classifier returning $h_i(\mathbf{x}) = 1$, and all the other classifiers returning $h_j(\mathbf{x}) = 0$, assuming, again, that each class can be linearly separated from the other classes.

Modifying the Training Data To exhibit this behavior, however, the individual classifiers need to be properly trained. Here, one can rely on the algorithms that have been described in the previous sections. The only additional "trick" is that the engineer needs to modify the training data accordingly.

Fig. 4.4 Converting a 4-class problem into four 2-class problems

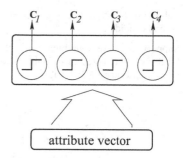

Table 4.5 A 4-class training set, T, converted to 4 binary training sets, $T_1 \ldots T_4$

T		T_1		T_2		T_3		T_4	
e_1	C_2	e_1	0	e_1	1	e_1	0	e_1	0
e_2	C_1	e_2	1	e_2	0	e_2	0	e_2	0
e_3	C_3	e_3	0	e_3	0	e_3	1	e_3	0
e_4	C_4	e_4	0	e_4	0	e_4	0	e_4	1
e_5	C_2	e_5	0	e_5	1	e_5	0	e_5	0
e_6	C_4	e_6	0	e_6	0	e_6	0	e_6	1

Table 4.5 illustrates the principle. On the left is the original training set, T, where each example is labeled with one of the four classes. On the right are four "derived" sets, T_1 through T_4, each consisting of the same six examples which, however, have been re-labeled so that an example that in the original set, T, represents class C_i is labeled with $c(\mathbf{x}) = 1$ in T_i and with $c(\mathbf{x}) = 0$ in all other sets.

The Need for a Master Classifier The training sets, T_i, are presented to a learner which induces from each of them a linear classifier dedicated to the corresponding class. This is not the end of the story, though. The training examples may poorly represent the classes, they may be corrupted by noise, and even the requirement of linear separability may not be satisfied. As a result of these complications, the induced classifiers may "overlap" each other in the sense that two or more of them will respond to the same example, $\mathbf{x}$, with $h_i(\mathbf{x}) = 1$, leaving the wrong impression that $\mathbf{x}$ belongs to more than one class. This is why a "master classifier" is needed, its task being to choose from the returned classes the one most likely to be correct.

This is not difficult. The reader will recall that a linear classifier labels $\mathbf{x}$ as positive if the weighted sum of $\mathbf{x}$'s attribute values exceeds zero: $\Sigma_{i=0}^{n} w_i x_i > 0$. This sum (usually different in each of the classifiers that have returned $h_i(\mathbf{x}) = 1$) can be interpreted as the amount of evidence in support of the corresponding class. The master classifier then simply gives preference to the class whose binary classifier delivered the highest $\Sigma_{i=0}^{n} w_i x_i$.

A Numeric Example The principle is illustrated in Table 4.6. Here, each of the rows represents a different class (with the total of four classes). Of course, each

Table 4.6 Illustration of the master classifier's behavior: choosing the example's class from several candidates

Suppose we have four binary classifiers (the i-th classifier used for the i-th class) defined by the weights listed in the table below. How shall the master classifier label example $\mathbf{x} = (x_1, x_2, x_3, x_4) = (1, 1, 1, 0)$?

| Class | Classifier | | | | | $\sum_{i=0}^{n} w_i x_i$ | $h(\mathbf{x})$ |
	w_0	w_1	w_2	w_3	w_4		
C_1	-1.5	1	0.5	-1	-5	-1	0
C_2	0.5	1.5	-1	3	1	4	1
C_3	1	-2	4	-1	0.5	2	1
C_4	-2	1	1	-3	-1	-3	0

The rightmost column tells us that two classifiers, C_2 and C_3, return $h(\mathbf{x}) = 1$. From these, C_2 is supported by the higher value of $\sum_{i=0}^{n} w_i x_i$. Therefore, the master classifier labels $\mathbf{x}$ with C_2.

classifier has a different set of weights, each weight represented by one column in the table. When an example is presented, its attribute-values are in each classifier multiplied by the corresponding weights. We observe that in the case of two classifiers, C_2 and C_3, the weighted sums are positive, $\sum_{i=0}^{n} w_i x_i > 0$, which might mean that both classifiers return $h(\mathbf{x}) = 1$. Since each example is supposed to be labeled with one and only one class, we need a master classifier to make a decision. In this particular case, the master classifier gives preference to C_2 because this classifier's weighted sum is greater than that of C_3.

A Practical Limitation A little disclaimer is in place here. This method of employing linear classifiers in multi-class domains is reliable only if the number of classes is moderate, say, 3–5. In domains with many classes, the "derived" training sets, T_i, will be imbalanced in the sense that most examples will have $c(\mathbf{x}) = 0$ and only a few $c(\mathbf{x}) = 1$. As we will learn in Sect. 10.2, imbalanced training sets tend to cause difficulties in noisy domains unless appropriate measures have been taken.

What Have You Learned?

To make sure you understand this topic, try to answer the following questions. If you have problems, return to the corresponding place in the preceding text.

- When trying to use N linear classifiers in an N-class domain, how will you create the training sets, T_i, for the induction of the individual binary classifiers?
- How can an example's class be decided upon in a situation where two or more binary classifiers return $h(\mathbf{x}) = 1$?

4.5 Polynomial Classifiers

It is now time to abandon the requirement that the positive examples be linearly separable from the negative; because fairly often, they are not. Not only can the linear separability be destroyed by noise. The very shape of the region occupied by one of the classes can make it impossible to separate it by a linear decision surface. Thus in the training set shown in Fig. 4.5, no linear classifier ever succeeds in separating the two classes, a feat that can only be accomplished by a non-linear curve such as the parabola shown in the picture.

Non-linear Classifiers The point having been made, we have to ask how to induce the more sophisticated *non-linear classifiers*. There is no doubt that they exist. For instance, math teaches us that *any* n-dimensional function can be approximated to arbitrary precision with a *polynomial* of a sufficiently high order. Let us therefore take a look at how to use—and induce—the polynomials for our classification purposes.

Polynomials of the Second Order The good news is that the coefficients of polynomials can be induced by the same techniques that we have used for linear classifiers. Let us explain how.

For the sake of clarity, we will begin by constraining ourselves to simple domains with only two boolean attributes, x_1 and x_2. The second-order polynomial function is then defined as follows:

$$w_0 + w_1 x_1 + w_2 x_2 + w_3 x_1^2 + w_4 x_1 x_2 + w_5 x_2^2 = 0 \qquad (4.12)$$

The expression on the left is a sum of terms that all have one thing in common: a weight, w_i, that multiplies a product $x_1^k x_2^l$. In the first term, we have $k + l = 0$, because $w_0 x_1^0 x_2^0 = w_0$; next come the terms with $k + l = 1$, concretely, $w_1 x_1^1 x_2^0 = w_1 x_1$ and $w_2 x_1^0 x_2^1 = w_1 x_2$; and the sequence ends with the three terms that have $k + l = 2$: specifically, $w_3 x_1^2$, $w_4 x_1^1 x_2^1$, and $w_5 x_2^2$. The thing to remember is that the expansion of the second-order polynomial stops when the sum of the exponents reaches 2.

Fig. 4.5 In some domains, no linear classifier can separate the positive examples from the negative. Only a *non-linear classifier* can do so

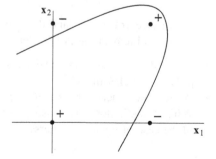

Of course, some of the weights can be $w_i = 0$, rendering the corresponding terms "invisible" such as in $7 + 2x_1x_2 + 3x_2^2$ where the coefficients of x_1, x_2, and x_1^2 are zero.

Polynomials of the r-th Order   More generally, the r-th order polynomial (still in a two-dimensional domain) will consist of a sum of weighted terms, $w_i x_1^k x_2^l$, such that $k + l = j$, where $j = 0, 1, \ldots r$.

The reader will easily make the next step and write down the general formula that defines the r-th order polynomial for domains with more than two attributes. A hint: the sum of the exponents in any single term never exceeds r.

Converting Them to Linear Classifiers   Whatever the polynomial's order, and whatever the number of attributes, the task for machine learning is to find weights that make it possible to separate the positive examples from the negative. The seemingly unfortunate circumstance that the terms are non-linear (the sum of the exponents sometimes exceeds 1) is easily removed by *multipliers*, devices that return the logical conjunction of inputs: 1 if *all* inputs are 1; and 0 if *at least one* input is 0. With their help, we can replace each product of attributes with a new attribute, z_i, and thus re-write Eq. (4.12) in the following way:

$$w_0 + w_1z_1 + w_2z_2 + w_3z_3 + w_4z_4 + w_5z_5 = 0 \tag{4.13}$$

This means, for instance, that $z_3 = x_1^2$ and $z_4 = x_1 \cdot x_2$. Note that this "trick" has transformed the originally non-linear problem with two attributes, x_1 and x_2, into a linear problem with five newly created attributes, z_1 through z_5.

Figure 4.6 illustrates the situation where a second-order polynomial is used in a domain with three attributes.

Since the values of z_i in each example are known, the weights can be obtained without any difficulties using *perceptron learning* or WINNOW. Of course, we must not forget that these techniques will find the solution only if the polynomial of the chosen order is indeed capable of separating the two classes.

What Have You Learned?

To make sure you understand this topic, try to answer the following questions. If you have problems, return to the corresponding place in the preceding text.

- When do we need non-linear classifiers? Specifically, what speaks in favor of polynomial classifiers?
- Write down the mathematical expression that defines a polynomial classifier. What "trick" allows us to use here the same learning techniques that were used in the case of linear classifiers?

The second-order polynomial function over three attributes is defined by the following function:

$$0 = w_0 + w_1 x_1 + w_2 x_2 + w_3 x_3 + w_4 x_1^2 + w_5 x_1 x_2 + w_6 x_1 x_3$$
$$+ w_7 x_2^2 + w_8 x_2 x_3 + w_9 x_3^2$$

Using the multipliers, we obtain the following:

$$0 = w_0 + w_1 z_1 + w_2 z_2 + w_3 z_3 + w_4 z_4 + w_5 z_5 + w_6 z_6 + w_7 z_7 + w_8 z_8 + w_9 z_9$$

Below is the schema of the whole "device" with multipliers. Before reaching the linear classifier, each signal z_i is multiplied by the corresponding weight, w_i.

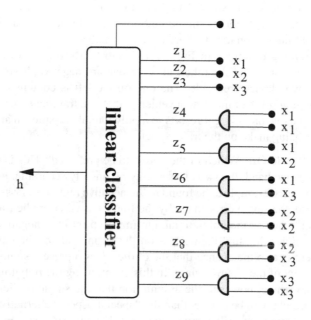

Fig. 4.6 A polynomial classifier can be converted into a linear classifier with the help of multipliers that pre-process the data

4.6 Specific Aspects of Polynomial Classifiers

To be able to use a machine-learning technique with success, the engineer must understand not only its strengths, but also its limitations, shortcomings, and pitfalls. In the case of polynomial classifiers, there are a few that deserve our attention. Let us briefly discuss them.

Fig. 4.7 The two classes are linearly separable, but noise has caused one negative example to be mislabeled as positive. The high-order polynomial on the *right overfits* the data, ignoring the possibility of noise

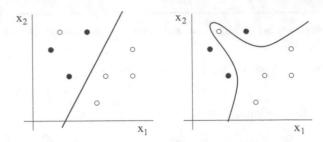

Overfitting Polynomial classifiers tend to *overfit* noisy training data. Since the problem of overfitting is encountered also in other machine-learning paradigms, we have to discuss its essence in some detail. For the sake of clarity, we will abandon the requirement that all attributes should be boolean; instead, we will rely on two-dimensional continuous domains that are easy to visualize.

The eight training examples in Fig. 4.7 fall into two groups. In one of them, all examples are positive; in the other, all save one are negative. Two attempts at separating the two classes are made. The one on the left is content with a linear classifier, shrugging off the minor inconvenience that one training example remains misclassified. The one on the right resorts to a polynomial classifier in an attempt to avoid any error being make on the training set.

An Inevitable Trade-Off Which of the two is to be preferred? This is not an easy question. On the one hand, the two classes may be linearly separable, and the only cause for one positive example to be found in the negative region is class-label noise. If this is the case, the single error made by the linear classifier on the training set is inconsequential, whereas the polynomial, cutting deep into the negative area, will misclassify examples that find themselves on the wrong side of the curve. On the other hand, there is also the chance that the outlier *does* represent some legitimate, even if rare, aspect of the positive class. In this event, using the polynomial will be justified. On the whole, however, the assumption that the single outlier is nothing but noise is more likely to be correct than the "special-aspect" alternative.

A realistic training set will contain not one, but quite a few, perhaps many examples that seem to find themselves in the wrong part of the instance space. And the inter-class boundary that our classifier attempts to model may indeed be curved, though *how much* curved is anybody's guess. The engineer may lay aside the linear classifier as too crude an approximation, and opt instead for the greater flexibility offered by the polynomials. This said, a very-high-order polynomial will avoid any error even in a very noisy training set—and then fail miserably on future data. The ideal solution is often somewhere between the extremes of linear classifiers and high-order polynomials. The best choice can be determined experimentally.

The Number of Weights The total number of the weights to be trained depends on the length of the attribute vector, and on the order of the polynomial. A simple analysis reveals that, in the case of n attributes and the r-th order polynomial, the number is determined by the following combinatorial expression:

$$N_W = \binom{n+r}{r} \tag{4.14}$$

Of course, N_W will be impractically high for large values of n. For instance, even for the relatively modest case of $n = 100$ attributes and a polynomial's order $r = 3$, the number of weights to be trained is $N_W = 176{,}851$ (103 choose 3). The computational costs thus incurred are not insurmountable for today's computers. What is worse is the danger of overfitting noisy training data; the polynomial is simply too flexible. The next paragraphs will tell us *how much* flexible.

Capacity The trivial domain in Fig. 4.1 consisted of four examples. Given that each of them can be labeled as either positive or negative, we have $2^4 = 16$ different ways of assigning labels to this training set. Of these sixteen, only two represent a situation where the two classes cannot be linearly separated—in this domain, linear inseparability is a rare event. But how typical is this situation in the more general case of m examples described by n attributes'? What are the chances that a random labeling of the examples will result in linearly separable classes?

Mathematics has found a simple guideline to be used in domains where n is "reasonably high" (say, ten or more attributes): if the number of examples, m, is less than twice the number of attributes ($m < 2n$), the probability that a random distribution of the two labels will result in linear separability is close to 100%. Conversely, this probability is almost zero when $m > 2n$. In this sense, "the *capacity* of a linear classifier is twice the number of attributes."

This result applies also to polynomial classifiers. The role of attributes is here played by the terms, z_i, obtained by the multipliers. Their number, N_W, is obtained by Eq. (4.14). We have seen that N_W can be quite high—and this makes the capacity high, too. In the case of $n = 100$ and $r = 3$, the number of weights is 176,851. This means that the third-order polynomial can separate the two classes (regardless of noise) as long as the number of examples is less than 353,702.

What Have You Learned?

To make sure you understand this topic, try to answer the following questions. If you have problems, return to the corresponding place in the preceding text.

- Elaborate on the term, "overfitting" and explain why this phenomenon (and its consequences) is in polynomial classifiers difficult to avoid.
- What is the upper bound on the number of weights to be trained in a polynomial of the r-th order in a domain that has n attributes?
- What is the *capacity* of the linear or polynomial classifier? What does capacity tell us about linear separability?

4.7 Numerical Domains and Support Vector Machines

Now that we have realized that polynomials do not call for new machine-learning algorithms, we can return to linear classifiers, a topic we have not yet exhausted. Time has come to abandon our self-imposed restriction to boolean attributes, and to start considering also the possibility of the attributes being continuous. Can we then still rely on the two training algorithms described above?

Perceptron in Numeric Domains In the case of *perceptron learning*, the answer is easy: yes, the same weight-modification formula can be used. Practical experience shows, however, that it is good to make all attribute values fall into the unit interval, $x_i \in [0, 1]$.

Let us repeat here, for the reader's convenience, the weight-adjusting formula:

$$w_i = w_i + \eta[c(\mathbf{x}) - h(\mathbf{x})]x_i \qquad (4.15)$$

While the learning rate, η, and the difference between the real and the hypothesized class label, $[c(\mathbf{x}) - h(\mathbf{x})]$, have the same meaning and impact as before, what has changed is the role of x_i. In the case of boolean attributes, the value of x_i decided whether or not the weight should change. Here, however, it rather says *how much* the weight should be affected: more in the case of higher attribute values.

The Multiplicative Rule In the case of WINNOW, too, essentially the same learning formula can be used as in the binary-attributes case:

$$w_i = w_i \alpha^{c(\mathbf{x}) - h(\mathbf{x})} \qquad (4.16)$$

This said, one has to be careful about *when* to apply the formula. Previously, one modified only the weights of attributes with $x_i = 1$. Now that the attribute values come from a continuous domain, some modification is needed. One possibility is the following rule:

"Update weight w_i only if the value of the i-th attribute is $x_i \geq 0.5$."

Let us remark that both of these algorithms (perceptron learning and WINNOW) usually find a relatively low-error solution even if the two classes are not linearly separable—for instance, in the presence of noise.

Which Linear Classifier Is Better? At this point, however, another important question needs to be discussed. Figure 4.8 shows three linear classifiers, each perfectly separating the positive training examples from the negative. Knowing that "good behavior" on the training set does not guarantee high performance in the future, we have to ask: which of the three is likely to score best on future examples?

The Support Vector Machine Mathematicians who studied this problem found an answer. When we take a look at Fig. 4.8, we can see that the dotted-line classifier all but touches the nearest examples on either side; we say that its *margin* is small.

Conversely, the margin is greater in the case of the solid-line classifier: the nearest positive example on one side of the line, and the nearest example on the other of the line are much farther than in the case of the other classifier. As it turns out, the greater the margin, the higher the chances that the classifier will do well on future data.

The technique of the *support vector machines* is illustrated in Fig. 4.9. The solid line is the best classifier. The graph shows also two thinner lines, parallel to the classifier, each at the same distance. The reader can see they pass through the examples nearest to the classifier. These examples are called *support vectors* (because, after all, each example is a vector of attributes).

The task for machine learning is to identify the support vectors that maximize the margin. The concrete mechanisms for finding the optimum set of support vectors exceed the ambition of an introductory text. The simplest technique would simply try all possible *n*-tuples of examples, and measure the margin implied by each such choice. This, however, is unrealistic in domains with many examples. Most of the time, therefore, engineers rely on some of the many software packages available for free on the internet.

What Have You Learned?

To make sure you understand this topic, try to answer the following questions. If you have problems, return to the corresponding place in the preceding text.

- Can *perceptron learning* and WINNOW be used in numeric domains? How?
- Given that there are infinitely many linear classifiers capable of separating the positive examples from the negative (assuming such separation exists), which of them can be expected to give the best results on future data?
- What is a *support vector*? What is meant by the *margin* to be maximized?

Fig. 4.8 Linearly separable classes can be separated in infinitely many different ways. Question is, which of the classifiers that are perfect on the training set will do best on future data

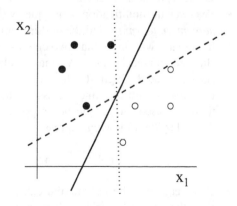

Fig. 4.9 The technique of the *support vector machine* looks for a separating hyperplane that has the maximum *margin*

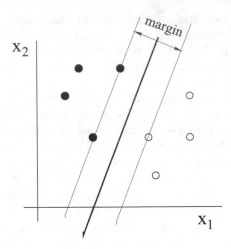

4.8   Summary and Historical Remarks

- Linear and polynomial classifiers define a *decision surface* that separates the positive examples from the negative. Specifically, *linear* classifiers label the examples according to the sign of the following expression:

$$w_0 + w_1 x_1 + \ldots w_n x_n$$

The concrete behavior is determined by the weights, w_i. The task for machine learning is to find appropriate values for these weights.

- The learning techniques from this chapter rely on the principle of "learning from mistakes." The training examples are presented, one by one, to the learner. Each time the learner misclassifies an example, the weights are modified. When the entire training set has been presented, one *epoch* has been completed. Usually, several epochs are needed.

- Two weight-modification techniques were considered here: the additive rule of *perceptron learning*, and the multiplicative rule of WINNOW.

- In domains with more than two classes, one can consider the use of a specialized classifier for each class. A "master classifier" then chooses the class whose classifier had the highest value of $\Sigma_{i=0}^{n} w_i x_i$.

- In domains with non-linear class boundaries, *polynomial* classifiers can sometimes be used. A second-order polynomial in a two-dimensional domain is defined by the following expression:

$$w_0 + w_1 x_1 + w_2 x_2 + w_3 x_1^2 + w_4 x_1 x_2 + w_5 x_2^2$$

- The weights of the polynomial can be found by the same learning algorithms as in the case of linear classifiers, provided that the non-linear terms (e.g., $x_1 x_2$)

have been replaced (with the help of *multipliers*) by newly created attributes such as $z_3 = x_1^2$ or $z_4 = x_1 x_2$.

- Polynomial classifiers are prone to *overfit* noisy training data. This is explained by the excessive flexibility caused by the very high number of trainable weights.
- The potentially best class-separating hyperplane (among the infinitely many candidates) is identified by the technique of the *support vector machines (SVM)* that seek to maximize the distance of the nearest positive and the nearest negative example from the hyperplane.

Historical Remarks The principle of *perceptron learning* was developed by Rosenblatt [81], whereas WINNOW was proposed and analyzed by Littlestone [54]. The question of the capacity of linear and polynomial classifiers was analyzed by Cover [18]. The principle of Support Vector Machines was invented by Vapnik [93] as one of the consequences of the Computational Learning Theory which will be the subject of Chap. 7.

4.9 Solidify Your Knowledge

The exercises are to solidify the acquired knowledge. The suggested thought experiments will help the reader see this chapter's ideas in a different light and provoke independent thinking. Computer assignments will force the readers to pay attention to seemingly insignificant details they might otherwise overlook.

Exercises

1. Write down equations for linear classifiers to implement the following functions:

   - At least two out of the boolean attributes $x_1, \ldots, x_5$ are *true*
   - At least three out of the boolean attributes $x_1, \ldots, x_6$ are *true*, and at least one of them is *false*.

2. Return to the examples from Table 4.2. Hand-simulate the perceptron learning algorithm's procedure, starting from a different initial set of weights than the one used in the table. Try also a different learning rate.
3. Repeat the same exercise, this time using WINNOW. Do not forget to introduce the additional "attributes" for what in perceptrons were the negative weights.
4. Write down the equation that defines a third-order polynomial in two dimensions. How many multipliers (each with up to three inputs) would be needed if we wanted to train the weights using the perceptron learning algorithm?

Give It Some Thought

1. How can induction of linear classifiers be used to identify irrelevant attributes? Hint: try to run the learning algorithm on different subsets of the attributes, and then observe the error rate achieved after a fixed number of epochs.
2. Explain in what way it is true that the 1-NN classifier applied to a pair of examples (one positive, the other negative) in a plane defines a linear classifier. Invent a machine learning algorithm that uses this observation for the induction of linear classifiers. Generalize the procedure to n-dimensional domains.
3. When is a linear classifier likely to lead to better classification performance on independent testing examples than a polynomial classifier?
4. Sometimes, a linearly separable domain becomes linearly non-separable on account of the class-label noise. Think of a technique capable of removing such noisy examples. Hint: you may rely on an idea we have already encountered in the field of k-NN classifiers.

Computer Assignments

1. Implement the perceptron learning algorithm and run it on the following training set where six examples (three positive and three negative) are described by four attributes:

x_1	x_2	x_3	x_4	Class
1	1	1	0	pos
0	0	0	0	pos
1	1	0	1	pos
1	1	0	0	neg
0	1	0	1	neg
0	0	0	1	neg

Observe that the linear classifier fails to reach zero error rate because the two classes are not linearly separable.

2. Create a training set consisting of 20 examples described by five binary attributes, $x_1, \ldots, x_5$. Examples in which at least three attributes have values $x_i = 1$ are labeled as positive, all other examples are labeled as negative. Using this training set as input, induce a linear classifier using perceptron learning. Experiment with different values of the learning rate, η. Plot a function where the horizontal axis represents η, and the vertical axis represents the number of example-presentations needed for the classifier to correctly classify all training examples. Discuss the results.

3. Use the same domain as in the previous assignment (five boolean attributes, and the same definition of the positive class). Add to each example N additional boolean attributes whose values are determined by a random-number generator. Vary N from 1 to 20. Observe how the number of example-presentations needed to achieve the zero error rate depends on N.

4. Again, use the same domain, but add attribute noise by changing the values of randomly selected examples (while leaving class labels unchanged). Observe what minimum error rate can then be achieved.

5. Repeat the last three assignments for different sizes of the training set, evaluating the results on (always the same) testing set of examples that have not been seen during learning.

6. Repeat the last four assignments, using WINNOW instead of the perceptron learning. Compare the results in terms of the incurred computational costs. These costs can be measured by the number of epochs needed to converge to the zero error rate on the training set.

7. Define a domain with three numeric attributes with values from the unit interval, $[0, 1]$. Generate 100 training examples, labeling as positive those for which the expression $1 - x_1 + x2 + x3$ is positive. Use the "perceptron learning algorithm" modified so that the following versions of the weight-updating rule are used:

   (a) $w_i = w_i + \eta[c(\mathbf{x}) - h(\mathbf{x})]x_i$
   (b) $w_i = w_i + \eta[c(\mathbf{x}) - h(\mathbf{x})]$
   (c) $w_i = w_i + \eta[c(\mathbf{x}) - \sum w_i x_i]x_i$
   (d) $w_i = w_i + \eta[c(\mathbf{x}) - \sum w_i x_i]$

   Will all of them converge to zero error rate on the training set? Compare the speed of conversion.

8. Create a training set where each example is described by six boolean attributes, $x_i, \ldots, x_6$. Label each example with one of the four classes defined as follows:

   (a) C_1: at least five attributes have $x_i = 1$.
   (b) C_2: three or four attributes have $x_i = 1$.
   (c) C_3: two attributes have $x_i = 1$.
   (d) C_4: one attribute has $x_i = 1$.

   Use perceptron learning, applied in parallel to each of the four classes.
   As a variation, use different numbers of irrelevant attributes, varying their number from 0 to 20. See if the zero error rate on the training set can be achieved.
   Record the number of false positives and the number of false negatives observed on an independent testing set.
   Design an experiment showing that the performance of K binary classifiers, connected in parallel as in Fig. 4.4, will decrease if we increase the number of classes. How much is this observation pronounced in the presence of noise?

9. Run induction of linear classifiers on selected boolean domains from the UCI repository[3] and compare the results.
10. Experimenting with selected domains from the UCI repository, observe the impact of the learning rate, η, on the convergence speed of the perceptron learning algorithm.
11. Compare the behavior of linear and polynomial classifiers. Observe how the former wins in simple domains, and the latter in highly non-linear domains.

---

[3]www.ics.uci.edu/~mlearn/MLRepository.html.

Chapter 5
Artificial Neural Networks

Polynomial classifiers can model decision surfaces of any shape; and yet their practical utility is limited because of the easiness with which they overfit noisy training data, and because of the sometimes impractically high number of trainable parameters. Much more popular are *artificial neural networks* where many simple units, called *neurons*, are interconnected by weighted links into larger structures of remarkably high performance.

The field of neural networks is too rich to be covered in the space we have at our disposal. We will therefore provide only the basic information about two popular types: multilayer perceptrons and radial-basis function networks. The chapter describes how each of them classifies examples, and then describes some elementary mechanisms to induce them from training data.

5.1 Multilayer Perceptrons as Classifiers

Throughout this chapter, we will suppose that all attributes are continuous. Moreover, it is practical (though not strictly necessary) to assume that they have been normalized so that their values always fall in the interval $[-1, 1]$.

Neurons The function of a *neuron*, the basic unit of a multilayer perceptron, is quite simple. A weighted sum of signals arriving at the input is subjected to a *transfer function*. Several different transfer functions can be used; the one that is preferred in this chapter is the so-called *sigmoid*, defined by the following formula where Σ is the weighted sum of inputs:

$$f(\Sigma) = \frac{1}{1 + e^{-\Sigma}} \tag{5.1}$$

© Springer International Publishing AG 2017
M. Kubat, *An Introduction to Machine Learning*,
DOI 10.1007/978-3-319-63913-0_5

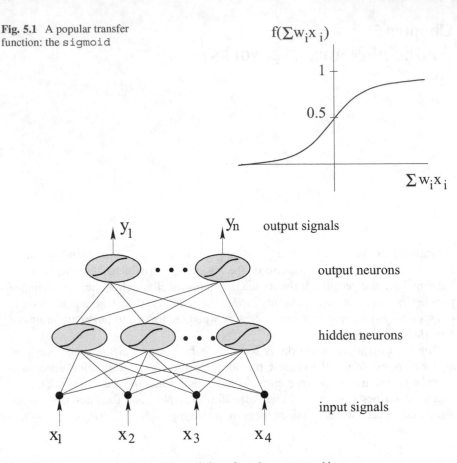

Fig. 5.1 A popular transfer function: the `sigmoid`

Fig. 5.2 An example neural network consisting of two interconnected layers

Figure 5.1 shows the curve representing the transfer function. The reader can see that $f(\Sigma)$ grows monotonically with the increasing value of Σ but is destined never to leave the open interval $(0, 1)$ because $f(-\infty) = 0$, and $f(\infty) = 1$. The vertical axis is intersected at f(0)=0.5. We will assume that each neuron has the same transfer function.

Multilayer Perceptron The neural network in Fig. 5.2 is known as the *multilayer perceptron*. The neurons, represented by ovals, are arranged in the *output layer* and the *hidden layer*.[1] For simplicity, we will consider only networks with a single hidden layer while remembering that it is quite common to employ two such layers, even three, though rarely more than that. Another simplification is that we have omitted the zero-th weights (providing for each neuron its trainable bias) so as not to clutter the picture.

---

[1]When we view the network from above, the hidden layer is obscured by the output layer.

While there is no communication between neurons of the same layer, adjacent layers are fully interconnected. Importantly, each neuron-to-neuron link is associated with a *weight*. The weight of the link from the j-th hidden neuron to the i-th output neuron is denoted as $w_{ji}^{(1)}$, and the weight of the link from the k-th attribute to the j-th hidden neuron as $w_{kj}^{(2)}$. Note that the first index always refers to the link's "beginning"; the second, to its "end."

Forward Propagation When we present the network with an example, $\mathbf{x} = (x_1, \ldots, x_n)$, its attribute values are passed along the links to the neurons. The values x_k being multiplied by the weights associated with the links, the j-th hidden neuron receives as input the weighted sum, $\sum_k w_{kj}^{(2)} x_k$, and subjects this sum to the sigmoid, $f(\sum_k w_{kj}^{(2)} x_k)$. The i-th output neuron then receives the weighted sum of the values coming from the hidden neurons and, again, subjects it to the transfer function. This is how the i-th output is obtained. The process of propagating in this manner the attribute values from the network's input to its output is called *forward propagation*.

Each class is assigned its own output neuron, and the value returned by the i-th output neuron is interpreted as the amount of evidence in support of the i-th class. For instance, if the values obtained at three output neurons are $\mathbf{y} = (0.2, 0.6, 0.1)$, the classifier will label the given example with the second class because 0.6 is greater than both 0.2 and 0.1.

In essence, this kind of two-layer perceptron calculates the following formula where f is the sigmoid transfer function (see Eq. (5.1)) employed by the neurons; $w_{kj}^{(2)}$ and $w_{ji}^{(1)}$ are the links leading to the hidden and output layers, respectively, and x_k are the attribute values of the presented example:

$$y_i = f\left(\sum_j w_{ji}^{(1)} f\left(\sum_k w_{kj}^{(2)} x_k\right)\right) \tag{5.2}$$

A Numeric Example The principle of forward propagation is illustrated by the numeric example in Table 5.1. At the beginning, the attribute vector $\mathbf{x}$ is presented. Before reaching the neurons in the hidden layer, the attribute values are multiplied by the corresponding weights, and the weighted sums are passed on to the sigmoid functions. The results ($h_1 = 0.32$ and $h_2 = 0.54$) are then multiplied by the next layer of weights, and forwarded to the output neurons where they are again subjected to the sigmoid function. This is how the two output values, $y_1 = 0.66$ and $y_2 = 0.45$, have been obtained. The evidence supporting the class of the "left" output neuron is higher than the evidence supporting the class of the "right" output neuron. The classifier therefore chooses the left neuron's class.

Universal Classifier Mathematicians have been able to prove that, with the right choice of weights, and with the right number of the hidden neurons, Eq. (5.2) can approximate with arbitrary accuracy any realistic function. The consequence of this so-called *universality theorem* is that the *multilayer perceptron* can in principle be used to address just about any classification problem. What the theorem *does not*

Table 5.1 Example of forward propagation in a *multilayer perceptron*

Task. Forward-propagate $\mathbf{x} = (x_1, x_2) = (0.8, 0.1)$ through the network below.

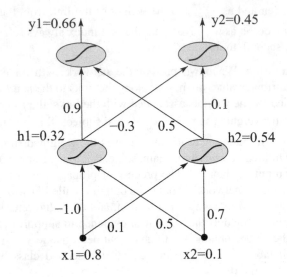

Solution.

inputs of hidden-layer neurons:

$$z_1^{(2)} = 0.8 \times (-1.0) + 0.1 \times 0.5 = -0.75$$
$$z_2^{(2)} = 0.8 \times 0.1 + 0.1 \times 0.7 = 0.15$$

outputs of hidden-layer neurons:

$$h_1 = f(z_1^{(2)}) = \tfrac{1}{1+e^{-(-0.75)}} = 0.32$$
$$h_2 = f(z_2^{(2)}) = \tfrac{1}{1+e^{-0.15}} = 0.54$$

inputs of output-layer neurons:

$$z_1^{(1)} = 0.32 \times 0.9 + 0.54 \times 0.5 = 0.56$$
$$z_2^{(1)} = 0.32 \times (-0.3) + 0.54 \times (-0.1) = -0.15$$

outputs of output-layer neurons:

$$y_1 = f(z_1^{(1)}) = \tfrac{1}{1+e^{-0.56}} = 0.66$$
$$y_2 = f(z_2^{(1)}) = \tfrac{1}{1+e^{-(-0.15)}} = 0.45$$

tell us, though, is how many hidden neurons are needed, and what the individual weight values should be. In other words, we know that the solution exists, yet there is no guarantee we will ever find it.

What Have You Learned?

To make sure you understand this topic, try to answer the following questions. If you have problems, return to the corresponding place in the preceding text.

- Explain how the example described by a vector of continuous attributes is forward-propagated through the *multilayer perceptron*. How is the network's output interpreted?
- What is the *transfer function*? Write down the formula defining the *sigmoid* transfer function, and describe its shape.
- What is the *universality theorem*? What does it tell us, and what does it *not* tell us?

5.2 Neural Network's Error

Let us defer to a later chapter the explanation of a technique to train the *multilayer perceptron* (to find its weights). Before we can address this issue, it is necessary to prepare the soil by taking a closer look at the method of classification, and at the way the accuracy of this classification is evaluated.

Error Rate Suppose a training example, $\mathbf{x}$, with known class, $c(\mathbf{x})$, has been presented to an existing version of the *multilayer perceptron*. The forward propagation step establishes the label, $h(\mathbf{x})$. If $h(\mathbf{x}) \neq c(\mathbf{x})$, an error has been made. This may happen to some other examples, too, and we want to know *how often* this happens. We want to know the *error rate*—which is for a given set of examples obtained by dividing the number of errors by the number of examples. For instance, if the classifier misclassifies 30 out of 200 examples, the error rate is $30/200 = 0.15$.

This, however, fails to give the full picture of the network's classification performance. What the error rate neglects to reflect is the sigmoid function's ability to measure the *size* of each error.

An example will clarify the point. Suppose we have two different networks to choose from, each with three output neurons corresponding to classes denoted by C_1, C_2, and C_3. Let us assume that, for some example $\mathbf{x}$, the first network outputs $\mathbf{y}1(\mathbf{x}) = (0.5, 0.2, 0.9)$ and the second, $\mathbf{y}2(\mathbf{x}) = (0.6, 0.6, 0.7)$. This means that both will label $\mathbf{x}$ with the third class, $h1(\mathbf{x}) = h2(\mathbf{x}) = C_3$. If the correct answer is $c(\mathbf{x}) = C_2$, both have erred, but the error does not appear to be the same. The reader will have noticed that the first network was "very sure" about the class being C_3 (because 0.9 is clearly greater than the other two outputs, 0.5 and 0.2), whereas the second network was less certain, the differences of the output values (0.6, 0.6, and 0.7) being so small as to give rise to the suspicion that C_3 has won merely by chance. Due to its weaker commitment to the incorrect class, the second network is somehow less wrong than the first.

This is the circumstance that can be captured by a more appropriate error function, the *mean square error (MSE)*.

Target Vector Before proceeding to the definition of the mean square error, however, we must introduce yet another important concept, the *target vector* which, too, depends on the concrete example, **x**. Let us denote it by $\mathbf{t}(\mathbf{x})$. In a domain with m classes, the target vector, $\mathbf{t}(\mathbf{x}) = (t_1(\mathbf{x}), \dots, t_m(\mathbf{x}))$, consists of m binary numbers. If the example belongs to the i-th class, then $t_i(\mathbf{x}) = 1$ and all other elements in this vector are $t_j(\mathbf{x}) = 0$ (where $j \neq i$). For instance, suppose the existence of three different classes, C_1, C_2, and C_3, and let **x** be known to belong to C_2. In the ideal case, the second neuron should output 1, and the two other neurons should output 0.[2] The target is therefore $\mathbf{t}(\mathbf{x}) = (t_1, t_2, t_3) = (0, 1, 0)$.

Mean Square Error The mean square error is defined using the differences between the elements of the output vector and the target vector:

$$MSE = \frac{1}{m} \sum_{i=1}^{m} (t_i - y_i)^2 \qquad (5.3)$$

When calculating the network's *MSE*, we have to establish for each output neuron the difference between its output and the corresponding element of the target vector. Note that the terms in the parentheses, $(t_i - y_i)$, are squared to make sure that negative differences are not subtracted from positive ones.

Returning to the example of the two networks mentioned above, if the target vector is $\mathbf{t}(\mathbf{x}) = (0, 1, 0)$, then these are the mean square errors:

$MSE_1 = \frac{1}{3}[(0 - 0.5)^2 + (1 - 0.2)^2 + (0 - 0.9)^2)] = 0.57$
$MSE_2 = \frac{1}{3}[(0 - 0.6)^2 + (1 - 0.6)^2 + (0 - 0.7)^2] = 0.34$

As expected, $MSE_2 < MSE_1$, which is in line with our intuition that the second network is "less wrong" on **x** than the first network.

What Have You Learned?

To make sure you understand this topic, try to answer the following questions. If you have problems, return to the corresponding place in the preceding text.

---

[2]More precisely, the outputs will only *approach* 1 and 0 because the sigmoid function is bounded by the *open* interval, $(0, 1)$.

- In what sense do we say that the traditional error rate does not provide enough information about a neural network's classification accuracy?
- Explain the difference between a neural network's output, the example's class, and the target vector.
- Write down the formulas defining the *error rate* and the *mean square error*.

5.3 Backpropagation of Error

In the *multilayer perceptron*, the parameters to affect the network's behavior are the sets of weights, $w_{ji}^{(1)}$ and $w_{kj}^{(2)}$. The task for machine learning is to find for these weights such values that will optimize the network's classification performance. Just like in the case of linear classifiers, this is achieved by *training*. This section is devoted to one popular technique capable of accomplishing this task.

The General Scenario In principle, the procedure is the same as in the previous chapter. At the beginning, the weights are *initialized* to small random numbers, typically from the interval $(-0.1, 0.1)$. After this, the training examples are presented, one by one, and each of them is forward-propagated to the network's output. The discrepancy between this output and the example's target vector then tells us how to modify the weights (see below). After the weight modification, the next example is presented. When the last training example has been reached, one *epoch* has been completed. In *multilayer perceptrons*, the number of epochs needed for successful training is much greater than in the case of linear classifiers: it can be thousands, tens of thousands, even more.

The Gradient Descent Before we proceed to the concrete formulas for weight adjustment, we need to develop a better understanding of the problem's nature. Figure 5.3 will help us. The vertical axis represents the mean square error, expressed as a function of the network's weights (plotted along the horizontal axes). For graphical convenience, we assume that there are only two weights. This, of course, is unrealistic to say the least. But if we want an instructive example, we simply cannot afford more dimensions—we can hardly visualize ten dimensions, can we? The message we want to drive home at this point is that the error function can be imagined as a kind of a "landscape" whose "valleys" represent the function's *local minima*. The deepest of them is the *global minimum*, and this is what the training procedure is, in the ideal case, expected to reach; more specifically, it should find the set of weights corresponding to the global minimum.

A quick glance at Fig. 5.3 tells us that any pair of weights defines for the given training example a concrete location on the landscape, typically somewhere on one of the slopes. Any weight change will then result in different coordinates along the horizontal axes, and thus a different location on the error function. Where exactly this new location is, whether "up" or "down" the slope, will depend on how much, and in what direction, each of the weights has changed. For instance, it may be that increasing both w_1 and w_2 by 0.1 will lead only to a minor reduction of the mean square error; whereas increasing w_1 by 0.3 and w_2 by 0.1 will reduce it considerably.

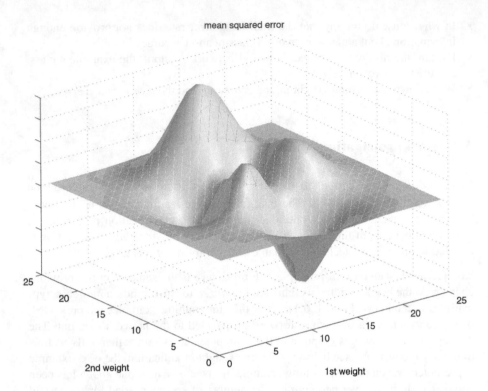

Fig. 5.3 For a given example, each set of weights implies a certain mean square error. Training should reduce this error as quickly as possible

In the technique discussed here, we want weight changes that will bring about the *steepest descent* along the error function. Recalling the terminology from Chap. 1, this is a job for *hill-climbing* search. The best-known technique used to this end in *multilayer perceptrons* is *backpropagation of error*.

Backpropagation of Error The specific weight-adjusting formulas can be derived from Eq. (5.2) by finding the function's gradient. However, as this book is meant for practitioners, and not for mathematicians, we will skip the derivation, and focus instead on explaining the learning procedure's behavior.

To begin with, it is reasonable to assume that the individual neurons differ in their contributions to the overall error. Some of them "spoil the game" more than the others. If this is the case, the reader will agree that the links leading to these neurons should undergo greater weight changes than the links leading to less offending neurons.

Fortunately, each neuron's amount of "responsibility" for the overall error can be easily obtained. Generally speaking, the concrete choice of formulas depends on what transfer function has been used. In the case of the sigmoid (see Eq. (5.1)), the responsibility is calculated as follows:

Output-layer neurons: $\delta_i^{(1)} = y_i(1 - y_i)(t_i - y_i)$

Here, $(t_i - y_i)$ is the difference between the i-th output and the corresponding target value. This difference is multiplied by $y_i(1 - y_i)$, a term whose minimum value is reached when $y_1 = 0$ or $y_i = 1$ (a "strong opinion" as to whether $\mathbf{x}$ should or should not be labeled with the i-th class); the term is maximized when $y_i = 0.5$, in which case the "opinion" can be deemed neutral. Note that the sign of $\delta_i^{(1)}$ depends only on $(t_i - y_i)$ because $y_i(1 - y_i)$ is always positive.

Hidden-layer neurons: $\delta_j^{(2)} = h_j(1 - h_j) \sum_i \delta_i^{(1)} w_{ji}$

The responsibilities of the hidden neurons are calculated by "backpropagating" the output-neurons' responsibilities obtained in the previous step. This is the role of the term $\sum_i \delta_i^{(1)} w_{ji}$. Note that each $\delta_i^{(1)}$ (the responsibility of the i-th output neuron) is multiplied by the weight of the link connecting the i-th output neurons to the j-th hidden neuron. The weighted sum is multiplied by, $h_j(1 - h_j)$, essentially the same term as the one used in the previous step, except that h_j has taken the place of y_i.

Weight Updates Now that we know the responsibilities of the individual neurons, we are ready to update the weights of the links leading to them. Similarly to *perceptron learning*, an additive rule is used:

output-layer neurons: $w_{ji}^{(1)} := w_{ji}^{(1)} + \eta \delta_i^{(1)} h_j$

hidden-layer neurons: $w_{kj}^{(2)} := w_{kj}^{(2)} + \eta \delta_j^{(2)} x_k$

The extent of weight correction is therefore determined by $\eta \delta_i^{(1)} h_j$ or $\eta \delta_j^{(2)} x_k$. Two observations can be made. First, the neurons' responsibilities, $\delta_i^{(1)}$ or $\delta_j^{(2)}$, are multiplied by η, the *learning rate* which, theoretically speaking, should be selected from the unit interval, $\eta \in (0, 1)$; however, practical implementations usually rely on smaller values, typically less than 0.1. Second, the terms are also multiplied by $h_j \in (0, 1)$ and $x_k \in [0, 1]$, respectively. The correction is therefore quite small, but its effect is relative. If the added term's value is, say, 0.02, then smaller weights, such as $w_{ij}^{(1)} = 0.01$, will be affected more significantly than greater weights such as $w_{ij}^{(1)} = 1.8$.

The whole training procedure is summarized by the pseudocode in Table 5.2. The reader will benefit also from taking a closer look at the example given in Table 5.3 that provides all the necessary details of how the weights are updated in response to one training example.

What Have You Learned?

To make sure you understand this topic, try to answer the following questions. If you have problems, return to the corresponding place in the preceding text.

- Why is the training technique called "backpropagation of error"? What is the rationale behind the need to establish the "neurons' responsibilities"?

Table 5.2 *Backpropagation of error* in a neural network with one hidden layer

---

1. Present example **x** to the input layer and propagate it through the network.
2. Let $\mathbf{y} = (y_1, \ldots y_m)$ be the output vector, and let $\mathbf{t(x)} = (t_1, \ldots t_m)$ be the target vector.
3. For each output neuron, calculate its responsibility, $\delta_i^{(1)}$, for the network's error:

   $\delta_i^{(1)} = y_i(1 - y_i)(t_i - y_i)$
4. For each hidden neuron, calculate its responsibility, $\delta_j^{(2)}$, for the network's error. While doing so, use the responsibilities, $\delta_i^{(1)}$, of the output neurons as obtained in the previous step.

   $\delta_j^{(2)} = h_j(1 - h_j) \sum_i \delta_i^{(1)} w_{ji}$
5. Update the weights using the following formulas, where η is the learning rate:

   output layer: $w_{ji}^{(1)} := w_{ji}^{(1)} + \eta \delta_i^{(1)} h_j$; h_j: the output of the j-th hidden neuron
   hidden layer: $w_{kj}^{(2)} := w_{kj}^{(2)} + \eta \delta_j^{(2)} x_k$; x_k: the value of the k-th attribute

6. Unless a termination criterion has been satisfied, return to step 1.

- Discuss the behaviors of the formulas for the calculation of the responsibilities of the neurons in the individual layers.
- Explain the behaviors of the formulas used to update the weights. Mention some critical aspects of these formulas.

5.4 Special Aspects of Multilayer Perceptrons

Limited space prevents us from the detailed investigation of the many features that make the training of *multilayer perceptrons* more art than science. To do justice to all of them, another chapter at least the size of this one would be needed. Still, the knowledge of certain critical aspects is vital if the training is ever to succeed. Let us therefore briefly survey some of the more important ones.

Computational Costs Backpropagation of error is computationally expensive. Upon the presentation of an example, the responsibility of each individual neuron has to be calculated, and the weights then modified accordingly. This has to be repeated for all training examples, usually for many epochs. To get an idea of the real costs of all this, consider a network that is to classify examples described by 100 attributes, a fairly realistic case. If there are 100 hidden neurons, then the number of weights in this lower layer is $100 \times 100 = 10^4$. This then is the number of weight changes after each training example. Note that the upper-layer weights can be neglected, in these calculations, as long as the number of classes is small. For instance, in a domain with three classes, the number of upper-layer weights is $100 \times 3 = 300$ which is much less than 10^4.

Suppose the training set consists of 10^5 examples, and suppose that the training will continue for 10^4 epochs. In this event, the number of weight-updates will be $10^4 \times 10^5 \times 10^4 = 10^{13}$. This looks like a whole lot, but many applications are even more demanding. Some fairly ingenious methods to make the training more efficient have therefore been developed. These, however, are outside the scope of our interest here.

Target Values Revisited For simplicity, we have so far assumed that each target value is either 1 or 0. This may not be the best choice. For one thing, these values

Table 5.3 Illustration of the backpropagation of error

Task. In the neural network below, let the transfer function be $f(\Sigma) = \frac{1}{1+e^{-\Sigma}}$. Using backpropagation of error (with $\eta = 0.1$), show how the weights are modified after the presentation of the following example: $[\mathbf{x}, \mathbf{t}(\mathbf{x})] = [(1, -1), (1, 0)]$

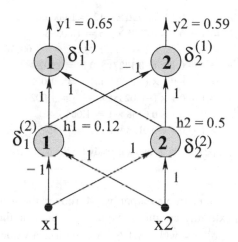

Forward propagation.

The picture shows the state after forward propagation when the signals leaving the hidden and the output neurons have been calculated as follows:

- $h_1 = \frac{1}{1+e^{-(-2)}} = 0.12$

  $h_2 = \frac{1}{1+e^0} = 0.5$

  $y_1 = \frac{1}{1+e^{-(0.12+0.5)}} = 0.65$

  $y_2 = \frac{1}{1+e^{-(-0.12+0.5)}} = 0.59$

(the solution continues on the next page)

(continued)

Table 5.3 (continued)

Backpropagation of error (cont. from the previous page)

The target vector being $\mathbf{t}(\mathbf{x}) = (1, 0)$, and the output vector $\mathbf{y} = (0.65, 0.59)$, the task is to establish each neuron's responsibility for the output error. Here are the calculations for the output neurons:

- $\delta_1^{(1)} = y_1(1 - y_1)(t_1 - y_1) = 0.65(1 - 0.65)(1 - 0.65) = 0.0796$
  $\delta_2^{(1)} = y_2(1 - y_2)(t_2 - y_2) = 0.59(1 - 0.59)(0 - 0.59) = -0.1427$

Using these values, we calculate the responsibilities of the hidden neurons. Note that we will first calculate (and denote by Δ_1 and Δ_2) the weighted sums, $\sum_i \delta_i^{(1)} w_{ij}^{(1)}$, for each of the two hidden neurons.

- $\Delta_1 = \delta_1^{(1)} w_{11}^{(1)} + \delta_2^{(1)} w_{12}^{(1)} = 0.0796 \times 1 + (-0.1427) \times (-1) = 0.2223$
  $\Delta_2 = \delta_1^{(1)} w_{21}^{(1)} + \delta_2^{(1)} w_{22}^{(1)} = 0.0796 \times 1 + (-0.1427) \times 1 = -0.0631$
  $\delta_1^{(2)} = h_1(1 - h_1)\Delta_1 = 0.12(1 - 0.12) \times 0.2223 = -0.0235$
  $\delta_2^{(2)} = h_2(1 - h_2)\Delta_2 = 0.5(1 - 0.5) \times (-0.0631) = 0.0158$

Once the responsibilities are known, the weight modifications are straightforward:

- $w_{11}^{(1)} = w_{11}^{(1)} + \eta \delta_1^{(1)} h_1 = 1 + 0.1 \times 0.0796 \times 0.12 = 1.00096$
  $w_{21}^{(1)} = w_{21}^{(1)} + \eta \delta_1^{(1)} h_2 = 1 + 0.1 \times 0.0796 \times 0.5 = 1.00398$
  $w_{12}^{(1)} = w_{12}^{(1)} + \eta \delta_2^{(1)} h_1 = -1 + 0.1 \times (-0.1427) \times 0.12 = -1.0017$
  $w_{22}^{(1)} = w_{22}^{(1)} + \eta \delta_2^{(1)} h_2 = 1 + 0.1 \times (-0.1427) \times 0.5 = 0.9929$
- $w_{11}^{(2)} = w_{11}^{(2)} + \eta \delta_1^{(2)} x_1 = -1 + 0.1 \times (-0.0235) \times 1 = -1.0024$
  $w_{21}^{(2)} = w_{21}^{(2)} + \eta \delta_1^{(2)} x_2 = 1 + 0.1 \times (-0.0235) \times (-1) = 1.0024$
  $w_{12}^{(2)} = w_{12}^{(2)} + \eta \delta_2^{(2)} x_1 = 1 + 0.1 \times 0.0158 \times 1 = 1.0016$
  $w_{22}^{(2)} = w_{22}^{(2)} + \eta \delta_2^{(2)} x_2 = 1 + 0.1 \times 0.0158 \times (-1) = 0.9984$

The weights having been updated, the network is ready for the next example.

---

can never be reached by a neuron's output, y_i. Moreover, the weight changes in the vicinity of these two extremes are miniscule because the calculation of the output-neuron's responsibility, $\delta_i^{(1)} = y_i(1 - y_i)(t_i - y_i)$, returns a value very close to zero whenever y_i approaches 0 or 1. Finally, we know that the classifier chooses the class whose output neuron has returned the highest value. The individual neuron's output precision therefore does not matter much; more important is the comparison with the other outputs. If the forward propagation results in $\mathbf{y} = (0.9, 0.1, 0.2)$, then the example is bound to be labeled with the first class (the one supported by $y_i = 0.9$), and this decision will not be swayed by minor weight changes.

In view of these arguments, more appropriate values for the target are recommended: for instance, $t_i(\mathbf{x}) = 0.8$ if the example belongs to the i-th class, and $t_i(\mathbf{x}) = 0.2$ if it does not. Suppose there are three classes, C_1, C_2, and C_3, and suppose $c(\mathbf{x}) = C_1$. In this case, the target vector will be defined as $\mathbf{t}(\mathbf{x}) = (0.8, 0.2, 0.2)$. Both 0.8 and 0.2 find themselves in regions of relatively high sensitivity of the sigmoid function, and as such will eliminate most of the concerns raised in the previous paragraph.

Local Minima Figure 5.3 illustrated the main drawback of the gradient-descent approach when adopted by *multilayer perceptron* training. The weights are changed in a way that guarantees descent along the steepest slope. But once the bottom of a *local minimum* has been reached, there is nowhere else to go—which is awkward: after all, the ultimate goal is to reach the *global minimum*. Two things are needed here: first, a mechanism to recognize the local minimum; second, a method to recover from having fallen into one.

One possibility of timely identification of local minima during training is to keep track of the mean square error, and to sum it up over the entire training set at the end of each epoch. Under normal circumstances, this sum tends to go down from one epoch to another. Once it seems to have reached a plateau where hardly any error reduction can be observed, the learning process is suspected of being trapped in a local minimum.

Techniques to overcome this difficulty usually rely on adaptive learning rates (see below), and on adding new hidden neurons (see Sect. 5.5). Generally speaking, the problem is less critical in networks with many hidden neurons. Also, local minima tend to be shallower, and less frequent, if all weights are very small, say, from the interval $(-0.01, 0.01)$.

Adaptive Learning Rate While describing backpropagation of error, we assumed that the user-set learning rate, η, was a constant. This, however, is rarely the case in realistic applications. Very often, the training starts with a large η which then gradually decreases in time. The motivation is easy to guess. At the beginning, greater weight changes reduce the number of epochs, and they may even help the learner to "jump over" some local minima. Later on, however, this large η might lead to "overshooting" the global minimum, and this is why its value should be decreased. If we express the learning rate as a function of time, $\eta(t)$, where t tells us which epoch the training is now going through, then the following negative-exponential formula will gradually reduce the learning rate (α is the slope of the negative exponential, and $\eta(0)$ is the learning rate's initial value):

$$\eta(t) = \eta(0)e^{-\alpha t} \tag{5.4}$$

It should perhaps be noted that some advanced weight-changing formulas are capable of reflecting "current tendencies." For instance, it is quite popular to implement "momentum": if the last two weight changes were both positive (or both negative), it makes sense to increase the weight-changing step; if, conversely, a positive change was followed by a negative change (of vice versa), the step should be reduced so as to prevent overshooting.

Overtraining *Multilayer perceptrons* share with polynomial classifiers one unpleasant property. Theoretically speaking, they are capable of modeling *any* decision surface, and this makes them prone to overfitting the training data. The reader remembers that overfitting typically means perfect classification of noisy training examples, which is inevitably followed by disappointing performance in the future.

For small *multilayer perceptrons*, this problem is not as painful; they are not flexible enough to overfit. But as the number of hidden neurons increases, the network gains in flexibility, and overfitting can become a real concern. However, as we will learn in the next section, this does *not* mean that we should always prefer small networks!

There is a simple way to discover whether the training process has reached the "overfitting" stage. If the training set is big enough, we can afford to leave aside some 10–20% examples. These will never be used for backpropagation of error; rather, after each epoch, the sum of mean square errors on these withheld examples is calculated. At the beginning, the sum will tend to go down, but only up to a certain moment; then it starts growing again, alerting the engineer that the training has begun to overfit the data.

What Have You Learned?

To make sure you understand this topic, try to answer the following questions. If you have problems, return to the corresponding place in the preceding text.

- What do you know about the computational costs of the technique of backprop-agation of error?
- Explain why this section recommended the values of the target vector to be chosen from {0.8, 0.2} instead of from {1,0}.
- Discuss the problem of local minima. Why do they represent such a problem for training? How can we reduce the danger of getting trapped in one?
- What are the benefits of an adaptive learning rate? What formula has been recommended for it?
- What do you know about the danger that the training of a neural network might result in overfitting the training data?

5.5 Architectural Issues

So far, one important question has been neglected: how many hidden neurons to use? If there are only one or two, the network will lack flexibility; not only will it be unable to model a complicated decision surface; the training of this network will be prone to get stuck in a local minimum. At the other extreme, using thousands of neurons will not only increase computational costs because of the need to train so many neurons. The network will be more flexible than needed. As a result, it will easily overfit the data. As usual, in situations of this kind, some compromise has to be found.

Performance Versus Size Suppose you decide to run the following experiment. The available set of pre-classified examples is divided into two parts, one for training, the other for testing. Training is applied to several neural networks, each

Fig. 5.4 The error rate
measured on testing examples
depends on the number of
neurons in the hidden layer

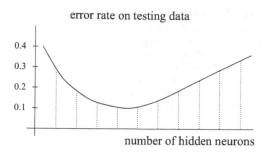

with a different number of hidden neurons. The networks are trained until no reduction of the training-set error rate is observed. After this, the error rate on the testing data is measured.

Optimum Number of Neurons When plotted in a graph, the results will typically look something like the situation depicted in Fig. 5.4. Here, the horizontal axis represents the number of hidden neurons; the vertical axis, the error rate measured on the testing set. Typically, the error rate will be high in the case of very small networks because these lack adequate flexibility, and also suffer from the danger of getting stuck in local minima. These two weaknesses can be mitigated if we increase the number of hidden neurons. As shown in the graph, the larger networks then exhibit lower error rates. But then, networks that are too large are vulnerable to overfitting. This is why, after a certain point, the testing-set error starts growing again (the right tail of the graph).

The precise shape of the curve depends on the complexity of the training data. Sometimes, the error rate is minimized when the network contains no more than 3–5 hidden neurons. In other domains, however, the minimum is reached only when hundreds of hidden neurons are used. Yet another case worth mentioning is the situation where the training examples are completely noise-free. In a domain of this kind, overfitting may never become an issue, and the curve's right tail may not grow at all.

Search for Appropriate Size The scenario described above is too expensive to be employed in practical applications. After all, we have no idea whether we will need just a few neurons, or dozens of them, or hundreds, and we may have to re-run the computationally intensive training algorithm a great many times before being able to establish the ideal size. Instead, we would like to have at our disposal a technique capable of finding the appropriate size more efficiently.

One such technique is summarized by the pseudocode in Table 5.4. The idea is to start with a very small network that only has a few hidden neurons. After each epoch, the learning algorithm checks the sum of the mean square errors observed on the training set. This sum of errors is likely to keep decreasing with the growing number of epochs—but only up to a certain point. When this is reached, the network's performance no longer improves, either because of its insufficient flexibility that makes correct classification impossible, or because it "fell" into a local minimum.

Fig. 5.5 When the
training-set mean square error
(MSE) does not seem to go
down, further improvement
can be achieved by adding
new hidden neurons. In this
particular case, this has been
done twice

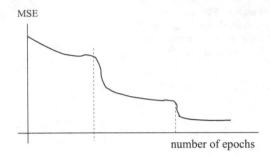

When this is observed, a few more neurons with randomly initialized weights are
added, and the training is resumed.

Usually, the added flexibility makes further error reduction possible. In the
illustration from Fig. 5.5, the error rate "stalled" on two occasions; adding new
hidden neurons then provided the necessary new flexibility.

What Have You Learned?

To make sure you understand this topic, try to answer the following questions. If
you have problems, return to the corresponding place in the preceding text.

- Discuss the shape of the curve from Fig. 5.4. What are the shortcomings of very
  small and very large networks?
- Explain the algorithm that searches for a reasonable size of the hidden layer.
  What main difficulties does it face?

5.6   Radial-Basis Function Networks

The behavior of an output neuron in *multilayer perceptrons* is similar to that of
a linear classifier. As such, it may be expected to fare poorly in domains with

Table 5.4 Gradual search for a good size of the hidden layer

1. At the beginning, use only a few hidden neurons, say, five.
2. Train the network until the mean square error no longer seems to improve.
3. At this moment, add a few (say, three) neurons to the hidden layer, each with randomly initialized weights, and resume training.
4. Repeat the previous two steps until a termination criterion has been satisfied. For instance, this can be the case when the new addition does not result in a significant error reduction, or when the hidden layer exceeds a user-set maximum size.

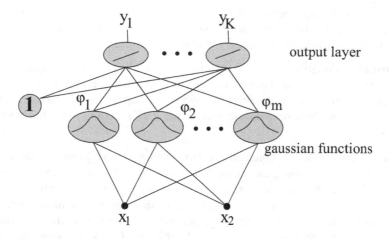

Fig. 5.6 Radial-basis function network

classes that are not linearly separable. In the context of the neural networks, however, this limitation is not necessarily hurtful. Thing is, the original examples have been transformed by the sigmoid functions in the hidden layer. Consequently, the neurons in the output layer deal with new "attributes," those obtained by this transformation. In the process of training, these transformed examples (the outputs of the hidden neurons) may become linearly separable so that the output-layer neurons can separate the two classes without difficulties.

The Alternative There is another way of transforming the attribute values; by using the so-called *radial-basis function*, RBR, as the transfer function employed by the hidden-layer neurons. This is the case of the network depicted in Fig. 5.6. An example presented to the input is passed through a set of neurons that each return a value denoted as φ_j.

Radial-Basis Function, RBF In essence, this is based on the normal distribution that we already know from Chap. 2. Suppose that the attributes describing the examples all fall into some reasonably sized interval, say $[-1, 1]$. For a given variance, σ^2, the following equation defines the n-dimensional gaussian surface centered at $\boldsymbol{\mu}_j = [\mu_{j1}, \ldots \mu_{jn}]$:

$$\varphi_j(\mathbf{x}) = \exp\{-\frac{\sum_{i=1}^{n}(x_i - \mu_{ji})^2}{2\sigma^2}\} \tag{5.5}$$

In a sense, $\varphi_j(\mathbf{x})$ measures similarity between the example vector, $\mathbf{x}$, and the gaussian center, $\boldsymbol{\mu}_j$: the larger the distance between the two, the smaller the value of $\varphi_j(\mathbf{x})$. If $\mathbf{x}$ is to be classified, the network first redescribes it as $\boldsymbol{\varphi}(\mathbf{x}) = [\varphi_1(\mathbf{x}), \ldots, \varphi_m(\mathbf{x})]$. The output signal of the i-th output neuron is then calculated as $y_i = \sum_{j=0}^{m} w_{ji}\varphi_j(\mathbf{x})$, where w_{ji} is the weight of the link from the j-th hidden neuron to the i-th output neuron (the weights w_{0i} are connected to a fixed $\varphi_0 = 1$). This output signal being interpreted as the amount of evidence supporting the i-th class, the example is labeled with the i-th class if $y_i = \max_k(y_k)$.

Output-Layer Weights It is relatively simple to establish the output-layer *weights*, w_{ij}. Since there is only one layer of weights to be trained, we can just as well rely on the *perceptron learning* algorithm described in Chap. 4, applying it to examples whose descriptions have been transformed by the RBF functions in the hidden-layer neurons.

Gaussian Centers It is a common practice to identify the gaussian centers, μ_j, with the individual training examples. If the training set is small, we can simply use one hidden neuron per training example. In many realistic applications, though, the training sets are much bigger, which can mean thousands of hidden neurons, or even more. Realizing that such large networks are unwieldy, many engineers prefer to select for the centers only a small subset of the training examples. Very often, a random choice is enough. Another possibility to identify groups of "similar" vectors, and then use for each RBF neuron the center of one group. To identify such groups of similar vectors is a task for so-called *cluster analysis* that will discussed in great detail in Chap. 14.

RBF-Based Support Vector Machines The RBF neurons transform the original example description into a new vector that consists of the values, $\phi_1, \ldots \phi_m$. Most of the time, this transformation increases the chances that the examples thus transformed will be linearly separable. This makes it possible to subject them to a linear classifier whose weights are trained by *perceptron learning*.

This is perhaps an appropriate place to mention that it is also quite popular to apply to the transformed examples the idea of the *support vector machine* introduced in Sect. 4.7. In this event, the resulting machine-learning tool is usually referred to as RBF-based SVM. Especially in domains where the boundary between the classes is highly non-linear, this classifier is more powerful than the plain (linear) SVM.

Computational Costs RBF-network training consists of two steps. First, the centers of the radial-basis functions (the Gaussians) are selected. Second, the output-neurons' weights are obtained by training. Since there is only one layer to train, the process is computationally much less intensive than the training in *multilayer perceptrons*.

What Have You Learned?

To make sure you understand this topic, try to answer the following questions. If you have problems, return to the corresponding place in the preceding text.

- Explain the principle of the radial-basis function network. In what aspects does it differ from the *multilayer perceptron*?

- How many weights need to be trained in a radial-basis function network? How can the training be accomplished?
- What are the possibilities for the construction of the gaussian hidden layer?

5.7 Summary and Historical Remarks

- The basic unit of a *multilayer perceptron* is a *neuron*. The neuron accepts a weighted sum of inputs, and subjects this sum to a *transfer function*. Several different transfer functions can be used. The one chosen in this chapter is the *sigmoid* defined by the following equation where Σ is the weighted sum of inputs:

$$f(\Sigma) = \frac{1}{1 + e^{-\Sigma}}$$

Usually, all neurons use the same transfer function.

- The simple version of the *multilayer perceptron* described in this chapter consists of one output layer and one hidden layer of neurons. Neurons in adjacent layers are fully interconnected; but there are no connections between neurons in the same layer. An example presented to the network's input is *forward-propagated* to its output, implementing, in principle, the following function:

$$y_i = f(\sum_j w_{ji}^{(1)} f(\sum_k w_{kj}^{(2)} x_k))$$

Here, $w_{ji}^{(1)}$ and $w_{kj}^{(2)}$ are the weights of the output and the hidden neurons, respectively, and f is the sigmoid function.

- The training of *multilayer perceptron* is accomplished by a technique known as *backpropagation of error*. For each training example, the technique first establishes each neuron's individual responsibility for the network's overall error, and then updates the weights according to these responsibilities.

Here is how the responsibilities are calculated:

output neurons: $\delta_i^{(1)} = y_i(1 - y_i)(t_i - y_i)$

hidden neurons: $\delta_j^{(2)} = h_j(1 - h_j) \sum_i \delta_i^{(1)} w_{ij}$

Here is how the weights are updated:

output layer: $w_{ji}^{(1)} := w_{ji}^{(1)} + \eta \delta_i^{(1)} h_j$

hidden layer: $w_{kj}^{(2)} := w_{kj}^{(2)} + \eta \delta_j^{(2)} x_k$

- Certain aspects of *backpropagation by error* have been discussed here. Among these, the most important are computational costs, the existence of local minima, adaptive learning rate, the danger of overfitting, and the problems of how to determine the size of the hidden layer.
- An alternative is the *radial-basis function* (RBF) network. For the transfer function at the hidden-layer neurons, the gaussian function is used. The output-layer neurons often use the step function (in principle, the linear classifier), or simply a linear function of the inputs.
- In RBF networks, each gaussian center is identified with one training example. If there are too many such examples, a random choice can be made. The gaussian's standard deviation is in this simple version set to $\sigma^2 = 1$,
- The output-layer neurons in RBF networks can be trained by perceptrons learning. Only one layer of weights needs to be trained.
- Sometimes, the idea of *support vector machine (SVM)* is applied to the outputs of hidden neurons. The resulting tool is then called the RBF-based SVM.

Historical Remarks Research of neural networks was famously delayed by the skeptical views formulated by Minsky and Papert [66]. The pessimism expressed by such famous authors was probably the main reason why an early version of neural-network training by Bryson and Ho [13] was largely overlooked, a fate soon to be shared by an independent successful attempt by Werbos [95]. It was only after the publication of the groundbreaking volumes by Rumelhart and McClelland [83], where the algorithm was independently re-invented, that the field of artificial neural networks became established as a respectable scientific discipline. The gradual growth of the multilayer perceptron was proposed by Ash [1]. The idea of radial-basis functions was first cast in the neural-network setting by Broomhead and Lowe [12].

5.8 Solidify Your Knowledge

The exercises are to solidify the acquired knowledge. The suggested thought experiments will help the reader see this chapter's ideas in a different light and provoke independent thinking. Computer assignments will force the readers to pay attention to seemingly insignificant details they might otherwise overlook.

Exercises

1. Return to the illustration of backpropagation of error in Table 5.3. Using only a pen, paper, and calculator, repeat the calculations for a slightly different training example: $\mathbf{x} = (-1, -1), t(\mathbf{x}) = (0, 1)$.

2. Hand-simulating backpropagation of error as in the previous example, repeat the calculation for the following two cases:

High output-layer weights: $w_{11}^{(1)} = 3.0, w_{12}^{(1)} = -3.0, w_{21}^{(1)} = 3.0, w_{22}^{(1)} = 3.0$

Small output-layer weights: $w_{11}^{(1)} = 0.3, w_{12}^{(1)} = -0.3, w_{21}^{(1)} = 0.3, w_{22}^{(1)} = 0.3$

Observe the relative changes in the weights in each case.

3. Consider a training set containing of 10^5 examples described by 1000 attributes. What will be the computational costs of training a *multilayer perceptron* with 1000 hidden neurons and ten output neurons for 10^5 epochs?

Give It Some Thought

1. How will you generalize the technique of backpropagation of error so that it can be used in a *multilayer perceptron* with more that one hidden layer?
2. Section 5.1 suggested that all attributes should be normalized here to the interval $[-1.0, 1.0]$. How will the network's classification and training be affected if the attributes are not so normalized? (hint: this has something to do with the sigmoid function).
3. Discuss the similarities and differences of the classification procedures used in radial-basis functions and in *multilayer perceptrons*.
4. Compare the advantages and disadvantages of radial-basis function networks in comparison with *multilayer perceptrons*.

Computer Assignments

1. Write a program that implements backpropagation of error for a predefined number of output and hidden neurons. Use a fixed learning rate, η.
2. Apply the program implemented in the previous task to some benchmark domains from the UCI repository.[3] Experiment with different values of η, and see how they affect the speed of convergence.
3. For a given data set, experiment with different numbers of hidden neurons in the *multilayer perceptron*, and observe how they affect the network's ability to learn.
4. Again, experiment with different numbers of hidden neurons. This time, focus on computational costs. How many epochs are needed before the network converges? Look also at the evolution of the error rate.
5. Write a program that will, for a given training set, create a radial-basis function network. For large training sets, select randomly the examples that will define the gaussian centers.
6. Apply the program implemented in the previous task to some benchmark domains from the UCI repository.

---

[3] www.ics.uci.edu/~mlearn/MLRepository.html.

Chapter 6
Decision Trees

The classifiers discussed in the previous chapters expect all attribute values to be presented at the same time. Such a scenario, however, has its flaws. Thus a physician seeking to come to grips with the nature of her patient's condition often has nothing to begin with save a few subjective symptoms. And so, to narrow the field of diagnoses, she prescribes lab tests, and, based on the results, perhaps other tests still. At any given moment, then, the doctor considers only "attributes" that promise to add meaningfully to her current information or understanding. It would be absurd to ask for all possible lab tests (thousands and thousands of them) right from the start.

The lesson is, exhaustive information often is not immediately available; it may not even be needed. The classifier may do better choosing the attributes one at a time, according to the demands of the situation. The most popular tool targeting this scenario is a *decision tree*. The chapter explains its classification behavior, and then presents a simple technique that induces a decision tree from data. The reader will learn how to benefit from tree-pruning, and how to convert the induced tree to a set of rules.

6.1 Decision Trees as Classifiers

The training set shown in Table 6.1 consists of eight examples described by three attributes and labeled as positive or negative instances of a given class. To simplify the explanation of the basic concepts, we will assume that all attributes are discrete. Later, when the underlying principles become clear, we will slightly generalize the approach to make it usable also in domains with continuous attributes or with mixed sets of continuous and discrete attributes.

Decision Tree Figure 6.1 shows a few example decision trees that are capable of dealing with the data from Table 6.1. The internal nodes represent attribute-value

© Springer International Publishing AG 2017
M. Kubat, *An Introduction to Machine Learning*,
DOI 10.1007/978-3-319-63913-0_6

Table 6.1 Eight training examples described by three symbolic attributes and classified as positive and negative examples of a given class

Example	crust size	shape	filling size	Class
e1	big	circle	small	**pos**
e2	small	circle	small	**pos**
e3	big	square	small	**neg**
e4	big	triangle	small	**neg**
e5	big	square	big	**pos**
e6	small	square	small	**neg**
e7	small	square	big	**pos**
e8	big	circle	big	**pos**

tests, the edges indicate how to proceed in the case of diverse test results, and the leafs[1] contain class labels. An example to be classified is first subjected to the test prescribed at the topmost node, the *root*. The result of this test then decides along which edge the example is to be sent down, and the process continues until a leaf node is reached. Once this happens, the example is labeled with the class associated with this leaf.

Let us illustrate the process using the tree from Fig. 6.1b. The root asks about the shape, each of whose three values is represented by one edge. In examples e1, e2, and e8, we find the value shape=circle which corresponds to the left edge. The example is sent down along this edge, ending in a leaf that labeled pos. This indeed is the class common to all these three examples. In e4, shape=triangle, and the corresponding edge ends in a leaf labeled neg—again, the correct class. Somewhat more complicated is the situation with examples e3, e5, e6, and e7 where shape=square. For this particular value, the edge ends not in a leaf, but only at a test-containing node, this one inquiring about the value of filling-size. In the case of e5 and e7, the value is big, which leads to a leaf labeled with pos. In the other two examples, e3 and e6, the value is small, and this sends them to a leaf labeled with neg.

We have shown that the decision tree from Fig. 6.1b identifies the correct class for all training examples. By way of a little exercise, the reader is encouraged to verify that the other trees shown in the picture are just as successful.[2]

Interpretability Comparing this classifier with those introduced earlier, we can see one striking advantage: *interpretability*. If anybody asks why example e1 is deemed positive, the answer is, "because its shape is circle." Other classifiers do not offer explanations of this kind. Especially the neural network is a real *black box*: when presented with an example, it simply returns the class and never offers any

---

[1]Both spellings are used: *leaves* and *leafs*. The latter is probably more appropriate because the "leaf" in question is supposed to be a data abstraction that has nothing to do with the original physical object.

[2]In as sense, the decision tree can be seen as a simple mechanism for data compression.

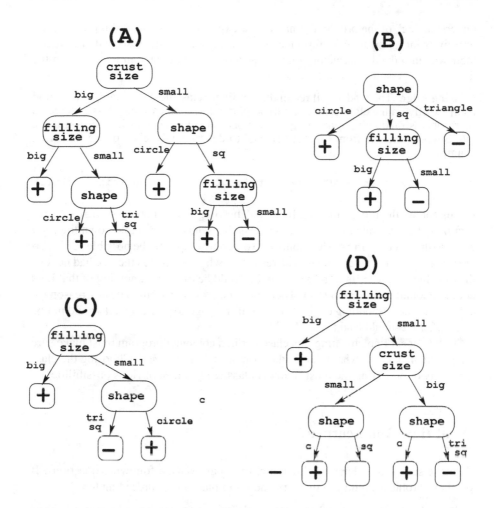

Fig. 6.1 Example decision trees for the "pies" domain. Note how they differ in size and in the order of tests. Each of them classifies correctly all training examples listed in Table 6.1, tri, sq, and c stand for `triangle`, `square`, and `circle`, respectively

insight as to why this particular label has been given preference over other labels. The situation is not much better in the case of Bayesian and linear classifiers. Only the *k-NN* classifier offers a semblance of a—rather rudimentary—argument. For instance, one can say that, "**x** should be labeled with pos because this is the class of the training example most similar to **x**." Such a statement, however, is a far cry from the explicit attribute-based explanation made possible by the decision tree.

One can go one step further and interpret a decision tree as a set of rules such as "if shape=square and filling-size=big, then the example belongs to class pos." A domain expert inspecting these rules may then decide whether they are intuitively appealing, and whether they agree with his or her "human understanding" of the problem at hand. The expert may even be willing to suggest improvements to the tree; for instance, by pointing out spurious tests that have found their way into the data structure only on account of some random regularity in the data.

Missing Edge The reader will recall that, in linear classifiers, an example may find itself exactly on the class-separating hyperplane, in which case the class is selected more or less randomly. Something similar occasionally happens in decision trees, too. Suppose the tree from Fig. 6.1a is used to determine the class of the following example:

$(\mathtt{crust-size} = \mathtt{small})\mathrm{AND}(\mathtt{shape} = \mathtt{triangle})\mathrm{AND}(\mathtt{filling-size} = \mathtt{small})$

Let us follow the procedure step by step. The root inquires about `crust-size`. Realizing that the value is `small`, the classifier sends the example down the right edge, to the test on `shape`. Here, only two outcomes appear to be possible: `circle` or `square`, but not `triangle`. The reason is, whoever created the tree had no idea that an object with `crust-size=small` could be triangular: nothing of that kind could be found in the training set. Therefore, there did not seem to be any reason for creating the corresponding edge. And even if the edge were created, it would not be clear where it should lead to.

The engineer implementing this classifier in a computer program must make sure the program "knows" what to do in the case of "missing edges." Choosing the class randomly or preferring the most frequent class are the most obvious possibilities.

What Have You Learned?

To make sure you understand this topic, try to answer the following questions. If you have problems, return to the corresponding place in the preceding text.

- Describe the mechanism that uses a decision tree to classify examples. Illustrate the procedure using the decision trees from Fig. 6.1 and the training examples from Table 6.1.
- What is meant by the statement that, "the decision tree's choice of a concrete class can be explained to the user"? Is something similar possible in the case of the classifiers discussed in the previous chapters?
- Under what circumstances can a decision tree find itself unable to determine an example's class? How would you handle the situation if you were the programmer?

6.2 Induction of Decision Trees

We will begin with a very crude induction algorithm. Applying it to a training set, we will realize that a great variety of alternative decision trees can be obtained. A brief discussion will convince us that, among these, the smaller ones are to be preferred. This observation will motivate an improved version of the technique, thus preparing the soil for the following sections.

Divide and Conquer Let us try our hand at creating a decision tree manually. Suppose we decide that the root node should test the value of shape. In the training set, three different outcomes are found: circle, triangle, and square. For each, the classifier will need a separate edge leading from the root. The first, defined by shape=circle, will be followed by examples $T_C = \{e1, e2, e8\}$; the second, defined by shape=triangle, will be followed by $T_T = \{e4\}$; and the last, defined by shape=square, will be followed by $T_S = \{e3, e5, e6, e7\}$. Each of the three edges will begin at the root and end in another node, either an attribute test or a leaf containing a class label.

Seeing that all examples in T_C are positive, we will let this edge point at a leaf labeled with pos. Similarly, the edge followed by the examples from T_T will point at a leaf labeled with neg. Certain difficulties will arise only in the case of the last edge because T_S is a mixture of both classes. To separate them, we need another test, say, filling-size, to be placed at the end of the edge. This attribute can acquire two values, small and big, dividing T_S into two subsets. Of these, $T_{S-S} = \{e3, e6\}$ is characterized by filling-size=small; the other, $T_{S-B} = \{e5, e7\}$, is characterized by filling-size=big. All examples in T_{S-S} are positive, and all examples in T_{S-B} are negative. This allows us to let both edges end in leafs, the former labeled with pos, the latter with neg. At this moment, the tree-building process can stop because each training example presented to the classifier thus created will reach a leaf.

The reader will have noticed that each node of the tree can be associated with a set of examples that pass through it or end in it. Starting with the root, each test divides the training set into disjoint subsets, and these into further subsets, and so on until each subset is "pure" in the sense that all its examples belong to the same class. This is why the approach is sometimes referred to as the *divide-and-conquer* technique.

Alternative Trees In the process thus described, the (rather arbitrary) choice of shape and filling-size resulted in the decision tree shown in Fig. 6.1b. To get used to the mechanism, the student is encouraged to experiment with alternatives such as placing at the root the test on crust-size or filling-size, and considering different options for the tests at the lower level(s). Quite a few other decision trees will thus be created—some of them depicted in Fig. 6.1.

That so many solutions can be found even in this very simple toy domain is a food for thought. Is there a way to decide which trees are better? So, an improved version of the divide-and-conquer technique should be able to arrive at a "good" tree *by design*, and not by mere chance.

The Size of the Tree The smallest of the data structures in Fig. 6.1 consists of two attribute tests; the largest, of five. Differences of this kind may have a strong impact on the classifier's behavior. Before proceeding to the various aspects of this phenomenon, however, let us emphasize that the number of nodes in the tree is not the only criterion of size. Just as important is the number of tests that have to be carried out when classifying an average example.

For instance, in a domain where `shape` is almost always `circle` or `triangle` (and only very rarely `square`), the average number of tests prescribed by the tree from Fig. 6.1b will only slightly exceed 1 because both `shape=circle` and `shape=triangle` immediately point at leafs with class labels. But if the prevailing `shape` is `square`, the average number tests approaches 2. Quite often, then, a bigger tree may result in fewer tests than a smaller one.

Small Trees Versus Big Trees There are several reasons why small decision trees are preferred. One of them is *interpretability*. A human expert finds it easy to analyze, explain, and perhaps even correct, a decision tree that consists of no more than a few tests. The larger the tree, the more difficult this is.

Another advantage of small decision trees is their tendency to dispose of *irrelevant* and *redundant* information. Whereas the relatively large tree from Fig. 6.1a employs all three attributes, the smaller one from Fig. 6.1b is just as good at classifying the training set—without ever considering `crust-size`. Such economy will come handy in domains where certain attribute values are expensive or time-consuming to obtain.

Finally, larger trees are prone to *overfit* the training examples. This is because the divide-and-conquer method keeps splitting the training set into smaller and smaller subsets, the number of these splits being equal to the number of attribute tests in the tree. Ultimately, the resulting training subsets can become so small that the classes may get separated by an attribute that only by chance—or noise—has a different value in the remaining positive and negative examples.

Induction of Small Decision Trees When illustrating the behavior of the divide-and-conquer technique on the manual tree-building procedure, we picked the attributes at random. When doing so, we observed that some choices led to smaller trees than others. Apparently, the attributes differ in how much information they convey. For instance, `shape` is capable of immediately labeling some examples as positive (if the value is `circle`) or negative (if the value is `triangle`); but `crust-size` cannot do so unless assisted by some other attribute.

Assuming that there is a way to measure the amount of information provided by each attribute (and such a mechanism indeed exists, see Sect. 6.3), we are ready to formalize the technique for induction of decision trees by a pseudocode. The reader will find it in Table 6.2.

Table 6.2 Induction of decision trees

Let T be the training set.

grow(T):

(1) Find the attribute, *at*, that contributes the maximum information about the class labels.
(2) Divide T into subsets, T_i, each characterized by a different value of *at*.
(3) For each T_i:
    If all examples in T_i belong to the same class, then create a leaf labeled with this class; otherwise, apply the same procedure recursively to each training subset: *grow(T_i)*.

What Have You Learned?

To make sure you understand this topic, try to answer the following questions. If you have problems, return to the corresponding place in the preceding text.

- Explain the principle of the divide-and-conquer technique for induction of decision trees.
- What are the advantages of small decision trees in comparison to larger ones?
- What determines the size of the decision tree obtained by the divide-and-conquer technique?

6.3 How Much Information Does an Attribute Convey?

To create a relatively small decision tree, the divide-and-conquer technique relies on one critical component: the ability to decide how much information about the class labels is conveyed by the individual attributes. This section introduces a mechanism to calculate this quantity.

Information Contents of a Message Suppose we know that the training examples are labeled as pos or neg, the relative frequencies of these two classes being p_{pos} and p_{neg}, respectively.[3] Let us select a random training example. How much information is conveyed by the message, "this example's class is pos"?

The answer depends on p_{pos}. In the extreme case where all examples are known to be positive, $p_{pos} = 1$, the message does not tell us anything new. The reader knows the example is positive even without being told so. The situation changes if

---

[3]Recall that the relative frequency of pos is the percentage (in the training set) of examples labeled with pos; this represents the probability that a randomly drawn example will be positive.

Table 6.3 Some values of
the information contents
(measured in bits) of the
message, "this randomly
drawn example is positive."

p_{pos}	$-\log_2 p_{pos}$
1.00	0 bits
0.50	1 bit
0.25	2 bits
0.125	3 bits

Note that the message is impossible for $p_{pos} = 0$

both classes are known to be equally represented so that $p_{pos} = 0.5$. Here, the guess is no better than a flipped coin, so the message *does* offer some information. And if a great majority of examples are known to be negative so that, say, $p_{pos} = 0.01$, then the reader is all but certain that the chosen example is going to be negative as well; the message telling him that this is not the case is unexpected. And the lower the value of p_{pos}, the more information the message offers.

When quantifying the information contents of such a message, the following formula has been found convenient:

$$I_{pos} = -\log_2 p_{pos} \tag{6.1}$$

The negative sign compensates for the fact that the logarithm of $p_{pos} \in (0, 1)$ is always negative. Table 6.3 shows the information contents for some typical values of p_{pos}. Note that the unit for the amount of information is 1 *bit*. Another comment: the base of the logarithm being 2, it is fairly common to write $\log p_{pos}$ instead of the more meticulous $\log_2 p_{pos}$.

Entropy (Average Information Contents) So much for the information contents of a single message. Suppose, however, that the experiment is repeated many times. Both messages will occur, "the example is positive," and "the example is negative," the first with probability p_{pos}, the second with probability p_{neg}. The average information contents of all these messages is then obtained by the following formulas where the information contents of either message is weighted by its probability (the T in the argument refers to the training set):

$$H(T) = -p_{pos} \log_2 p_{pos} - p_{neg} \log_2 p_{neg} \tag{6.2}$$

The attentive reader will protest that the logarithm of zero probability is not defined, and Eq. (6.2) may thus be useless if $p_{pos} = 0$ or $p_{neg} = 0$. Fortunately, a simple analysis (using limits and l'Hopital's rule) will convince us that, for p converging to zero, the expression $p \log p$ converges to zero, too, which means that $0 \cdot \log 0 = 0$.

$H(T)$ is called *entropy of T*. Its value reaches its maximum, $H(T) = 1$, when $p_{pos} = p_{neg} = 0.5$ (because $0.5 \cdot \log 0.5 + 0.5 \cdot \log 0.5 = 1$); and it drops to its minimum, $H(T) = 0$, when either $p_{pos} = 1$ or $p_{neg} = 1$ (because $0 \cdot \log 0 + 1 \cdot$

log $1 = 0$). By the way, the case with $p_{pos} = 1$ or $p_{neg} = 1$ is regarded as perfect regularity because all examples belong to the same class; conversely, the case with $p_{pos} = p_{neg} = 0.5$ is seen as a total lack of regularity.

Amount of Information Contributed by an Attribute The concept of entropy (lack of regularity) will help us deal with the main question: how much does the knowledge of the value of a discrete attribute, at, tell us about an example's class?

Let us remind ourselves that at divides the training set, T, into subsets, T_i, each characterized by a different value of at. Quite naturally, each subset will be marked by its own probabilities (estimated by relative frequencies) of the two classes, p_{ipos} and p_{ineg}. Based on the knowledge of these, Eq. (6.2) will give us the corresponding entropies, $H(T_i)$.

Now let $|T_i|$ be the number of examples in T_i, and let $|T|$ be the number of examples in the whole training set, T. The probability that a randomly drawn training example will be in T_i is estimated as follows:

$$P_i = \frac{|T_i|}{|T|} \tag{6.3}$$

We are ready to calculate the weighted average of the entropies of the subsets.

$$H(T, at) = \Sigma_i P_i \cdot H(T_i) \tag{6.4}$$

The obtained result, $H(T, at)$, is the entropy of a system where not only the class labels, but also the values of at are known for each training example. The amount of information contributed by at is then the difference between the entropy *before* at has been considered and the entropy *after* this attribute has been considered:

$$I(T, at) = H(T) - H(T, at) \tag{6.5}$$

It would be easy to prove that this difference cannot be negative; information can only be gained, never lost, by considering at. In certain rare cases, however, $I(T, at) = 0$, which means that no information has been gained, either.

Applying Eq. (6.5) separately to each attribute, we can find out which of them provides the maximum amount of information, and as such is the best choice for the "root" test in the first step of the algorithm from Table 6.2.

The procedure just described is summarized by the pseudocode in Table 6.4. The process starts by the calculation of the entropy of the system where only class percentages are known. Next, the algorithm calculates the information gain conveyed by each attribute. The attribute that offers the highest information gain is deemed best.

Illustration Table 6.5 shows how to use the mechanism for the selection of the most informative attribute in the domain from Table 6.1. At the beginning, the entropy, $H(T)$, of the system without attributes is established. Then, we observe that, for instance, the attribute shape divides the training set into three subsets. The

Table 6.4 The algorithm to find the most informational attribute

1. Calculate the entropy of the training set, T, using the percentages, p_{pos} and p_{neg}, of the positive and negative examples:

$$H(T) = -p_{pos} \log_2 p_{pos} - p_{neg} \log_2 p_{neg}$$

2. For each attribute, at, that divides T into subsets, T_i, with relative sizes P_i, do the following:

    (i)   calculate the entropy of each subset, T_i;
    (ii) calculate the average entropy: $H(T, at) = \Sigma_i P_i \cdot H(T_i)$;
    (iii) calculate information gain: $I(T, at) = H(T) - H(T, at)$

3. Choose the attribute with the highest value of information gain.

---

average of their entropies, $H(T, \texttt{shape})$, is calculated, and the difference between $H(T)$ and $H(T, \texttt{shape})$ gives the amount of information conveyed by this attribute. Repeating the procedure for `crust-size` and `filling-size`, and comparing the results, we realize the `shape` contributes more information than the other two attributes, and this is why we choose `shape` for the root test.

This, by the way, is how the decision tree from Fig. 6.1b was obtained.

What Have You Learned?

To make sure you understand this topic, try to answer the following questions. If you have problems, return to the corresponding place in the preceding text.

- What do we mean when we talk about the "amount of information conveyed by a message"? How is this amount determined, and what units are used?
- What is *entropy* and how does it relate to the frequency of the positive and negative examples in the training set?
- How do we use entropy when assessing the amount of information contributed by an attribute?

6.4   Binary Split of a Numeric Attribute

The entropy-based mechanism from the previous section requested that all attributes should be discrete. With a little modification, however, the same approach can be applied to continuous attributes as well. All we need is to convert them to boolean attributes.

Table 6.5 Illustration of the search for the most informative attribute

Example	crust size	shape	filling size	Class
e1	big	circle	small	**pos**
e2	small	circle	small	**pos**
e3	big	square	small	**neg**
e4	big	triangle	small	**neg**
e5	big	square	big	**pos**
e6	small	square	small	**neg**
e7	small	square	big	**pos**
e8	big	circle	big	**pos**

Here is the entropy of the training set where only class labels are known:

$$H(T) = -p_{pos} \log_2 p_{pos} - p_{neg} \log_2 p_{neg}$$
$$= -(5/8) \log(5/8) - (3/8) \log(3/8) = 0.954$$

Next, we calculate the entropies of the subsets defined by the values of shape:

$$H(\text{shape=square}) = -(2/4) \cdot \log(2/4) - (2/4) \cdot \log(2/4) = 1$$
$$H(\text{shape=circle}) = -(3/3) \cdot \log(3/3) - (0/3) \cdot \log(0/3) = 0$$
$$H(\text{shape=triangle}) = -(0/1) \cdot \log(0/1) - (1/1) \cdot \log(1/1) = 0$$

From these, we obtain the average entropy of the system where the class labels *and* the value of shape is known:

$$H(T, \text{shape}) = (4/8) \cdot 1 + (3/8) \cdot 0 + (1/8) \cdot 0 = 0.5$$

Repeating the same procedure for the other two attributes, we obtain the following:

$$H(T, \text{crust} - \text{size}) = 0.951$$
$$H(T, \text{filling} - \text{size}) = 0.607$$

These values give the following information gains:

$$I(T, \text{shape}) = H(T) - H(T, \text{shape}) = 0.954 - 0.5 = 0.454$$
$$I(T, \text{crust} - \text{size}) = H(T) - H(T, \text{crust} - \text{size}) = 0.954 - 0.951 = 0.003$$
$$I(T, \text{filling} - \text{size}) = H(T) - H(T, \text{filling} - \text{size}) = 0.954 - 0.607 = 0.347$$

We conclude that maximum information is contributed by shape.

Converting a Continuous Attribute to a Boolean One Let us denote the continuous attribute by x. The trick is to choose a threshold, θ, and then decide that if $x < \theta$, then the value of the newly created boolean attribute is *true*, and otherwise it is *false* (or vice versa).

Simple enough. But what concrete θ to choose? Surely there are many of them? Here is one possibility.

Suppose that x has a different value in each of the N training examples. Let us sort these values in ascending order, denoting by x_1 the smallest, and by x_N the highest. Any pair of neighboring values, x_i and x_{i+1}, then defines a threshold, $\theta_i = (x_i + x_{i+1})/2$. For instance, a four-example training set where x has values 3, 4, 6, and 9 leads us to consider $\theta_1 = 3.5, \theta_2 = 5.0$, and $\theta_3 = 7.5$. For each of these $N-1$ thresholds, we calculate the amount of information contributed by the boolean attribute thus defined, and then choose the threshold where the information gain is maximized.

Candidate Thresholds The approach just described deserves to be criticized for its high computational costs. Indeed, in a domain with a hundred thousand examples described by a hundred attributes (nothing extraordinary), the information contents of $10^5 \times 10^2 = 10^7$ different thresholds would have to be calculated. Fortunately, mathematicians have been able to prove that a great majority of these thresholds can just as well be ignored. This reduces the costs to a mere fraction.

The principle is illustrated in Table 6.6. In the upper part, 13 values of x are ordered from left to right, each labeled with the class (positive or negative) of the training example in which the value was found. And here is the rule: the best threshold never finds itself between values that are labeled with the same class. This means that it is enough to investigate the contributed information only for locations between values with opposite class labels. In the specific case shown in Table 6.6, only three *candidate thresholds*, θ_1, θ_2, and θ_3, need to be investigated (among the three, θ_1 is shown to be best).

The Root of a Numeric Decision Tree The algorithm summarized by the pseudocode in Table 6.7 determines the best attribute test for the root of a decision tree in a domain where all attributes are continuous. Note that the test consists of a pair, $[at_i, \theta_{ij}]$, where at_i is the selected attribute and θ_{ij} is the best threshold found for this attribute. If an example's value of the i-th attribute is below the threshold, $at_i < \theta_{ij}$, the left branch of the decision tree is followed; otherwise, the right branch is chosen.

What Have You Learned?

To make sure you understand this topic, try to answer the following questions. If you have problems, return to the corresponding place in the preceding text.

- Explain why this section suggested to divide the domain of a continuous attribute into two subdomains.
- What mathematical finding has reduced the number of thresholds that need to be investigated?

Table 6.6 Illustration of the search for the best threshold

The values of attribute x are sorted in ascending order. The candidate thresholds are located between values labeled with opposite class labels.

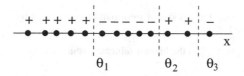

Here is the entropy of the training set, ignoring attribute values:

$$H(T) = -p_+ \log p_+ - p_- \log p_-$$
$$= -(7/13) \log(7/13) - (6/13) \log(6/13) = 0.9957$$

Here are the entropies of the training subsets defined by the three candidate thresholds:

$$H(x < \theta_1) = -(5/5) \log(5/5) - (0/5) \log(0/5) = 0$$
$$H(x > \theta_1) = -(2/8) \log(2/8) - (6/8) \log(6/8) = 0.8113$$

$$H(x < \theta_2) = -(5/10) \log(5/10) - (5/10) \log(5/10) = 1$$
$$H(x > \theta_2) = -(2/3) \log(2/3) - (1/3) \log(1/3) = 0.9183$$

$$H(x < \theta_3) = -(7/12) \log(7/12) - (5/12) \log(5/12) = 0.9799$$
$$H(x > \theta_3) = -(0/1) \log(0/1) - (1/1) \log(1/1) = 0$$

Average entropies associated with the individual thresholds:

$$H(T, \theta_1) = (5/13) \cdot 0 + (8/13) \cdot 0.8113 = 0.4993$$
$$H(T, \theta_2) = (10/13) \cdot 1 + (3/13) \cdot 0.9183 = 0.9811$$
$$H(T, \theta_3) = (12/13) \cdot 0.9799 + (1/13) \cdot 0 = 0.9045$$

Information gains entailed by the individual candidate thresholds:

$$I(T, \theta_1) = H(T) - H(T, \theta_1) = 0.9957 - 0.4993 = 0.4964$$
$$I(T, \theta_2) = H(T) - H(T, \theta_2) = 0.9957 - 0.9811 = 0.0146$$
$$I(T, \theta_3) = H(T) - H(T, \theta_3) = 0.9957 - 0.9045 = 0.0912$$

Threshold θ_1 gives the highest information gain.

Table 6.7 Algorithm to find the best numeric-attribute test

1. For each attribute at_i:

   (i) Sort the training examples by the values of at_i;
   (ii) Determine the candidate thresholds, θ_{ij}, as those lying between examples with opposite labels;
   (iii) For each θ_{ij}, determine the amount of information contributed by the boolean attribute thus created.

2. Choose the pair $[at_i, \theta_{ij}]$ that offers the highest information gain.

6.5 Pruning

Section 6.2 extolled the virtues of small decision trees: interpretability, removal of irrelevant and redundant attributes, reduced danger of overfitting. These were the arguments that motivated the use of information theory in the course of decision-tree induction; they also motivate the step that follows: *pruning*.

Fig. 6.2 A simple approach to pruning will replace a subtree with a leaf

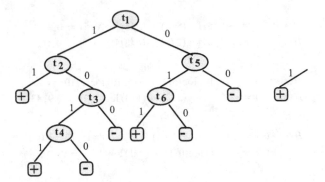

The Essence Figure 6.2 will help us explain the principle. On the left is the original decision tree whose six attribute tests are named $t_1, \ldots t_6$. On the right is a pruned version. Note that the subtree rooted in test t_3 in the original tree is in the pruned tree replaced with a leaf labeled with the negative class; and the subtree rooted in test t_6 is replaced with a leaf labeled with the positive class. The reader can see the point: pruning consists of replacing one or more subtrees with leafs, each labeled with the class most common among the training examples that reach—in the original classifier—the removed subtree.

This last idea sounds counterintuitive: the induction mechanism seeks to create a decision tree that scores zero errors on the training examples, but this perfection may be lost in the pruned tree! But the practically minded engineer is not alarmed. The ultimate goal is *not* to classify the training examples (their classes are known anyway). Rather, we want a tool capable of labeling *future* examples. And experience shows that this kind of performance is often improved by reasonable pruning.

Error Estimate Pruning is typically carried out in a sequence of steps: first replace with a leaf one subtree, then another, and so on, as long as the replacements appear to be beneficial according to some reasonable criterion. The term "beneficial" is meant to warn us that small-tree advantages should not be outbalanced by reduced classification performance.

Which brings us to the issue of *error estimate*. The principle is illustrated in Fig. 6.2. Let m be the number of training examples that reach test t_3 in the decision tree on the left. If we replace the subtree rooted in t_3 by a leaf (as happened in the tree on the right), some of these m examples may become misclassified. Denoting the number of these misclassified examples by e, we may be tempted to estimate the probability that an example will be misclassified at this leaf by the relative frequency: e/m. But admitting that small values of m may render this estimate problematic, we prefer the following formula where N is the total number of training examples:

$$E_{estimate} = \frac{e+1}{N+m} \tag{6.6}$$

The attentive reader may want to recall (or re-read) what Sect. 2.3 had to say about the difficulties of probability estimates of rare events.

Error Estimates for the Whole Tree Once again, let us return to Fig. 6.2. The tree on the left has two subtrees, one rooted at t_2, the other at t_5. Let m_2 and m_5 be the numbers of the training examples reaching t_2 and t_5, respectively; and let E_2 and E_5 be the error estimates of the two subtrees, obtained by Eq. (6.6). For the total of $N = m_2 + m_5$ training examples, the error rate of the whole subtree is estimated as the weighted average of the two subtrees:

$$E_R = \frac{m_2}{N}E_2 + \frac{m_5}{N}E_5 \tag{6.7}$$

Of course, in a situation with more than just two subtrees, the weighted average has to be taken over all of them. This should present no major difficulties.

As for the values of E_2 and E_5, these are obtained from the error rates of the specific subtrees, and these again from the error rates of their sub-subtrees, and so on, all the way down to the lowest level tests. The error-estimating procedure is a recursive undertaking.

Suppose that the tree to be pruned is the one rooted at t_3, which happens to be one of the two children of test t_2. The error estimate for t_2 is calculated as the weighted average of E_3 and the error estimate for the other child of t_2 (the leaf labeled with the positive class). The resulting estimate would then be combined with E_5 as shown above.

Post-pruning The term refers to the circumstance that the decision tree is pruned *after* it has been fully induced from data (an alternative is the subject of the next subsection). We already know that the essence is to replace a subtree with a leaf labeled with the class most frequent among the training examples reaching that leaf. Since there are usually several (or many) subtrees that can thus be replaced, a choice has to be made; and the existence of a choice means we need a criterion to guide our decision.

Here is one possibility. We know that pruning is likely to change the classifier's performance. One way to assess this change is to compare the error estimate of the decision tree after the pruning with that of the tree before the pruning:

$$D = E_{after} - E_{before} \tag{6.8}$$

From the available pruning alternatives, we choose the one where this difference is the smallest, D_{min}; but we carry out the pruning only if $D_{min} < c$, where c is a user-set threshold for how much performance degradation can be tolerated in exchange for the tree's compactness. The mechanism is then repeated, with the decision tree becoming smaller and smaller, the stopping criterion being imposed by the constant c. Thus in Fig. 6.2, the first pruning step might have removed the subtree rooted at t_3; and the second step, the subtree rooted at t_6. Here the procedure was stopped because any further attempt at pruning resulted in a tree whose error estimate increased too much: the difference between the estimated error of the final (pruned) tree and that of the original tree on the left of Fig. 6.2 exceeded the user's threshold: $D > c$.

The principle is summarized by the pseudocode in Table 6.8.

On-line Pruning In the divide-and-conquer procedure, each subsequent attribute divides the set of training examples into smaller and smaller subsets. Inevitably, the evidence supporting the choice of the tests at lower tree-levels will be weak. When a tree node is reached by only, say, two training examples, one positive and one negative, a totally irrelevant attribute may by mere coincidence succeed in separating the positive example from the negative. The only "benefit" to be gained from adding this test to the decision tree is training-set overfitting.

The motivation behind *on-line* pruning is to make sure this situation is prevented. Here is the rule: if the training subset is smaller than a user-set minimum, m, stop further expansion of the tree.

Impact of Pruning How far the pruning goes is controlled by two parameters: c in post-pruning, and m in on-line pruning. In both cases, higher values result in smaller trees.

Table 6.8 The algorithm for decision-tree pruning

---

c . . . a user-set constant

---

(1) Estimate the error rate of the original decision tree. Let its value be denoted by E_{before}.
(2) Estimate the error rates of the trees obtained by alternative ways of pruning the original tree.
(3) Choose the pruning after which the estimated error rate experiences minimum increase, $D_{min} = E_{before} - E_{after}$, but only if $D_{min} < c$.
(4) Repeat steps (2) and (3) as long as $E_{before} - E_{after} < c$.

The main reason why pruning tends to improve classification performance on future examples is that the removal of low-level tests, which have poor statistical support, usually reduces the danger of overfitting. This, however, works only up to a certain point. If overdone, a very high extent of pruning can (in the extreme) result in the decision being replaced with a single leaf labeled with the majority class. Such classifier is unlikely to be useful.

Figure 6.3 shows the effect that gradually increased pruning typically has on classification performance. Along the horizontal axis is plotted the extent of pruning as controlled by c or m or both. The vertical axis represents the error rate measured on the training set as well as on some hypothetical testing set (the latter consisting of examples that have *not* been used for learning, but whose class labels are known).

On the training-set curve, error rate is minimized when there is no pruning at all. More interesting, however, is the testing-set curve. Its shape is telling us that the unpruned tree usually scores poorly on testing data, which is explained by the unpruned tree's tendency to overfit the training set, a phenomenon that can be reduced by increasing the extent of pruning. Excessive pruning, however, will remove attribute tests that *do* carry useful information, and this will have a detrimental effect on classification performance.

By the way, the two curves can tell us a lot about the underlying data. In some applications, even very modest pruning will impair error rate on testing data; for instance, in a noise-free domain with a relatively small training set.

Another thing to notice is that the error rate on the testing set is almost always greater than the error rate on the training set.

What Have You Learned?

To make sure you understand this topic, try to answer the following questions. If you have problems, return to the corresponding place in the preceding text.

- What are the potential benefits of decision-tree pruning?
- How can we estimate the tree's error rate on future data? Write down the formula and explain how it is used.

Fig. 6.3 With the growing extent of pruning, error rate on the testing set usually drops, then starts growing again. Error rate on the training set usually increases monotonically

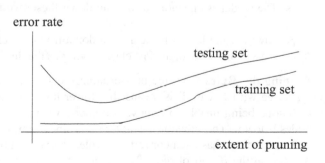

- Describe the principle of post-pruning and the principle of on-line pruning.
- What parameters control the extent of pruning? How do they affect the error rate on the training set, and how do they affect the error rate on the testing set?

6.6   Converting the Decision Tree into Rules

One of the advantages of decision trees in comparison with the other classifiers is their interpretability. Any sequence of tests along the path from the root to a leaf represents an *if-then* rule, and this rule explains why the classifier has labeled a given example with this or that class.

Rules Generated by a Decision Tree   The reader will find it easy to convert a decision tree to a set of rules. It is enough to notice that a leaf is reached through a series of edges whose specific choice is determined by the results of the attribute tests encountered along the way. Each leaf is thus associated with a concrete conjunction of test results.

For the sake of illustration, let us write down the complete set of rules for the pos class as obtained from the decision tree in Fig. 6.1a.

if   `crust-size=big AND filling-size=big` *then* pos
if   `crust-size=big AND filling-size=small AND shape=circle`
       then pos
if   `crust-size=small AND shape=circle` *then* pos
if   `crust-size=small AND (shape=square OR triangle)`
       `AND filling-size=big` *then* pos
else   neg

Note the *default class*, neg, in the last line. An example is labeled with the default class if all rules fail, which means that the value of the *if*-part of each rule is *false*. We notice that, in this two-class domain, we need to consider only the rules resulting in the pos class, the other class being the default option. We could have done it the other way round, considering only the rules for the neg class, and making pos the default class. This would actually be more economical because there are only two leafs labeled with the neg, and therefore only two corresponding rules. The reader is encouraged to write down these two rules by way of a simple exercise.

At any rate, the lesson is clear: in a domain with K classes, only the rules for $K - 1$ classes are needed, the last class serving as the default.

Pruning the Rules   The tree post-pruning mechanism described earlier replaced a subtree with a leaf. This means that lower-level tests were the first to go, the technique being unable to remove a higher-level node before the removal of the nodes below it. The situation is similar in on-line pruning.

Once the tree has been converted to rules, however, pruning gains in flexibility: *any* test in the *if*-part of *any* rule is a potential candidate for removal; and entire

Table 6.9 The algorithm for rule pruning

Re-write the decision tree as a set of rules.

Let c be a user-set constant controlling the extent of pruning

(1) In each rule, calculate the increase in error estimate brought about by the removal of individual tests.
(2) Choose those removals where this increase, D_{min} is smallest. Remove the tests, but only if $D_{min} < c$.
(3) In the set of rules, search for the weakest rules to be removed.
(4) Choose the default class.
(5) Order the rules according to their strengths (how many training examples they cover).

Table 6.10 Illustration of the algorithm for rule pruning

Suppose that the decision from the left part of Fig. 6.2 has been converted into the following set of rules (**neg** is the default label to be used when the if-parts of all rules are *false*).

$$t_1 \wedge t_2 \qquad\qquad \rightarrow \text{pos}$$
$$t_1 \wedge \neg t_2 \wedge t_3 \wedge t_4 \rightarrow \text{pos}$$
$$\neg t_1 \wedge t_5 \wedge t_6 \qquad \rightarrow \text{pos}$$
$$else \quad \text{neg}$$

Suppose that the evaluation of the tests in the rules has resulted in the conclusion that t_3 in the second rule and t_5 in the third rule can be removed without a major increase in the error estimate. We obtain the following set of modified rules.

$$t_1 \wedge t_2 \qquad\quad \rightarrow \text{pos}$$
$$t_1 \wedge \neg t_2 \wedge t_4 \rightarrow \text{pos}$$
$$\neg t_1 \wedge t_6 \qquad \rightarrow \text{pos}$$
$$else \quad \text{neg}$$

The next step can reveal that the second (already modified) rule can be removed without a major increase in the error estimate. After the removal, the set of rules will look as follows.

$$t_1 \wedge t_2 \qquad \rightarrow \text{pos}$$
$$\neg t_1 \wedge t_6 \quad \rightarrow \text{pos}$$
$$else \quad \text{neg}$$

This completes the pruning.

rules can be deleted. This is done by the rule-pruning algorithm summarized by the pseudocode in Table 6.9 and illustrated by the example in Table 6.10. Here, the initial set of rules was obtained from the (now familiar) tree in the left part of Fig. 6.2. The first pruning step removes those tests that do not appear to contribute much to the overall classification performance; the next step deletes the weakest rules.

We haste to admit, however, that the price for this added flexibility is a significant increase in computational costs.

What Have You Learned?

To make sure you understand this topic, try to answer the following questions. If you have problems, return to the corresponding place in the preceding text.

- Explain the mechanism that converts a decision tree to a set of rules. How many rules are thus obtained? What is the motivation behind such conversion?
- What is meant by the term, *default class*? How would you choose it?
- Discuss the possibilities of rule-pruning. In what sense can we claim that rule-pruning offers more flexibility than decision-tree pruning? What is the price for this increased flexibility?

6.7   Summary and Historical Remarks

- In decision trees, the attribute values are tested one at a time, the result of each test indicating what should happen next: either another attribute test, or a decision about the class label if a leaf has been reached. One can say that a decision tree consists of a set of partially ordered set of tests, each sequence of tests defining one branch in the tree terminated by a leaf.
- From a typical training set, many alternative decision trees can be created. As a rule, smaller trees are to be preferred, their main advantages being interpretability, removal of irrelevant and redundant attributes, and lower danger of overfitting noisy training data.
- The most typical procedure for induction of decision trees from data proceeds in a recursive manner, always seeking to identify the attribute that conveys maximum information about the class label. This approach tends to make the induced decision trees smaller. The "best" attribute is identified by simple formulas borrowed from information theory.
- An important aspect of decision-tree induction is pruning. The main motivation is to make sure that all tree branches are supported by sufficient evidence. Further on, pruning reduces the tree size which has certain advantages (see above). Two generic types of pruning exist. (1) In post-pruning, the tree is first fully developed, and then pruned. (2) In on-line pruning (which is perhaps a bit of a misnomer), the development of the tree is stopped once the training subsets used to determine the next attribute test become too small. In both cases, the extent of pruning is controlled by user-set parameters (denoted c and m, respectively).
- A decision tree can be converted to a set of rules that can further be pruned. In a domain with K classes, it is enough to specify the rules for $K - 1$ classes,

the remaining class becoming the *default class*. The rules are usually easier to interpret. Rule-pruning algorithms sometimes lead to more compact classifiers, though at significantly increased computational costs.

Historical Remarks   The idea behind decision trees was first put forward by Hoveland and Hund in the late 1950s. The work was later summarized in the book Hunt et al. [39] that reports experience with several implementations of their Concept Learning System (CLS). Friedman et al. [30] developed a similar approach independently. An early high point of the research was reached by Breiman et al. [11] where the system CART is described. The idea was then imported to the machine-learning world by Quinlan [75, 76]. Perhaps the most famous implementation is C4.5 from Quinlan [78]. This chapter was based on a simplified version of C4.5.

6.8   Solidify Your Knowledge

The exercises are to solidify the acquired knowledge. The suggested thought experiments will help the reader see this chapter's ideas in a different light and provoke independent thinking. Computer assignments will force the readers to pay attention to seemingly insignificant details they might otherwise overlook.

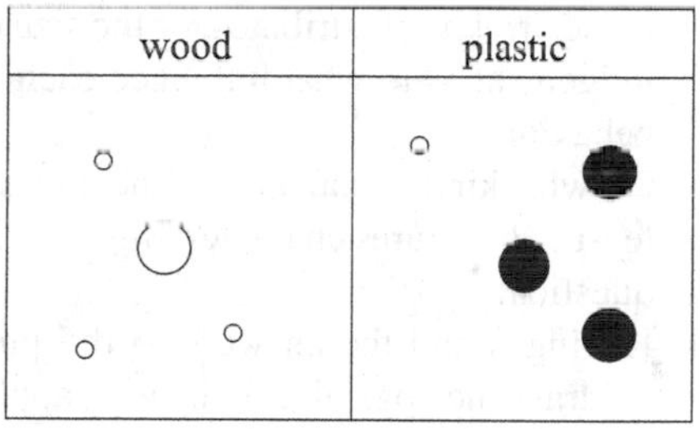

Fig. 6.4 Another example with wooden and plastic circles

Exercises

1. In Fig. 6.4, eight training examples are described by two attributes, `size` and `color`, the class label being the material: either `wood` or `plastic`.

   - What is the entropy of the training set when only the class labels are considered (ignoring the attribute values)?
   - Using the information-based mechanism from Sect. 6.3, decide which of the two attributes better predicts the class

2. Take the decision tree from Fig. 6.1a and remove from it the bottom-right test on `filling-size`. Based on the training set from Table 6.1, what will be the error estimate before and after this "pruning"?
3. Choose one of the decision trees in Fig. 6.1 and convert it to a set of rules. Pick one of these rules and decide which of its tests can be removed with the minimum increase in the estimated error.
4. Consider a set of ten training examples. Suppose there is a continuous attribute that has the following values: $3.6, 3.2, 1.2, 4.0, 0.8, 1.2, 2.8, 2.4, 2, 2, 1.0$. Suppose that the first five of these examples, and also the last one, are positive, all other examples being negative. What will be the best binary split of the range of this attribute's values?

Give It Some Thought

1. The baseline performance criteria used for the evaluation of decision trees are error rate and the size of the tree (the number of nodes). These, however, may not be appropriate in certain domains. Suggest applications where either the size of the decision tree or its error rate may be less important. Hint: Consider the costs of erroneous decisions and the costs of obtaining attribute values.
2. What are likely to be the characteristics of a domain where a decision tree clearly outperforms the baseline 1-NN classifier? Hint: Consider such characteristics as noise, irrelevant attributes, or the size of the training set; and then make your own judgement as to what influence each of them is likely to have on the classifier's behavior.
3. On what kind of data may a linear classifier do better than a decision tree? Give at least two features characterizing such data. Rely on the same hint as the previous question.
4. Having found the answers to the previous two questions, you should be able to draw the logical conclusion: applying to the given data both decision-tree induction and linear-classifier induction, what will their performance betray about the characteristics of the data?
5. The decision tree as described in this chapter gives only "crisp" yes-or-no decisions about the given example's class (in this sense, one can argue that Bayesian classifiers or multilayer perceptrons are more flexible). By way of mitigating this weakness, suggest a mechanism that would modify the decision-trees framework so as to give, for each example, not only the class label, but also the classifier's confidence in this class label.

Computer Assignments

1. Implement the baseline algorithm for the induction of decision trees and test its behavior on a few selected domains from the UCI repository.[4] Compare the results with those achieved by the k-NN classifier.
2. Implement the simple pruning mechanism described in this chapter. Choose a data file from the UCI repository. Run several experiments and observe how different extent of pruning affects the error rate on the training and testing sets.
3. Choose a sufficiently large domain from the UCI repository. Put aside 30% of the examples for testing. For training, use 10%, 20%, ... 70% of the remaining examples, respectively. Plot a graph where the horizontal axis gives the number of examples, and the vertical axis gives the computational time spent on the induction. Plot another graph where the vertical axis will give the error rate on the testing set. Discuss the obtained results.

---

[4] www.ics.uci.edu/~mlearn/MLRepository.html.

Chapter 7
Computational Learning Theory

As they say, nothing is more practical than a good theory. And indeed, mathematical models of learnability have helped improve our understanding of what it takes to induce a useful classifier from data, and, conversely, why the outcome of a machine-learning undertaking so often disappoints. And so, even though this textbook does not want to be mathematical, it cannot help introducing at least the basic concepts of the *computational learning theory*.

At the core of this theory is the idea of *PAC learning*, a paradigm that makes it possible to quantify learnability. Restricting itself to domains with discrete attributes, the first section of this chapter derives a simple expression that captures the mutual relation between the training-set size and the induced classifier's error rate. Some consequences of this formula are briefly discussed in the two sections that follow. For domains with continuous attributes, the so-called VC-dimension is then introduced.

7.1 PAC Learning

Perhaps the most useful idea contributed by computational learning theory is that of "probably approximate learning," sometimes abbreviated as *PAC learning*. Let us first explain the underlying principles and derive a formula that will then provide some useful guidance.

Assumptions and Definitions The analysis will be easier if we build it around a few simplifying assumptions. First, the training examples—as well as all future examples—are completely noise-free. Second, all attributes are discrete (none of them is continuous-valued). Third, the classifier acquires the form of a logical

© Springer International Publishing AG 2017
M. Kubat, *An Introduction to Machine Learning*,
DOI 10.1007/978-3-319-63913-0_7

expression of attribute values; the expression is *true* for positive examples and *false* for negative examples. And, finally, there exists at least one expression that correctly classifies all training examples.[1]

Each of the logical expressions can then be regarded as one hypothesis about what the classifier should look like. Together, all of the hypotheses form a *hypothesis space* whose size (the number of distinct hypotheses) is $|H|$. Under the assumptions listed above, $|H|$ is a finite number.

Inaccurate Classifier May Still Succeed on the Training Set Available training data rarely exhaust all subtleties of the underlying class; a classifier that labels correctly all training examples may still perform poorly in the future. The frequency of these mistakes is usually lower if we add more training examples because these additions may reflect aspects that the original data failed to represent. The rule of thumb is actually quite simple: the more examples we have, the better the induced classifier.

How many training examples will give us a fair chance of future success? To find the answer, we will first consider a hypothetical classifier whose error rate on the entire instance space is greater than some predefined ϵ. Put another way, the probability that this classifier will label correctly a randomly picked example is less than $1 - \epsilon$. Taking this reasoning one step further, the probability, P, that this imperfect classifier will label correctly m random examples is bounded by the following expression:

$$P \leq (1 - \epsilon)^m \tag{7.1}$$

And here is what this means: with probability $P \leq (1 - \epsilon)^m$, an entire training set consisting of m examples will be correctly classified by a classifier whose error rate actually exceeds ϵ. Of course, this probability is for a realistic m very low. For instance, if $\epsilon = 0.1$ and $m = 20$ (which is a very small set), we will have $P < 0.122$. If we increase the training-set size to $m = 100$ (while keeping the error rate bound by $\epsilon = 0.1$), then P drops to less than 10^{-4}. This is indeed small; but low probability is not the same as impossibility.

Eliminating Poor Classifiers Suppose that an error rate greater than ϵ is deemed unacceptable. What are the chances that a classifier with performance this poor is induced from the given training set, meaning that it classifies correctly all training examples?

The hypothesis space consists of $|H|$ classifiers. Let us consider the theoretical possibility that we evaluate all these classifiers on the m training examples, and then keep only those classifiers that have never made any mistake. Amongthese

---

[1]The reader will have noticed that all these requirements are satisfied by the "pies" domain from Chap. 1.

"survivors," some will disappoint in the sense that, while being error-free on the training set, their error rates on the entire instance space actually exceed ϵ. Let there be k such classifiers.

The concrete value of k cannot be established without evaluating each single classifier on the entire instance space. This being impossible, all we can say is that $k \leq |H|$, which is a somewhat better situation because $|H|$ is known in many realistic cases, or at least can be calculated.[2]

Let us re-write the upper bound on the probability that at least one of the k offending classifiers will be error-free on the m training examples.

$$P \leq k(1 - \epsilon)^m \leq |H|(1 - \epsilon)^m \qquad (7.2)$$

With this, we have established an upper bound on the probability that m training examples will succeed in eliminating all classifiers whose error rate exceeds ϵ.

To become useful, the last expression has to be modified. We know from mathematics that $1 - \epsilon < e^{-\epsilon}$, which means that $(1 - \epsilon)^m < e^{-m\epsilon}$. With this in mind, we are able to express the upper bound in an exponential form:

$$P \leq |H| \cdot e^{-m\epsilon} \qquad (7.3)$$

Suppose we want this probability to be lower than some user-set δ:

$$|H| \cdot e^{-m\epsilon} \leq \delta \qquad (7.4)$$

Taking the logarithm of both sides, and rearranging the terms, we obtain the formula that we will work with in the next few pages:

$$m > \frac{1}{\epsilon}(\ln |H| + \ln \frac{1}{\delta}) \qquad (7.5)$$

"Probably Approximately Correct" Learning The main reason we have gone through all the details of the derivation is that the reader thus gets a better grasp of the meanings and interpretations of the involved variables which may otherwise be a bit confusing. This is also why these variables are summarized in Table 7.1 for quick reference.

We have now reached a stage where we are able to define some important concepts. A classifier with error rate below ϵ is deemed *approximately correct*; and δ is the *probability* that this approximately correct classifier will be induced from m training examples (m being a finite number). Hence the name of the whole paradigm: *probably approximately correct* learning, or simply *PAC learning*. For the needs of this chapter, we will say that a class is *not PAC-learnable* if the number of examples

---

[2]Recall that in the "pies" domain from Chap. 1, the size of the hypothesis space was $|H| = 2^{108}$. Of these hypotheses, 2^{96} classified correctly the entire training set.

Table 7.1 The variables involved in our studies of PAC-learnability

m	...	The number of the training examples
$|H|$	...	The size of the hypothesis space
ϵ	...	The classifier's maximum permitted error rate
δ	...	The probability that a classifier with error rate greater than ϵ is error-free on the training set

needed to satisfy the given (ϵ, δ)-requirements is so high that we cannot expect a training set of this size ever to be available—or, if it *is* available, then the learning software will need impracticably long time (say, thousands of years) to induce from it the classifier.

Interpretation Inequality (7.5) specifies how many training examples, m, are needed if, with probability at least δ, a classifier with error rate lower than ϵ is to be induced. Note that this result does not depend on the concrete machine-learning technique, only on the size, $|H|$, of the hypothesis space defined by the given type of classifier.

An important thing to remember is that m grows linearly in $1/\epsilon$. For instance, if we strengthen the limit on the error rate from $\epsilon = 0.2$ to $\epsilon = 0.1$, we will need (at least in theory) twice as many training examples to have the same chance, δ, of success. At the same time, the reader should also notice that m is less sensitive to changes in δ, growing only logarithmically in $1/\delta$.

However, we must not forget that our derivation was something like a worst-case analysis. As a result, the bound it has given us is less tight than a perfectionist might desire. For instance, the derivation allowed the possibility that $k = |H|$, which is clearly too pessimistic for the vast majority of practical applications. Inequality (7.5) should therefore never be interpreted as telling the engineer how many training examples to use. Rather, it should be seen as a guideline that makes it possible to compare the learnability of alternative classifier types. We will pursue this idea in the next section.

What Have You Learned?

To make sure you understand this topic, try to answer the following questions. If you have problems, return to the corresponding place in the preceding text.

- What is the meaning of the four variables ($m, |H|, \epsilon$, and δ) used in Inequality (7.5)?
- What does the term *PAC learning* refer to? Under what circumstances do we say that a class is *not PAC-learnable*?
- Derive Inequality (7.5). Discuss the meaning and practical benefits of this result, and explain why it should only be regarded as a worst-case analysis.

7.2 Examples of PAC Learnability

Inequality (7.5) tells us how learnability, defined by the (ϵ, δ)-requirements, depends on the size of the hypothesis space. Let us illustrate this result on two concrete types of classifiers.

Conjunctions of Boolean Attributes: Hypothesis Space Suppose that all attributes are boolean, that all data are noise-free, and that an example's class is known to be determined by a logical conjunction of attribute values: if *true*, then the example is positive, otherwise it is deemed negative. For instance, the labels of the training examples may be determined by the following expression:

```
att1 = true AND att3 = false
```

This means that an example is labeled as pos if the value of the first attribute is *true* and the value of the third attribute is *false*, regardless of the value of any other attribute. An example that fails to satisfy these two conditions is labeled as neg.

The task for the machine-learning program is to find a conjunction that correctly labels all training examples. The set of all conjunctions permitted by our informal definition forms the hypothesis space, $|H|$. What is the size of this space?

In a logical conjunction of the kind specified above, each attribute is either *true* or *false* or irrelevant. This gives us three possibilities for the first attribute, times three possibilities for the second, and so on, times three possibilities for the last, n-th, attribute. The size of the hypothesis space is therefore $|H| = 3^n$.

Conjunctions of Boolean Attributes: PAC-Learnability Suppose that a training set consisting of noise-free examples is presented to some machine-learning program that is capable of inducing classifiers of the just-defined form.[3] To satisfy the last of the assumptions listed at the beginning of Sect. 7.1, we will assume that at least one of the logical conjunctions classifies correctly all training examples.

Since $\ln |H| = \ln 3^n = n \ln 3$, we can re-write Inequality (7.5) in the following form:

$$m > \frac{1}{\epsilon}(n \ln 3 + \ln \frac{1}{\delta}) \qquad (7.6)$$

With this, we have obtained a conservative lower bound on the number of training examples that are needed if our (ϵ, δ)-requirements are to be satisfied: with probability δ, the induced classifier (error-free on the training set) will exhibit error rate less than ϵ on the entire instance space. Note that the value of this expression grows linearly in the number of attributes, n. Theoretically speaking, then, if n is doubled, then twice as many training examples will be needed if, with probability limited by δ, classifiers with error rates above the predefined ϵ are to be weeded out.

---

[3]For instance, a variation of the hill-climbing search from Sect. 1.2 might be used to this end.

Any Boolean Function: Hypothesis Space Let us now investigate a broader class of classifiers, namely those defined by *any* boolean function, allowing for all three basic logical operators (AND, OR, and NOT) as well as any combination of parentheses. As before, we will assume that the examples are described by n boolean attributes, and that they are noise-free.

What is the size of this hypothesis space?

From n boolean attributes, 2^n different examples can be created. This defines the size of the instance space. For any subset of these 2^n examples, there exists at least one logical function that is *true* for all examples from this subset (labeling them as pos) and *false* for all examples from outside this subset (labeling them as neg). Two logical functions will be regarded as identical from the classification point of view if each of them labels any example with the same class; that is, if they never differ in their "opinion" about any example's class. The number of logical functions that are mutually distinct in their classification behavior then has to be the same as the number of the subsets of the instance space.

A set consisting of X elements is known to have 2^X subsets. Since our specific instance space consists of 2^n examples, the number of its subsets is 2^{2^n}—and this is the size of our hypothesis space:

$$|H| = 2^{2^n} \tag{7.7}$$

Any Boolean Function: PAC-Learnability Since $\ln |H| = \ln 2^{2^n} = 2^n \ln 2$, Inequality (7.5) acquires the following form:

$$m > \frac{1}{\epsilon}(2^n \ln 2 + \ln \frac{1}{\delta}) \tag{7.8}$$

We conclude that the lower bound on the number of the training examples that are needed if the (ϵ, δ)-requirements are to be satisfied grows here exponentially in the number of attributes, n. Such growth is prohibitive for any realistic value of n. For instance, even if we add only a single attribute, $n + 1$, the value of $\ln |H|$ will double because $\ln |H| = 2^{n+1} \ln 2$ which is twice a much as $2^n \ln 2$. And if we add ten attributes, $n + 10$, then the value of $\ln |H|$ increases a thousand times because $\ln |H| = 2^{n+10} \ln 2 = 2^{10} \cdot 2^n \ln 2 = 1024 \cdot 2^n \cdot \ln 2$.

This observation is enough to convince us that a classifier in this general form is *not PAC-learnable*.

A Word of Caution We must be careful not to jump to conclusions. As pointed out earlier, the derivation of Inequality (7.5)—a worst-case analysis of sorts—relied on quite a few simplifying assumptions that render the obtained bounds very conservative. In reality, the number of the training examples needed for the induction of a reliable classifier is much lower than that indicated by our "magic formula."

For the engineer seeking to choose the most appropriate learning technique, the main contribution of Inequality (7.5) is that it helps him compare PAC-learnability of classifiers constrained by different "languages." For instance, we have seen that a conjunction of attribute values can be learned from a reasonably sized training set, whereas a general concept defined by *any* boolean function usually cannot.

Some other corollaries of this analysis will be the subject of Sect. 7.3.

What Have You Learned?

To make sure you understand this topic, try to answer the following questions. If you have problems, return to the corresponding place in the preceding text.

- What is the size of a hypothesis space consisting of conjunctions of attribute values? Substitute this size to Inequality (7.5) and discuss the result.
- Can you find the value of $|H|$ for some other class of classifiers?
- Explain, in plain English, why a boolean function in its general form is not *PAC-learnable*.

7.3 Some Practical and Theoretical Consequences

The theoretical analysis from the first section of this chapter offers a broader perspective of the learning task as well as clues whose significance cannot be overstated. Their benefits range from purely intellectual satisfaction enjoyed by a theoretician to practical guidelines appreciated by the most down-to-earth engineer. Let us take a quick look at some of them.

Bias and Learnability Section 7.2 investigated *PAC-learnability* of classes whose descriptions are limited to conjunctions of attribute values. A constraint of this kind is called a *bias* for a specific type of class.

Allowing for some attributes to be ignored as irrelevant, we calculated the size of the hypothesis space thus defined as $|H| = 3^n$. If we strengthen the bias by insisting that every single attribute must be involved, we will reduce the size of the hypothesis space: $|H| = 2^n$. This is because every attribute in the class-describing expression is either *true* or *false*, whereas in the previous case, a third possibility ("ignored") was permitted.

Many other biases exist, some limiting the number of terms in the conjunction, others preferring a disjunction or some pre-specified combination of conjunctions and disjunctions, or imposing yet another constraint. What matters, for our discussion, is that each bias is likely to result in a different size of the hypothesis space. And, as we have seen, this size affects learnability.

No Learning Without a Bias In the absence of *any* bias, allowing for any boolean function, we have established that the value of $\ln |H|$ grows exponentially in the number of attributes, $\ln |H| = 2^n \ln 2$, which means that the class is in this most general form *not PAC-learnable*. The explanation is simple. The unconstrained hypothesis space is so vast that there is a reasonable chance that one may find a classifier that labels correctly the entire training set, and yet exhibits poor performance on future examples. Put bluntly, the induced classifier cannot be trusted.

It is in *this* sense that we say, with a grain of salt, that "there is no learning without a bias."

The thing to remember is that the machine-learning adventure can only succeed if the engineer constrains the hypothesis space by a meaningful bias. It stands to reason, however, that this bias should not be misleading. The reader will recall that our analysis assumed that the hypothesis space *does* contain the solution, and that the examples are noise-free.[4]

Occam's Razor Quite often, the engineer has an opportunity to choose from two or more biases. For instance, the class to be learned is described by a conjunction of attribute values, in which each single attribute plays a part. A weaker bias that permits the absence of some attributes from the conjunctions also includes the case where *zero* attributes are absent, and is therefore correct, too. In the former case, we have $\ln |H| = n \ln 2$; and in the latter, $\ln |H| = n \ln 3$, which is bigger. We thus have two correct biases, each defining a hypothesis space of a different size. Which of them to prefer?

The attentive reader no doubt already knows the answer. In Inequality (7.5), the number of training examples needed for successful learning depends on $\ln |H|$. A lower value of this term indicates that fewer examples are needed if we want to satisfy the given (ϵ, δ)-requirements. The engineer will thus benefit by choosing the bias whose hypothesis space is smaller.

By the way, scientists have been using a similar rule for centuries: in a situation where two different hypotheses can explain a certain phenomenon, it is assumed that the simpler one has a higher chance of success. The fact that this principle, *Occam's Razor*, has been named after a scholastic theologian who died in the fourteenth century indicates that the use of this rule pre-dates modern science. The word *razor* is here to emphasize that, when formulating a hypothesis, we better slice away all unnecessary information.

That mathematicians have now been able to find a formal proof of the validity of this principle in the field of machine learning, and even to quantify the scope of its utility, is a remarkable achievement.

Irrelevant and Redundant Attributes In the types of classes investigated in Sect. 7.2, the lower bound on the number of examples, m, depended on the number of attributes, n. For instance, in the case of conjunctions of attribute values, we now know that $\ln |H| = n \ln 3$; and the number of examples needed to satisfy given (ϵ, δ)-requirements thus grows linearly in n.

---

[4] Analysis of situations where these requirements are not satisfied would go beyond the scope of an introductory textbook.

The same result enables us to form an opinion about how a class learnability might be affected by the presence of irrelevant or redundant attributes. The higher the number of these attributes, the higher the value of n, which means that more training examples have to be provided if we want to satisfy the (ϵ, δ)-requirements. The lesson is clear: whenever we have a chance to identify (and remove) the less-then-useful attributes, we better do so.

These considerations also explain why it is so difficult to induce a good classifier in domains where only a tiny percentage of attributes carry the relevant information—as is the case in text classification.

What Have You Learned?

To make sure you understand this topic, try to answer the following questions. If you have problems, return to the corresponding place in the preceding text.

- What is meant by the statement, "there is no learning without a bias"? Conversely, under what circumstances will the engineer's bias tend to hurt the learning algorithm's chances of success?
- Explain the meaning of the term, *Occam's Razor*. In what way did mathematicians provide a solid ground for this—formerly philosophical—principle?
- What does Inequality (7.5) tell us about the role of irrelevant and redundant attributes?

7.4 VC-Dimension and Learnability

So far, we have only dealt with domains where the number of hypotheses is finite. Let us now turn our attention to domains where the hypothesis space is infinite, which is often inevitable in applications where at least some of the attributes are continuous-valued. In this situation, it would appear that comparing the learnability of two classes of classifiers is all but impossible because both of them have infinite $|H|$ anyway.

To begin with, the reader will agree that some classes of classifiers (say, polynomial) are more *flexible* than others (say, linear). Surely this flexibility is likely to have some impact on learnability? Indeed it does. Let us take a quick look at how theoreticians quantify learnability under this situation.

Shattered Training Set Consider the three examples in the two-dimensional space depicted on the left of Fig. 7.1. No matter how we distribute the positive and negative labels among them, we will always be able to find a linear classifier that separates the two classes (the classifiers are shown in the same picture). We say that this particular set of examples is "shattered" by the class of linear classifiers.

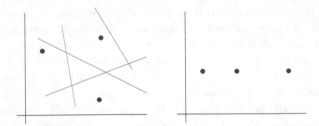

Fig. 7.1 The set of the three points on the *left* is "shattered" by a linear classifier. The set of the three points on the *right* is *not* shattered by a linear classifier because no straight line can separate the point in the middle from those on the sides

This, however, is not the case of the three points on the right. As we can see, these points find themselves all on the same line. If we label the one in the middle as positive and the other two as negative, no linear classifier will ever succeed in separating the two classes from each other. In other words, this particular set of examples is *not* shattered by the linear classifier.

A different class of classifiers will have a different power to shatter a given set of examples. For example, a parabola (a special kind of a quadratic, and thus polynomial function) will shatter the three aligned points on the right; it will even shatter four points that do not lie on the same line. And other classes of classifiers, such as high-order polynomials, will shatter any realistically sized set of examples.

Vapnik-Chervonenkis Dimension The Vapnik-Chervonenkis dimension (usually abbreviated as *VC-dimension*) of a given class of classifiers is defined as the size of the largest set of examples shattered by this class.

We have seen that, in a two-dimensional space, a linear classifier fails to shatter three points that are all on the same line, but that the linear classifier *does* shatter them if they do *not* lie on the same line. At the same time, four points, no matter how we arrange them in a plane, can always be labeled in a way that makes it impossible to separate positive examples from the negative linearly. Since the definition says, "the largest set of examples shattered by this class," we are bound to conclude that the VC-dimension of a linear classifier in a two-dimensional space is $VC_L = 3$.

The point to remember is that the value of the VC-dimension reflects—and quantifies—the geometrical properties of the given class of classifiers.

Learnability in Continuous Domains The concept of VC-dimension makes it possible for us to deal with learnability in continuous domains. Let us give here a slightly modified version of a famous theorem, omitting certain technicalities that are irrelevant for our specific needs:

If the VC-dimension, d, of a classifier class, H, is finite, then an error rate below ϵ can be achieved with confidence $1 - \delta$ if the target class is identical with some hypothesis $h \in H$, and if the number of the training examples, m, satisfies the following inequality:

Table 7.2 VC-dimensions for some hypothesis classes in R^n

Hypothesis class	VC-dimension
Hyperplane	$n+1$
Hypersphere	$n+2$
Quadratic	$\dfrac{(n+1)(n+2)}{2}$
r-Order polynomial	$\dbinom{n+r}{r}$

$$m \geq \max(\frac{4}{\epsilon} \log \frac{2}{\delta}, \frac{8d}{\epsilon} \log \frac{13}{\epsilon}) \qquad (7.9)$$

Note that this means that the lower bound on m is either $(\frac{4}{\epsilon} \log \frac{2}{\delta})$ or $(\frac{8d}{\epsilon} \log \frac{13}{\epsilon})$, whichever is greater.

The engineer interprets this result as telling him that he can then trust any classifier which correctly classifies the entire training set of size m, regardless of the machine-learning algorithm that has induced the classifier.

Note that the number of examples necessary for PAC learning grows linearly in the VC-dimension. This, however, is no reason to rejoice. Thing is, VC-dimensions of many realistic classes of classifiers have a way of growing very fast with the growing number of attributes—see below.

Some Example VC-Dimensions Table 7.2 lists some VC-dimensions for linear and polynomial classifiers. The reader may want to contemplate the inevitable trade-off: more complex classes of classifiers are more likely to contain the correct solution, but the number of training examples needed for success increases so dramatically that a classifier from this class cannot be deemed learnable.

Indeed, in the case of higher-order polynomials, the demands on the training-set size become all but prohibitive. For instance, the VC-dimension of a second-order polynomial in 100 dimensions (a fairly normal number of attributes) is as follows:

$$d = \frac{102 \cdot 101}{2 \cdot 1} = 5050$$

This is much larger than that of a linear classifier, but perhaps still acceptable. At any rate, the value is sufficiently high to make the first term in Inequality (7.9) small enough (by comparison) to be ignored.

If, however, we increase the polynomial's order to $r = 4$, the VC-dimension will become all but prohibitive:

$$d = \frac{104 \cdot 103 \cdot 102 \cdot 101}{4 \cdot 3 \cdot 2 \cdot 1} = 4{,}598{,}126$$

Polynomials of higher order therefore better be avoided. On the other hand, the VC-dimensions of neural networks and decision trees (see Chaps. 5 and 6) are known to be more affordable—which is why these classifiers are preferred.

A Word of Caution Just as in the case of Inequality (7.5), this one (Formula (7.9)) is the result of a worst-case analysis during which certain simplifying assumptions were made. The results therefore should not be interpreted as telling us how many examples are needed in any concrete application.

Importantly, in realistic applications, examples are not labeled arbitrarily. Since the examples of the same class are somehow similar to each other, they tend to be clustered together in the instance space.

What Have You Learned?

To make sure you understand this topic, try to answer the following questions. If you have problems, return to the corresponding place in the preceding text.

- How does the number of examples that are necessary for the (ϵ, δ)-requirements grow with the increasing VC-dimension, d?
- What is the VC-dimension of a linear classifier in the two-dimensional continuous space?
- What is the VC-dimension of an n-dimensional polynomial of the r-th order? Why does the high value of this VC-dimension mean that polynomial classifiers may not be a good choice?

7.5 Summary and Historical Remarks

- The *Computational Learning Theory* has been able to establish certain limits on the number of the training examples needed for successful classifier induction. These limits depend on two fundamental parameters: the threshold on error rate, ϵ, and the upper bound, δ, on the probability that m training examples will succeed in eliminating all classifiers whose error rate is above ϵ.
- If all attributes are discrete, the minimum number of examples needed to satisfy the (ϵ, δ)-requirements is determined by the size, $|H|$, of the given hypothesis space. Here is the classical formula:

$$m > \frac{1}{\epsilon}(\ln |H| + \ln \frac{1}{\delta}) \tag{7.10}$$

- In a domain where some attributes are continuous, the minimum number of examples needed to satisfy the (ϵ, δ)-requirements is determined by the so-called VC-dimension of the given class of classifiers. Specifically, if the value of the

VC-dimension is denoted by d, then the number of the needed examples is determined by the following inequality:

$$m \geq \max(\frac{4}{\epsilon} \log \frac{2}{\delta}, \frac{8d}{\epsilon} \log \frac{13}{\epsilon}) \tag{7.11}$$

- Inequalities (7.5) and (7.9) have been found by a worst-case analysis. In reality, therefore, much smaller training sets are usually sufficient for the induction of high-quality classifiers. The main benefit of these formulas is that they help us compare learnability in different machine-learning paradigms.

Historical Remarks The principles underlying the idea of PAC learning were proposed by Valiant [91]. The classical paper on what later came to be known as VC-dimension is Vapnik and Chervonenkis [94]; somewhat more elaborate version was later developed by Vapnik [92]. The idea to apply VC-dimension to learnability, and to the investigation of Occam's Razor is due to Blumer et al. [6]. Tighter bounds were later found, for instance, by Shawe-Taylor et al. [86]. VC-dimensions of linear and polynomial classifiers can be derived from the results published by Cover [18]. Readers interested in learning more about the *Computational Learning Theory* will greatly benefit from the excellent (even if somewhat older) book by Kearns and Vazirani [41].

7.6 Exercises and Thought Experiments

The exercises are to solidify the acquired knowledge. The ambition of the suggested thought experiments is to let the reader see this chapter's ideas in a different light and, somewhat immodestly, to provoke his or her independent thinking.

Exercises

1. Suppose that the instance space is defined by the attributes used in the "pies" domain from Chap. 1. Determine the size of the hypothesis space if the classifier is to be a conjunction of attribute values. Consider both cases: the one that assumes that some attributes might be ignored as irrelevant (or redundant), and the one that insists that all attributes must take part in the conjunction.
2. Return to the case of conjunctions of boolean attributes from Sect. 7.2. How many more examples will have to be used (in the worst-case analysis) if we change the required error rate from $\epsilon = 0.2$ to $\epsilon = 0.05$? Conversely, how will the size of the necessary training set be affected by changes in δ?
3. Again, consider the case where all attributes are boolean, and the classifier has the form of a conjunction of attribute values. What is the size of the hypothesis space

if the conjunction is permitted to involve exactly three attributes? For instance, here is one conjunction from this class:

```
att1 = true AND attr2 = false AND att3 = false
```

4. Consider a domain with $n = 20$ continuous-valued attributes. Calculate the VC-dimension for a classifier that has the form of a quadratic function, and compare it with that of a third-order polynomial.

   Second, suppose that the engineer has realized that half of the attributes are irrelevant. Having removed them, he now has $n = 10$. How will this reduction affect the VC-dimensions of the two classifiers?
5. Compare the *PAC-learnability* of the boolean function involving 8 attributes with the *PAC-learnability* of a quadratic classifier in a domain with 4 numeric attributes

Give It Some Thought

1. A certain role in Inequality (7.5) is played by δ, a term that quantifies the probability that a successful classifier will be induced from the given training set. Under what conditions can the impact of δ be neglected?
2. From the perspective of *PAC-learnability*, is there a difference between irrelevant and redundant attributes?
3. We have seen that a classifier is not *PAC-learnable* in the absence of a bias. The bias, however, many not be known. Suggest a learning procedure that would induce a classifier in the form of a boolean expression in this case. (Hint: consider two or more alternative biases.)
4. In the past, some machine-learning scientists considered the idea of converting continuous attributes into discrete ones by a process called *discretization*. By this they meant dividing the range of attribute values into intervals, each interval treated as a boolean attribute (the given numeric value either is or is not in the given interval).

   Suppose you are considering two ways of dividing the range $[0, 100]$. The first consists of two subintervals, $[0, 50]$, $[51, 100]$, and the second consists of ten equally sized subintervals: $[0, 10], \ldots [91, 100]$. Discuss the advantages and disadvantages of the two options from the perspective of *PAC-learnability*.

Chapter 8
A Few Instructive Applications

For someone wishing to become an expert on machine learning, mastering a handful of baseline techniques is not enough. Far from it. The world lurking behind a textbook's toy domains has a way of complicating things, frustrating the engineer with unexpected obstacles, and challenging everybody's notion of what exactly the induced classifier is supposed to do and why. Just as in any other field of technology, success is hard to achieve without a healthy dose of creativity.

Some practical experience helps, too, either your own, or at least of those who have tried and succeeded (or failed) before you. And this is what this chapter wants to offer. Using a few carefully selected case studies, it will acquaint you with some issues typically encountered in realistic applications—and with practical ways of dealing with them.

8.1 Character Recognition

The techniques from the previous chapters target recognition skills that are too difficult to be hard-coded in a computer program, but can be conveyed by means of pre-classified training examples. The ability to read text, even hand-written text, belongs to this category. Applications are legion: automated pre-processing of various forms, software for converting a text scribbled on a notepad into a format to be used by a text editor, and various newspaper-digitization programs. A mere generation or two ago, an undertaking of this kind was judged so ambitious as to be almost unrealistic; today, no one finds it extraordinary.

The Task To describe each character in a way that facilitates its recognition by a computer is not easy. To get the idea, just try to explain, in plain English, how to distinguish digit "5" from digit "6," or what constitutes the differencebetween a

© Springer International Publishing AG 2017
M. Kubat, *An Introduction to Machine Learning*,
DOI 10.1007/978-3-319-63913-0_8

Fig. 8.1 A simple way to convert the image of a hand-written character into a vector of 64 continuous attributes, each giving the mean intensity of the corresponding field

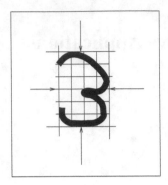

a 6−by−4 matrix of numeric attributes, each giving mean intensity of a field

hand-written "t" and "l." You will be surprised how difficult this is. And to convert the plain-English explanation to a code that a computer can understand is harder still.

This is why machine learning has been asked to help. If we prepare a collection of training examples, and describe them by pre-classified attribute vectors, then surely some of the techniques described in the previous chapters can induce the requisite classifier. The process is so simple as to seem almost trivial. Let us take a look at some of its crucial aspects.

Examples and Attribute Vectors To begin with, the engineer has to define the attributes to describe the examples. Figure 8.1 illustrates one possibility. The first step identifies a rectangular region in which the digit is located; the second divides this region into $6 \times 4 = 64$ equally sized fields, each characterized by a continuous attribute whose value gives the field's mean intensity. More ink means the field reflects less light, and thus results in a lower value of the attribute, and, conversely, the value is maximized if the field contains no ink at all. Of course, this is just the principle. There is no reason why there should be only 64 attributes. Indeed, the higher the level of detail the program is expected to discern, the smaller the size of the fields it needs to rely on, and thus the greater the number of attributes that will be used to describe the examples.

In a realistic application, the classifier will have to recognize not just one isolated character (such as the "3" in Fig. 8.1), but rather to "read" a whole text consisting of many such characters. This generalization, however, does not represent any major complication. As long as the individual characters can be isolated, and then treated separately, the same principle of describing each of them by an attribute vector can be applied.

Nowadays it is possible to download datasets of hundreds of thousands of hand-written characters, some described in the way presented above, some relying on other approaches.

Choosing the Classifier Now that we know how the training examples are going to be described, we are ready to proceed to the choice of the induction technique.

The first thing to be considered is that, in the attribute-vector obtained by the mechanism from Fig. 8.1, only a small percentage of the attributes (if any) are likely to be irrelevant or redundant, which means that this is not an issue to worry about (unless the number of attributes is increased way beyond those shown in Fig. 8.1). Another aspect guiding our choice of an appropriate machine-learning paradigm is the fact that we do not know whether the classes are linearly separable, and therefore hesitate to use linear classifiers. Finally, the classifiers need not be capable of offering explanations. If the intention is to read and convert to a text editor a long text, the user will hardly care to know the exact reason why a concrete character was classified as a "P" and not as a "D."

Based on these observations, the simple and easy-to-implement k-NN classifier looks like a good candidate. A cautious engineer will perhaps be concerned about the computational costs incurred in a domain with hundreds of thousands of training and testing examples. But as long as the number of attributes is moderate, these costs are unlikely to be prohibitive. From the practical point of view, a reasonable limit on what constitutes "prohibitive costs" will be determined by the computations associated with the isolation of the individual characters and the conversion of their images to attribute vectors. As long as these are comparable with the costs of classification (and they usually are), the classifier is affordable.

And indeed, the nearest-neighbor classifier is the most common choice, in this application, typically exhibiting an error rate of less than 2%. In some really illegible hand-writings, the error rate will of course be higher. But then, we should not be too harsh on the innocent machine, knowing as we do that even an experienced pharmacist finds it difficult to read certain hand-written prescriptions.

The Number of Classes In a domain where all characters are capitalized, the induced classifier is to discern 10 digits and 26 letters, which amounts to 36 classes. If both lowercase and uppercase letters are allowed, the total increases to $10 + 2 \times 26 = 62$ classes, and to these we may have to add special characters such as "?," "!," "$," and so on. This relatively high number of classes is not without consequences, and these deserve our attention.

The most immediate concern is the induced product's evaluation. Mere information about the error rate is somewhat inadequate here. Thus the performance of a classifier that correctly identifies 98% characters may appear good enough, even impressive; what this value fails to tell us, though, is how the errors are distributed. Typically, some characters will be correctly identified most of the time, while others pose difficulties—and as such, deserve further attention. For instance, certain pairs of similar characters tend to be mutually confused; the practically minded engineer then wants to know *which* pairs so as to mitigate the problem by providing additional training examples for the "difficult" classes.

Moreover, some letters will be less common than others. In a situation of this kind, it is known that the rarer classes get "overlooked" by the classifier-inducing algorithms unless special precautions have been taken. Section 10.2 will have more to say about this issue.

The Classifier Should Be Allowed to Reject an Example   To decipher a person's handwriting is far from easy. Certain letters are so ambiguous as to make the reader shrug his shoulders in despair. Yet the k-NN classifier from Chap. 3 is undeterred: it always finds a nearest neighbor, and then simply returns its class, no matter how arbitrary this class is.

Practical experience shows this circumstance to be harmful because the costs of getting the wrong class can be greater than the costs of not knowing the class at all. Thus in an automated reader of postal codes, an incorrectly read digit can result in the letter being sent to a wrong destination, which, in turn, may cause great delay in delivery. On the other hand, if the classifier does not give *any* answer, a human employee will have to do the reading. The costs of manual processing may then be lower than the costs of getting the wrong address.

We have convinced ourselves that the classifier should be implemented in a way that makes it possible to refuse to classify an example if the evidence supporting the winning class is insufficient. The simplest way of doing so in the context of the k-NN classifier is to require a certain minimum margin between the number of votes supporting the winner and the number of votes supporting the runner-up. For instance, if the winning class in a 7-NN classifier receives only four votes as compared to the three votes supporting another class, the example is rejected as ambiguous.

Something similar is easy to accomplish also in some other paradigms such as the Bayesian classifiers or neural networks: the classifier simply compares the probabilities (or output signals) of the two most likely classes, and rejects this example if the difference does not exceed a predefined threshold.

Error Rate Versus Rejection Rate   A classifier that rejects ambiguous examples will surely reduce its error rate; on the other hand, excessive reluctance to classify will not be beneficial, either. What if *all* examples are rejected? The error rate then drops to zero—and yet the user will question the tool's practical utility.

The lesson is, the engineer needs to consider the trade-off between the rejection rate and the error rate. True enough, increasing the former is likely to reduce the latter; but overdoing it may render the classifier useless.

What Have You Learned?

To make sure you understand this topic, try to answer the following questions. If you have problems, return to the corresponding place in the preceding text.

- Explain how to describe hand-written characters by attribute vectors.
- What aspects did we consider (and why) when looking for the most appropriate machine-learning tool to be used here?
- What is the immediate consequence of the relatively high number of classes in this domain? Why is here the error rate unable to give the full picture?

- Why should the classifier be allowed to refuse to classify certain examples? Discuss the trade-off between error rate and rejection rate; comment on the interplay between performance and utility.

8.2 Oil-Spill Recognition

Figure 8.2 shows a radar image of the sea surface as taken by a satellite-born device. Against the grayish background, the reader can see several dark regions of the most varied characteristics: small or large, sharp or barely discernible, and of all possible shapes. What interests us here primarily is the sharp and elongated object in the vicinity of the upper-right corner: an oil spill, illegally dumped by a tanker's captain who has chosen to get rid of the residues in its bilges in the open sea rather than doing so in a specially designed terminal as required by the law. As such, this particular oil spoil is of great interest to the Coast Guard.

For all relevant sea-surface areas (close to major ports, for example), satellites take many of such "snapshots" and send them down to collaborating ground stations. Here, experts pore over these images in search for signs of illegal oil spills. Whenever they detect one, an airplane is dispatched and verifies the suspicion by a spectrometer (which is unavailable to the satellite), collects evidence, and perhaps even identifies the perpetrator.

Unfortunately, human experts are expensive—and not always available. They may be on holidays, on a sick leave, or not present for any number of other reasons. Besides, in view of the high number of images, the work is tedious, and prone to human errors. This is why the idea came up to develop a computer program to automate the oil-spill recognition process.

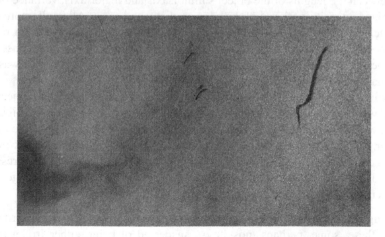

Fig. 8.2 A radar image of a sea surface. The "wiggly" elongated dark region in the upper-right corner represents environmental hazard: an oil spill

The picture shown in Fig. 8.2 has been selected out of many, the main criterion being its rare clarity. Indeed, the oil spill it contains is so different from the other dark regions that even an untrained eye will easily recognize it as such. Even so, the reader will find it difficult to specify the oil-spill's distinguishing features in a manner that can be used in a computer program. In the case of more realistic objects, the task will be even more difficult. At any rate, to hard-code the oil-spill recognition ability is quite a challenge.

Again, machine learning got its chance, the intention being to let the machine develop the requisite skills automatically, by induction from training examples. The general scenario of the adventure can be summarized into the following steps.

1. collect a set of radar images containing oil spills;
2. use some image-processing software capable of identifying, in these images, the dark regions of interest;
3. ask an expert to label the oil spills as positive examples, and the other dark regions (so-called "look-alikes") as negative examples;
4. describe all examples by attribute vectors, and let a machine-learning program induce the requisite classifier from the training set thus obtained.

As in the previous application, we will try to glean some useful lessons by taking a closer look at certain critical aspects of this undertaking.

Attributes and Class Labels   State-of-the-art image-processing techniques easily discover dark regions in an image. For these to be used by the machine-learning tool, we need to describe them by attributes that are hoped to capture those aspects that distinguish spills from "look-alikes." Preferably, their values should be unaffected by the given object's size and orientation.

The attributes that have been used in this project include the region's mean intensity, average edge-gradient (which quantifies the sharpness of the edges), the ratio between the lengths of the object's minor-axis and major-axis, variance of the background intensity, variance of the edge gradient, and so on. All in all, more than forty such attributes were selected in a rather ad hoc manner. Which of them would really be useful was hard to tell because experts were unable to reach consensus about the attributes' respective relevance and redundancy. The final choice was left to the machine-learning software.

Labeling the training examples with classes was not any easier. The objects in Fig. 8.2 were easy to categorize; in other images, they were much more ambiguous. On many occasions, the best the expert could say was, "yes, this looks like an oil spill" or, "I rather doubt this is what we are looking for." The correctness of the selected class labels was thus uncertain, resulting in class-label noise. The presence of class-label noise reduces our expectations: the classifier can be only as good as the data that have been used during its induction.

Choosing the Classifier   The examples were described by some forty attributes. Among these, some, perhaps most, were suspected of being either irrelevant or redundant. This is an important circumstance; the reader will recall that the presence of irrelevant and redundant attributes makes some classifiers underperform. In this particular project, the problem was side-stepped by first inducing a decision tree, and then eliminating all attributes that never appeared in any of the tree's tests. This was quite logical. After all, the choice of which attributes to include in the tree has been made based on the attributes' information contents. Since information contents of irrelevant attributes is low, this method of identifying them is quite reliable. Also redundant attribute can thus be eliminated, at least to some extent.

When the k-NN classifier was applied to examples described by attributes that "survived" this decision-tree-based elimination process, the classification performance turned out to be acceptable. The oil-spill problem was thus solved with a very simple machine-learning technique—which is good: having a choice, the engineer is always well advised to give preference to the simpler tool.

The decision to use the k-NN classifier was also driven by another important consideration. Since this is typical of many realistic applications, let us discuss it in some detail.

Cost of Errors Performance evaluation is here not as straightforward as it was in the previous chapters. For one thing, error rate can be a fairly misleading indicator in domains where each type of error carries a different penalty.

This is manifestly the case in the oil-spill adventure. Here, a false positive results in an aircraft being unnecessarily dispatched to a suspicious though "innocent" region, and this means a waste of time and resources (note that the costs are not only financial, but also moral because these kind of failures tend to undermine the user's trust in the classifier). On the other hand, a false negative means an undetected environmental hazard whose consequences (financial, environmental, and political) are hard to predict. In view of all this, the reader will agree that the two types of cost are of such a different nature that it is almost impossible to compare them. This, of course, makes it difficult to specify the project's goal.

Experiments on pre-classified testing data indicated that most of the errors made by the induced classifier were of the false-positive kind (i.e., false alarms). As for false negatives, these were relatively rare. But then: was this good, or should this observation be taken as a signal of a need to modify the classifier in order to change the ratio of the two types of errors?

The question cannot be answered in isolation from the application's momentary needs. Financial constraints may force the user occasionally to accept the risk of environmental hazard, simply because the budget can no longer tolerate false alarms. The user then wants to reduce the frequency of false positives even if this means to pay the price of an increased number of undetected oil spills (false negatives).

However, the situation will change in more prosperous times when the user who does not want to miss an oil spill is prepared to accept higher frequency of false positives for the sake of making false negatives rare.

Leaning Toward One or the Other Class In view of these trade-offs, it is necessary to give the user the opportunity to adjust the classifier's behavior so as to modify the frequency of one or the other type of error.

As already mentioned, this project relied on the k-NN classifier where this requirement is easy to satisfy: the trick consists in manipulating the margin between the number of votes supporting either of the two classes. For the sake of illustration, suppose the 7-NN classifier is used. Here, the number of false positives can be reduced if we instruct the classifier to label as positive only those examples where, say, at least five of the nearest neighbors are positive; any example that fails to satisfy this condition is deemed negative. Conversely, if we desire to lower the frequency of false negatives, we tell the classifier to return the positive label whenever, say, at least three of the nearest neighbors are positive. The user's preference for either type of error is thus expressed in terms of the number of votes that are necessary for the example to be labeled as positive.

What Have You Learned?

To make sure you understand this topic, try to answer the following questions. If you have problems, return to the corresponding place in the preceding text.

- How did the engineers deal with the fact that many attributes were redundant or irrelevant? What did they identify these less-then-useful attributes?
- What can be said about the reliability of the class labels in the training set? What does it mean for the classifier's expected performance?
- Discuss the respective costs, in this domain, of the two types of error: false positive versus false negative. Can they be compared using the same units? Why does the user need a mechanism to reduce one type of error at the cost of the other?
- Explain the essence of the mechanism that enables the k-NN classifier to increase or reduce either of the two errors.

8.3   Sleep Classification

Throughout the night, we go through different sleep stages such as deep, shallow, or rapid-eye movements (REM, this is when we dream). To identify these stages in a concrete sleeping subject, advanced instrumentation is used: an electrooculogram to record eye movements, an electromyogram to record muscle contractions, and contact electrodes attached to the scalp to record the brain's neural signals. Based on the readings of all these instruments, a medical expert can decide what stage the sleeping subject is in at any given moment, and can even draw a *hypnogram* such as the one shown in Fig. 8.3.

Note that the deepest sleep stage occurs here only three times during the 8-h sleep, and that it usually does not last long. Note also the move stage; this occurs, for instance, when the subject turns from one side to another, moves the head or an arm, and the like.

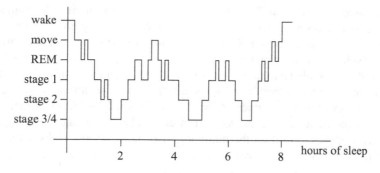

Fig. 8.3 An example hypnogram that records the sleep stages experienced by a subject during an 8-h sleep

Why Is It Important Medical practice needs to be able to recognize specific sleep stages. A case in point is the so-called *sudden infant death syndrome* (SIDS): an infant dies without any apparent cause. A newborn suspected of being in danger has to be watched, in a hospital, so that resuscitation can be started immediately. Fortunately, SIDS is known almost always to occur during the REM stage. This means that it is not necessary to watch the patient all the time, but only during this period of increased risk. For instance, a device capable of recognizing the onset of the REM stage might alert the nurse that more attention might be needed during the next five of so minutes.

The hypnogram, in turn, is a useful diagnostic tool because the distribution of the sleep stages during the night may indicate specific neural disorders such as epilepsy.

Why Machine Learning To draw the hypnogram manually is a slow and tedious undertaking, easily taking 3–5 h of a highly qualified expert's time. Moreover, the expert is not always available. This is why efforts have been made to develop a computer program capable of identifying the individual sleep stages based on observed data, and, hopefully, even to draw the hypnogram.

To be able to help, the computer needs a description of the individual stages. Such description, however, is difficult to obtain. Medical experts rely on skills obtained in the course of long training, and they use features and indications that are too subjective to be converted to a computer program.

This motivated an attempt to induce the classifier from pre-classified data. Specifically, the data-acquisition process divided the 8-h sleep period into 30-s samples, each of them treated as a separate training example. All in all, a few hours' sleep thus provided hundreds of training examples. Diverse instruments than provide data to describe each 30-s sample.

Attributes and Classes Again, the first task was to remove attributes suspected of being irrelevant or redundant. In the previous application, oil-spill recognition, this removal was motivated by the intention to increase the performance of the k-NN classifier (which is known to be sensitive to their presence). In sleep classification, another reason comes to the fore: the physician wants to minimize the number of measurement devices attached to the sleeping subject. Not only does their presence make the subject feel uncomfortable; they also disturb the sleep, and thus interfere with the results of the sleep analysis.

As for the class labels, these are even less reliable than in the oil-spills domain. The differences between "neighboring" (similar) sleep stages are so poorly defined that any two experts rarely agree on more than 70–80% of the class labels. No wonder that the training set contains a lot of class-label noise, and the low quality of the data imposes a limit on the minimum error rate that any realistic classifier induced from the data can achieve.

The Classifier and Its Performance The classifier employed in this particular case combined decision trees with a neural network in a manner whose details are unimportant of our needs here. Suffice it to say that the classifier's accuracy on independent data indeed achieved those 70–80% observed in human experts, which means that the natural performance limits have been reached. It is perhaps worth noting that plain decision trees were a few percent weaker than that.

This said, it is important to understand that classification accuracy does not give the full picture of the classifier's behavior (similarly as in the OCR domain from Sect. 8.1). For one thing, the classifier correctly recognized some of the seven classes most of the time, while failing on others. Particularly disappointing was its treatment of the REM state. Here, classification accuracy was in the range 90–95%, which, at first sight, looked good enough. However, closer inspection of the training data revealed that only less than 10% of all examples belonged to the REM class; this means that a comparable classification accuracy could be achieved by a classifier that says, "there is not a single REM example"—and yet this is hardly what the engineers hoped to achieve.

We realize that this way of measuring performance is not without its limitations, and that other criteria have to be found. These indeed exist. They will be discussed in Chap. 11.

Improving Classification Performance by Post-processing The accuracy of the hypnogram can be improved by post-processing whose nature relies on the domain's logic. Indeed, several rules of thumb can be used here. For instance, the deepest sleep (stage 3/4) is unlikely to occur right after the REM stage, and stage 2 does not happen after move. Also, the hypnogram can be "smoothed out" by the removal of any stage lasting only one 30-s period. Applying such rules in the course of post-processing makes it possible to improve the hypnogram's informational value.

The lesson is worth remembering. In domains where the examples are ordered in time, the classes of the individual examples may not be independent of those preceding or following them. In this event, post-processing can improve the classifier's performance.

Proper Use of the Induced Classifier The original idea was to induce the classifier from examples obtained from a few subjects, and then to classify future data using this induced classifier instead of the much more expensive "manual" classification.

This ambition, however, turned out to be unrealistic. Practical experience showed that no "universal classifier" of sleep data could be obtained in this manner: a classifier induced from one person's data could not be used to draw a hypnogram for another person without serious degradation in classification performance.[1]

This does not mean that machine learning is in this domain totally disqualified. Far from it. However, the user needs to modify his or her expectations: instead of being universal, the classifier will be induced separately for each subject. The expert's efforts are still significantly reduced. The following three-step scenario is indicated:

1. The expert determines the class labels of a subset of the available examples, thus creating a training set.
2. From this training set, a classifier is induced.
3. The induced classifier is used to classify the remaining examples, thus saving the expert's time.

What Have You Learned?

To make sure you understand this topic, try to answer the following questions. If you have problems, return to the corresponding place in the preceding text.

- Why was it important to minimize the number of attributes in this domain? How were the relevant attributes identified?
- The sleep-classification domain has seven classes. What does it mean for an engineer trying to evaluate the induced classifier's performance?
- In the hypnogram, the examples are ordered in time. How can this circumstance be exploited in data post-processing?
- In what manner can machine learning reduce the burden imposed on someone who seeks to classify available data?

8.4 Brain–Computer Interface

The muscle-controlling commands are issued at specific locations of *motor cortex*, a relatively well-understood region of the cortex. Even the brain of many totally paralyzed patients is fully capable of generating these commands; unfortunately, the information fails to reach the muscles. The fact that these signals can be detected by contact electrodes inspired a fantastic idea: can these signals, properly recorded and interpreted, be used to control a cursor on a computer screen? If yes, then there

---

[1]One can speculate that a different set of attributes might perhaps make this possible; the case study reported here did not attempt to do so.

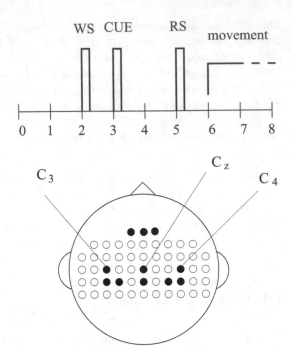

Fig. 8.4 Organization of the experiment. After a "ready" signal (WS) comes the CUE ("left" vs. "right"). Once RS appears on the screen, the subject waits 1 s, then presses the button indicated by the CUE

Fig. 8.5 Arrangement of electrodes on the subject's scalp. Only some of the electrodes are actually used (*highlighted*)

might be a way for the patient to communicate with the outside world, in spite of being unable to speak, in spite of being unable even to move his or her pupils.

The exact nature of the motor commands is too complicated to be expressed by a mathematical formula. But we can record the signals, describe them by attributes, and label with the concrete motor commands. If we do so, we have a training set from which a machine-learning program can induce the classifier.

The Training Examples In the specific case to be described in this section, only two classes were considered: left and right. The training and testing examples were obtained in the course of the following experiment. A subject was sitting in front of a computer monitor. On the desk in front of him was a wooden board with two buttons, one to be pressed by the left index finger, the other by the right index finger, according to instructions displayed on the monitor (Fig. 8.4).

The whole scenario followed the time-line shown in the upper part of Fig. 8.5. The contact electrodes attached to the scalp (see the bottom part of Fig. 8.5) record the intensity of the neural signals during a certain period of time before the button is pressed. The signals from each electrode are characterized by five numeric attributes, each representing the power of the signal over a specific time interval (a fraction of a second).

As indicated in the picture, only some (typically 11) of the electrodes were actually used. For 11 electrodes, each represented by 5 attributes, this amounted to 55 attributes. All in all, the subject "produced" a few hundred examples with both classes equally represented. Since it was always known which button was actually pressed, the training set was free of class-label noise. As for the attributes, the experimenters suspected many of them to be irrelevant.

Classifier Induction and Its Limitations Several machine learning paradigms were experimented with, including multilayer perceptrons and nearest-neighbor classifiers with attribute-value ranges normalized so as to avoid scaling-related problems. Relevant attributes were selected by a decision tree, similarly as in the applications discussed in the previous sections.

Typical error rates of these induced classifiers on testing data were in the range 20–30%, depending on the concrete subject. It is quite possible that better accuracy could not be achieved: the information provided by the electrodes might have been insufficient for this kind of experiment.

Importantly, just as in the sleep-classification domain, the classifier could only be applied to the subject from whose training data it has been induced. All attempts to induce a "general classifier" (to be used on any future subject) failed—the error rates were so high as to make the classifications process look random.

Testing the Classifier's Utility As in the previous domain, the error rate measured on independent data does not give the whole picture; the classifier's practical utility depends on how it is actually employed. In our case, the real question is: will the classifier succeed in sending the cursor to the correct location?

The question was answered by an experiment illustrated in Fig. 8.6. The subject is still sitting in front of a computer monitor, with electrodes attached to the scalp. However, the board with the two buttons has been removed. On the monitor, we can see two rectangles, one on the left and one on the right. In the middle is a cursor in the shape of a little cross.

When instructed to move the cursor, say, to the right, the subject only *imagined* he was pushing the right button. The electrodes recorded the neural signals and passed them to the classifier. Based on these, the classifier selected the side (the class, left or right) to which the cursor is to be moved. The movement was very slow so as to give the subject the opportunity to "correct" the wrong direction. To make this possible, the neural signals were sampled repeatedly, each sample becoming one example to be passed on to the classifier. Suppose the subject's intention is to move the cursor to the right. It is possible that the first sample will be misclassified as

Fig. 8.6 The classifier's task is to move the cross into either the *left box* or the *right box*, according to the subject's "thoughts"

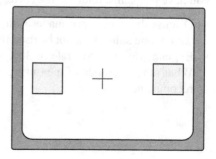

left, sending the cursor in the wrong direction; but if the following samples are correctly classified as right, the cursor will, after all, land in the correct rectangle.

What Error Rate Is Acceptable?   Under the scenario just described, an error rate of 30% (which seems quite high) turned out to be fully acceptable. Even when the cursor occasionally did start in the wrong direction (on account of one "example" being misclassified), there was still the chance of correcting the mistake before the cursor reached its destination. As a result, the cursor almost always landed in the correct box, even if in a somewhat hesitating manner in which it moved back and forth before finding the right direction.

An impartial observer, uninformed about how the classifier was actually used, rarely noticed anything was going wrong. The occasional errors on the part of the classifier (manifested by the cursor's back-and-forth movements) were sometimes perceived as a sign of indecisiveness on the part of the subject, and not of the classifier's imperfection.

Only when the classifier's error rate dropped well below those 30% did the cursor miss its target frequently enough to cause disappointment. One can speculate that a better measure of the classifier's performance would in this case be the average time needed for moving the cursor.

An Honest Historical Remark The classifier's performance differed from one person to another. On quite a high percentage of the subjects, the induced classifiers exhibited an almost prohibitively high error rate. Also the fact that the classifier had to be trained on the same patient on which it was supposed to be used was perceived as a shortcoming. Further on, two classes are not enough. At the very least, also up and down classes would be needed. On these, the experimental results were less impressive. In the end, other methods of communication with paralyzed patients came to be studied. Though successful as a machine-learning exercise, the project did not satisfy the expectations of the medical domain.

What Have You Learned?

To make sure you understand this topic, try to answer the following questions. If you have problems, return to the corresponding place in the preceding text.

- Discuss the experience made in this application: a classifier induced from the data of one subject cannot be used to classify examples in another subject.
- Explain why the error rate of 30% was in this domain still deemed acceptable. What other methods of measuring performance, in this application would you recommend?

8.5 Medical Diagnosis

A physician attempting to find the cause of her patient's complaints behaves a bit like a classifier: based on certain attributes (the patient's symptoms, results of laboratory tests), she suggests a diagnosis—and a treatment. No wonder that, in the early days of machine learning, many believed medical diagnosis to be its natural, almost straightforward application.

Some Results The optimism was motivated by early studies such as the one whose results are summarized in Table 8.1. The numbers give the classification accuracy achieved by Bayesian classifiers, decision trees, and physicians on the same testing set—patients described on attribute vectors. The four domains differ in the number of classes and in the difficulty of the task (e.g., noise in the data, relevance and reliability of the available information).

The values indeed seem to confirm machine-learning's ability to induce classifiers capable of competing with human experts; indeed, they seem to outperform them. In the multi-class `primary tumor` domain, the margin is perhaps unimpressive, but in all the other domains, the machine learning classifiers appear to be clearly better. It is only a pity that the table does not tell us more about the variation in the results. At least standard deviations would help, here, but there are other ways how to ascertain whether the results are statistically reliable—Chaps. 11 and 12 will have more to say about the methods to be used here.

Are the Results Encouraging? The results seem impressive, but we must not jump to conclusions. The first question to ask is whether the comparison was fair. Since the examples were described by attributes, the data available to the machine-learning tool, and to the induced classifier, could hardly be the same as used by the physician who could rely also on subjective information that might not have been available to the machine. In this respect, the human enjoyed a certain advantage, and this makes the machine learning's results all the more impressive.

On the other hand, the authors of the study admit that the physicians participating in the study were not necessarily top specialists in the given field, and machine learning thus outperformed only general practitioners. Another aspect worth our attention is that this study was conducted in the 1980s when the diagnostic tools were less sophisticated than those in use today. It is hard to tell who will benefit more from modern laboratory tests: human experts or machine learning?

Table 8.1 Classification accuracy of two classifiers compared to the performance of physicians in four medical domains

	Bayesian classifier	Decision tree	General practitioner
Primary tumor	0.49	0.44	0.42
Breast cancer	0.78	0.77	0.64
Thyroid	0.70	0.73	0.64
Rheumatology	0.67	0.61	0.56

Apart from Classification, Also Explanation Is Needed For medical diagnosis, classification accuracy is not enough; more accurately, it is not enough for the diagnosis to be correct. A patient will hardly agree with a surgery if the only argument supporting this recommendation is, "this is what the machine says." The statement that "the machine is on average a few percentage points better than a physician" is unlikely to convince the patient, either.

What a reasonable person wants is a convincing explanation of why the surgery is to be preferred over a more conservative treatment. In the case of the domains from Table 8.1, the doctor is usually able to itemize what evidence supports the concrete diagnosis and why, and what treatment options are recommended in this particular case and why. Unfortunately, the baseline machine learning techniques are not very good at explaining their decisions (perhaps with the exception of decision trees). Software for automated reasoning would be needed, here. This, however, is beyond the scope of this book.

The Need for Measuring Confidence There is another problem to be aware of. Most of the classifiers induced by the techniques treated in the previous chapters only give the final verdict; for instance, "the example belongs to class 77." The physician (as well as the patient) needs to know more. For instance, is the recommended class absolutely certain or is it only just a little more likely than some other class? And if so, which other classes should be considered?

In other words, medical domains usually call for classifiers capable of telling us how confident they are in the returned class. For example, such classifier should say that "the example belongs to C_1 with probability 0.8, and to C_5 with probability 0.7." There probabilities are explicitly calculated by Bayesian classifiers, so these may be a good choice; similar information can be obtained also from neural networks. Both of these approaches, however, are rather unable to offer explanations.

On the other hand, decision trees *are* capable of offering some kind of explanation. As for the confidence, this is only possible with additional functions that we have not treated here.

Cultural Barriers There is one last reason why the encouraging results of the early applications of machine learning to medical diagnosis failed to inspire followers. With only minor injustice, they can be called *cultural barriers*. In a way, they are understandable. Medical doctors surely did not like to be told how easily they would be replaced by computers. No wonder that, in the early days of machine learning, many of them were not eager to collaborate on the development of the requisite software.

This, of course, is a misunderstanding. Machine learning is not meant to replace a human expert. At its very best, is only offer advice; the final diagnostic decision will always be the responsibility (even in legal terms) of the physician treating the patient. Still, the value of the advice should not be underestimated. For instance, the classifier can alert the doctor that additional, previously unsuspected, disorders my accompany the current diagnosis (the patient often suffers from more than just one problem at a time), and it can even point out the need to take specific additional laboratory tests.

What Have You Learned?

To make sure you understand this topic, try to answer the following questions. If you have problems, return to the corresponding place in the preceding text.

- As for the results mentioned in this section, why was it impossible to conclude from them that the induced classifiers outperformed the human expert?
- Apart from classification accuracy, what is needed in medical diagnosis?
- Discuss the limitations of machine-based diagnosis. Suggest a reasonable way of realistic application of machine learning in medical domains.

8.6   Text Classification

Suppose you have a vast collection of text documents, and you need to decide which of them is relevant to a specific topic of interest. If there are really a great many of them, there is no way you can do so manually. However, taking inspiration for some of the above-described applications, you choose a manageable subset of these documents, read them, assign the class labels to them (positive versus negative), and induce a classifier that will then be used to classify the rest.

Attributes A common way to describe a document is by the frequency of the words it contains. Each attribute represents one word, and its value represents the frequency with which it appears in the text. For instance, in a document that is 983 words long, the term "classifier" may appear five times, in which case its frequency is $5/983 = 0.005$.

Since the vocabulary typically contains tens of thousands of words, the attribute vectors will be very long. And of course, since only a small subset of the vocabulary finds its way into the document, most attribute values will be zero. This being somewhat unwieldy, simple applications prefer to work with only a subset of the whole vocabulary, say, 1000 most common words.

Class Labels The class labels for the training examples may be difficult to determine. Even if we want only something very easy such as to decide, for each document, if it deals with computer science, whether an example clearly belongs into this class may not be easy to decide because some documents are more relevant to this category than others. For instance, a very clear-cut case will be a scientific paper dealing with algorithm analysis; on the other hand, a less relevant document will only mention in passing that some algorithms are more expensive than others; and yet another document will be an article from a popular magazine.

One possibility to deal with this circumstance is to "grade" the class labels. For instance, if the document is most certainly relevant, the class is 2; if it is only somewhat relevant, the class is 1; and only if it is totally irrelevant, the class is 0. The user can then decide what level of relevance is needed for the given application.

Typical Observations The great number of attributes makes the induction computationally expensive. Thus in the case of decision trees, where one has to calculate the information gain separately for each attribute, this means that tens of thousands of attributes need to be investigated for each of the tests, and these calculations can take a lot of time, especially if the number of examples is high. Upon some thought, the reader will agree that Bayesian classifiers and neural networks are quite expensive, too.

One can suggest that induction will be cheaper in the case of the k-NN classifier or a linear classifier. Here, however, another difficulty comes to the fore: for each given topic, the vast majority of the attributes will be irrelevant, a circumstance that reduces the utility of both of these paradigms.

The Problem of Multi-Label Examples The applications discussed in the previous sections assumed that each example is labeled with one and only one class. Here, however, each document can belong to two or more (sometimes many) classes at the same time.

The most commonly used way of dealing with the situation is to induce a different classifier for each class. Whenever a future document's class is needed, the document's attribute vector is submitted to all of these classifiers in parallel; some of them will return "1" (meaning the document belongs to the corresponding class), others will return a "0."

This, however, makes the adventure even more expensive because of the need to induce hundreds of classifiers. And given that the induction of each of these classifiers is expensive in its own right, the result can be prohibitive.

Moreover, the classes are not independent of each other. At the very least, some are subclasses or others, giving rise to a whole hierarchy of classes. This means that their corresponding classifiers perhaps should not be induced in a manner totally independent of the induction of the other classifiers.

What Have You Learned?

To make sure you understand this topic, try to answer the following questions. If you have problems, return to the corresponding place in the preceding text.

- How are examples typically described in text-classification domains? Why are the attribute vectors so long?
- What did this section mean by "graded class labels"?
- What makes induction of text classifiers expensive?
- Discuss the problem of multi-label examples.

8.7 Summary and Historical Remarks

- The number of examples that can be used for learning depends on the particular application. In some domains, examples are abundant; for instance, if they can be automatically extracted from a database. In others, they may be rare and expensive, as was the case of the oil spill domain.
- We may be totally in the dark as to which attributes really matter. This was the case of the oil spill domain where the shape and other characteristics could be described by a virtually unlimited number of attributes obtained by image-processing techniques.
- In some domains, the really important attributes are not known or cannot be obtained at reasonable costs. Inevitably then, the induction has to use those attributes that are available, even if the performance of the classifier thus induced will be diminished.
- In the brain–computer-interface domain, it turned out to be enough that a majority of the decisions were correct: in this event, the cursor landed in the intended rectangle. We can see that the maximum classification accuracy may not be strictly necessary.
- In domains with more than just two classes, the error rate does not give the full picture of the classifier's behavior. Quite often, some classes will be almost always perfectly recognized while others will pose problems. It is therefore important to know not only the average performance, but also the performance on each individual class.
- In some domains, we want to be able to explain the decision. High classification accuracy is not enough. This is the case of medical diagnosis.
- The costs of false positives can be different from the costs of false negatives. In the oil-spill domain, it was difficult to express these costs in monetary terms.
- Many applications are interested in minimizing error rate. Sometimes, however, error rate fails to give the full picture.
- Another important requirement: the possibility that the user be able to tweak a parameter that will trade false positives for false negatives (see the oil spill domain)

Historical Remarks The results from Table 8.1 are taken from Kononenko et al. [48] who refer there to an older project of theirs. The oil-spill project was reported in Kubat et al. [52]. The sleep classification task is addressed by Kubat et al. [50], and the experience with using machine learning in brain–computer interface was published by Kubat et al. [53]. The character-recognition problem has been addressed for decades by the field of Computer Vision; the first major text systematically addressing the issue from the machine-learning perspective seems to be Mori et al. [69]. The text-classification task was first studied by Lewis and Gale [55].

A minor remark is in place, here. Of these six applications, the author of this book participated in two. His motivation for including his own work was not to stun the reader with his incredible scholarship and inestimable contribution to the field of machine learning. Far from it. The truth is, if you work on a project for quite some time, you develop not only personal attachment to it, but also certain deeper understanding that makes this particular application more appropriate for instruction purposes than those which you only read about in the literature.

8.8   Exercises and Thought Experiments

The exercises are to solidify the acquired knowledge. The ambition of the suggested thought experiments is to let the reader see this chapter's ideas in a different light and, somewhat immodestly, to provoke his or her independent thinking.

Give It Some Thought

1. Suppose that you know that the correctness of some class labels in your training set is not certain. Would you recommend that these "unreliable" examples be removed from the training set? In your considerations, do not forget that some of the pre-classified examples will be used as testing examples to assess the classification performance of the induced classifier.
2. Discuss the reasons why application-domain users may be reluctant to accept machine-learning tools. Suggest ways to eliminate, or at least diminish, their suspicions and concerns.
3. Section 8.2 mentioned a simple mechanism by which the k-NN classifier can manipulate the two kinds of error (false negatives versus false positives). Suggest a mechanism that a Bayesian classifier or a neural network can use to the same end. How would you go about implementing this mechanism in a linear classifier?
4. In more than one of the domains discussed in this chapter, it was necessary to reduce the number of irrelevant and/or redundant attributes. In the projects reported here, decision trees were used. Suggest another possibility that would use a technique from one of the previous chapters. Discuss its advantages and disadvantages.
5. Suppose you wanted to implement a program that would decide whether a given position in the tic-tac-toe game (see Fig. 8.7) is winning. What attributes would you use? How would you collect the training examples? What can you say about expected noise in the data thus collected? What classifier would you use? What difficulties are to be expected?

Fig. 8.7 In tic-tac-toe, the goal is to achieve three *crosses* (or *circles*) in a row or a column or on a diagonal

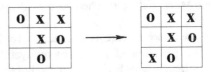

Computer Assignments

1. Some realistic data sets for machine-learning experimentation can be found on the website of the National Institute of Standards and Technology (NIST). Find this web site, then experiment with some of these domains.
2. Go to the web and find a website about the demography of the 50 states in the U.S. Identify an output variable whose value will be deemed positive if it is above the U.S. average and negative otherwise. Each state thus constitutes an example. Based on the information provided by the website, identify the attributes to describe the examples. From the data thus obtained, induce a classifier to predict the output variable's value.

Chapter 9
Induction of Voting Assemblies

A popular way of dealing with difficult problems is to organize a brainstorming session in which specialists from different fields share their knowledge, offering diverse points of view that complement each other to the point where they may inspire innovative solutions. Something similar can be done in machine learning, too. A group of classifiers is created in a way that makes each of them somewhat different. When they vote about the recommended class, their "collective wisdom" often compensates for each individual's imperfections.

This chapter deals with mechanisms for the induction of such sets of classifiers from data. The reasons behind the high performance of these "classifier assemblies" are explained on the simplest approach known as *bagging*. Building on these foundations, we proceed to the more sophisticated *boosting* algorithms and their variations, including the so-called *stacking* approach.

9.1 Bagging

For the sake of simplicity, we will limit ourselves to the field of two-class domains where each example is either positive or negative. As always, the ultimate classifier is to be induced from a set of pre-classified training examples.

The Underlying Principle The approach known under the name of *bagging*[1] induces a group of classifiers. When presented with an example, the classifiers are used in parallel, each offering an opinion as to which class the example should be labeled with. A "master classifier" collects this information, and then chooses the label that has received more votes.

---

[1] The name is an acronym of *booststrap aggregation*.

© Springer International Publishing AG 2017
M. Kubat, *An Introduction to Machine Learning*,
DOI 10.1007/978-3-319-63913-0_9

Table 9.1 The algorithm of *Bagging*

Input: the training set, T, and the user's choice of the induction technique

1. Using random sampling with replacement, create from T several training subsets, $T_1, T_2, \ldots T_n$. Each subset consists of the same number of examples.
2. From each T_i, induce classifier C_i (for $i = 1 \ldots, n$).
3. When presented with a new example, submit it to all C_i's in parallel and let each classifier, C_i, offer its recommendation for the example's class.
4. A "master classifier" decides which of the two classes received more votes.

Assuming that each of the participating classifiers represents a somewhat different aspect of the recognition task, the classifier group (sometimes called a "voting assembly") is expected to outperform any individual.

Induction of the Voting Assembly The principle of *bagging* is summarized in Table 9.1. The idea is to take the original training set, T, and to create from it a certain number of training subsets, $T_1, \ldots T_n$, each of the same size.

Once the subsets T_i have been created, a machine-learning algorithm induces a classifier, C_i, separately from each of them. For this, any of the induction techniques from the previous chapters can be used. However, the baseline version of *bagging* assumes that the same technique (say, induction of decision trees with the same user-set parameters) is always used.

Bootstrapping Let us now explain how the training examples for T_i are selected. Each example has the same chance of being picked. Once it *has* been included in T_i, it is "returned" to T, by which we mean that it will participate in the selection of examples for T_{i+1} with the same probability as any other example. For a training set, T, consisting of N examples, the selection is repeated n times in a process known as *bootstrapping*.

An example can appear in T_i more than once and, conversely, some examples will fail to appear in T_i. This means that each T_i consists of N examples (with duplicities), but each of these training subsets is different. As a result, each of the induced classifiers will focus on a different aspect of the learning problem.

Why It Works Figure 9.1 explains how the voting can help reduce the error rate. Three classifiers are considered. If the errors are rare, there is a chance that each classifier will err on different examples. In this event, each example will be misclassified at most once, the other two class labels being correct. An individual's occasional mistake will thus be "outvoted" (and thus corrected) by the others.

Of course, the situation will rarely be so convenient. If an example is misclassified by two out of the three classifiers, the voting will result in the wrong class. One can stipulate, however, that this unfavorable situation might be improved if the number of classifiers is increased.

Fig. 9.1 The three classifiers were asked to classify the same 17 examples. By doing so each erred on three examples—different for each classifier. The reader can see that these errors can be eliminated by voting

● ... a correctly classified example

○ ... an incorrectly classified example

The most important lesson to be drawn is that *bagging* works well if each classifier tends to err on different examples.

Some Observations Practical experience shows that *bagging* achieves good results if the error rate of each individual classifier is low. Then, with sufficiently large number of classifiers, chances are high that an individual's errors will be corrected by the others, similarly as in Fig. 9.1.

This, however, is not necessarily the case when the error rates of the individual classifiers are high. Up to a certain point, these errors are still corrected if there are really a great many classifiers because then, by the law of large numbers, each aspect of the underlying recognition problem is likely to be represented. The only thing to criticize is that perhaps too many classifiers are then needed.

Conceptually, however, the approach involves too much randomness. Figure 9.1 leads us to believe that the same effect might be achieved if the classifiers were induced not independently from each other, but rather in a way that might increase their ability to complement each other. Such mechanisms will be the subject of the following sections.

What Have You Learned?

To make sure you understand this topic, try to answer the following questions. If you have problems, return to the corresponding place in the preceding text.

- What makes us believe that a group of voting classifiers will outperform an individual? Under what circumstances will the scheme fail?
- How are the individual classifiers induced in the *bagging* approach?
- What is meant by *bootstrapping*?

9.2   Schapire's Boosting

Although the *bagging* approach often achieves impressive results, it suffers from a serious shortcoming: the voting classifiers have all been induced independently of each other from randomly selected data. One would surmise that a smarter—and perhaps more successful—approach should rely on a mechanism that makes the classifiers complement each other. For instance, this can be done by inducing each of them from training examples that are perceived as difficult by the other classifiers. *Schapire's boosting* was invented with this idea in mind.

Induction of Three Mutually Complementing Classifiers Suppose that a random subset, $T_1 \in T$, of m training examples has been created. These are used to induce the first classifier, C_1. When testing this classifier on the entire training set, T, we will observe that it misclassifies a certain number of examples.

Suppose we now create another training subset, $T_2 \in T$. Let it consist of m examples selected in a manner that ensures that the previously induced C_1 classifies correctly 50% of them, misclassifying the remaining 50%. This means that T_2 is so difficult for C_1 that the classifier will not outperform a flipped coin. From the training subset thus created, the second classifier, C_2, is induced.

The two classifiers, C_1 and C_2, having been induced each from different examples, will inevitably differ in how they label certain instances. A tie-breaker is therefore needed. To this end, a third training subset, T_3, is created, consisting only of examples on which C_1 and C_2 differ. From this third subset, T_3, the third classifier, C_3, is induced.

The principle is summarized by the pseudocode in Table 9.2. When an example is presented, the master classifier collects the labels recommended by the three classifiers, and then returns the class that has received more votes.

Ideally, each of the training sets, T_i, is of the same size, m.

Recursive Implementation The principle just described can be implemented recursively. Figure 9.2 shows how. The resulting triplet of classifiers (in the dotted rectangle) is treated as a single classifier. In the next step, a new training subset, T_4, is created in a manner that ensures that the triplet's error rate on T_4 is 50%. From

Table 9.2 The algorithm of *Schapire's boosting*

Input: the training set, T, and the user's choice of the induction technique

---

1. Create a random training subset, T_1, and induce from it classifier C_1.
2. Create a training subset T_2 in a manner that makes sure that C_1 scores 50% on it. Induce from T_2 classifier C_2.
3. Create T_3 such that C_1 and C_2 disagree on each of the examples it contains. Induce from T_3 classifier C_3.
4. For classification, use plain majority voting.

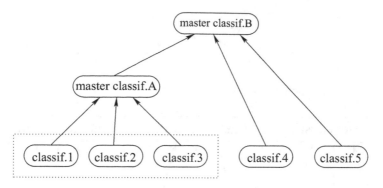

Fig. 9.2 Recursive application of *Schapire's boosting*. Master classifier A combines the votes of classifiers 1–3. Then, master classifier B combines the votes of master classifier A with those of classifiers 4 and 5

these, classifier C_4 is induced. Finally, training subset T_5 is created from examples on which the triplet and C_4 differ; from these, classifier C_5 is induced. Figure 9.2 also shows how the voting in the resulting scheme is hierarchically organized.

It is important to understand the logic of the voting procedure. If classifiers 1 through 3 all return 0, master classifier A returns 0, too; however, this result can still be outvoted if both classifier 4 and classifier 5 return 1. The whole structure thus may return 1 even if three out of the five participating classifiers return 0.

Note that the total number of classifiers induced using this single level of recursion is $3 + 2 = 5$. Of course, the principle can be repeated, resulting in $5 + 2 = 7$ classifiers, and so on. Assuming N_R levels of recursion, the total number of participating classifiers is $2N_R + 3$.

Performance Considerations Suppose that each of the induced classifiers has an error rate below a certain c. It has been proved that the voting triplet's error rate is less than $3\epsilon^2 - 2\epsilon^3$, which is always smaller than ϵ. For instance, if $\epsilon = 0.2$, then $3\epsilon^2 - 2\epsilon^3 = 2 \cdot 0.04 - 2 \cdot 0.008 = 0.104$. And if $\epsilon = 0.1$, then $3\epsilon^2 - 2\epsilon^3 = 0.03 - 0.002 = 0.028$.

Put another way, *Schapire's boosting* brings about an improvement over the performance of the individual classifiers. But if the first voting triplet achieves (see the previous paragraph) $3\epsilon^2 - 2\epsilon^3 = 0.104$, it may be difficult to achieve an equally low error rate with the other two classifiers (4 and 5). One possibility to handle this situation is to create each of them (classifier 4 and classifier 5) as a triplet in its own right as indicated in Fig. 9.3. In this case, the total number of classifiers participating in N_R levels of recursion will be 3^{N_R}.

The thing to remember is that each added level of recursion reduces the error rate. It is therefore theoretically possible to reduce the error rate all the way down to zero. Practice, however, is different—for reasons addressed in the next paragraph.

Limitations The main difficulty is how to find the right examples for each of the subsequent training subsets. These, we already know, have to satisfy certain criteria.

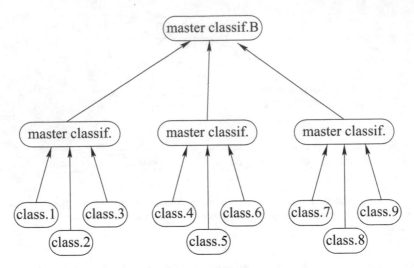

Fig. 9.3 Another approach to recursive application of *Schapire's boosting*

Recall that, to induce the second classifier, C_2, we need to create a training subset, T_2, that has been chosen in a way that makes the previous classifier, C_1, classify correctly only 50% of them.

This may be easier said than done. For instance, if the entire training set that we have at our disposal consists of 100 examples, and the first classifier's error rate is 10%, then we have only 10 misclassified examples, to which 10 correctly classified examples are added to create T_2; this means that the size of T_2 cannot exceed 20.

In the same spirit, we may find it impossible to find m examples that satisfy the requirements for the third training subset, T_3, which, as we know, should consist only of examples for which C_1 and C_2 give the opposite class labels.

And even if we do succeed in creating all the three training subsets, we may find it impossible to apply the same principle recursively. To be able to do this, we would need an almost inexhaustible (and certainly not affordable) source of examples—a luxury rarely available.

What Have You Learned?

To make sure you understand this topic, try to answer the following questions. If you have problems, return to the corresponding place in the preceding text.

- How does *Schapire's boosting* create the training subsets from which to induce the individual classifiers?

- Explain the two ways of implementing the principle recursively. How many classifiers are induced in each case if N_R levels of recursion are used? Explain also the voting mechanism that reaches the final classification.
- When compared to the error rate of the individual classifiers, what is the error rate of the final classification?
- Explain the practical limitation of *Schapire's boosting*: the problem with finding enough examples.

9.3 Adaboost: Practical Version of Boosting

The main weakness of *bagging* is the randomness of its behavior: each of the constituting classifiers is induced in total isolation from all the other classifiers. In *Schapire's boosting*, the randomness is minimized, but we have noticed another complaint: in a realistic setting, it is often impossible to find the right number of the training examples that satisfy the conditions for the second and the third classifier. And so perhaps the most practical approach to boosting is that of *Adaboost* where the examples are chosen from a probabilistic distribution that is gradually modified in a way that makes the whole "assembly" focus on those aspects that appear to be difficult.

The General Approach Similarly to Schapire's approach, *Adaboost* creates the classifiers one by one, each of them from a different training subset whose composition depends on the behavior of the previous classifiers. There is a major difference, though. Whereas *Schapire's boosting* selects the examples according to certain precisely defined conditions, *Adaboost* chooses them probabilistically. This means that each example has a certain probability of being drawn, the probabilities of all examples summing up to 1. The mechanism is implemented in a way that makes sure that examples that were repeatedly misclassified by the previous classifiers will get a higher chance of being included in the training subset for the induction of the next classifier.

Another difference is the number of classifiers. Unlike Schapire's triplets, this approach typically relies on a great number of classifiers. Further on, the final decision is not achieved by the plain voting used in *bagging*. Instead, the so-called *weighted majority voting* scheme is used (see later).

Probabilistic Selection of Training Examples We have mentioned that, for the individual training subsets, T_i, the examples are selected from the original set, T, probabilistically. Here is a simple way of implementing this principle. Suppose the i-th example's probability is $p(\mathbf{x_i}) = 0.1$. A random-number generator is asked for a number between 0.0 and 1.0. If the returned number is from the interval $[0.0, 0.1]$, then the example is chosen; otherwise, it is not chosen.

At the beginning, when the first training set, T_1, is being created, each example has the same chance of being selected. Put more technically, if T consists of m examples, then the probability of each example is $p = 1/m$.

For each of the next subsets, T_i, the probabilities of the individual examples are modified based on the observed behavior of the previous classifier, C_{i-1}. The idea is to make sure that examples misclassified by the previous classifiers should get a higher chance of being included in T_i. This will focus the next classifier, C_i, on those aspects that the previous classifiers found difficult.

Modifying the Probabilities for the i-th Training Set   To begin with, all m training examples have the same probability, $p = 1/m$, of being selected. Using this probability, the first training subset, T_1, is created. From T_1, the first classifier, C_1 is induced. After this, the probabilities of the training examples are modified according to the classifier's behavior. More specifically, the probability of examples that have been correctly classified by C_1 will be reduced, thus increasing the future chances of examples misclassified by C_1.

The concrete way in which the probabilities are modified according to the i-th classifier's behavior is shown in Table 9.3. First, *Adaboost* calculates the i-th classifier's overall error, ϵ_i, as the weighted sum of errors on the whole original training set: this is obtained simply by summing up the probabilities of the misclassified examples. Once the weighted sum has been obtained, it is used to reduce the probability of those examples that have been correctly classified by C_i. To be more specific, each such probability is multiplied by the term, $\beta_i = \epsilon_i/(1-\epsilon_i)$.

Normalizing the Probabilities   After this, the probabilities are normalized to make sure that their values sum up to 1 as required by the theory of probability. The easiest way to normalize them is to divide each probability by the sum of all probabilities. For instance, suppose that the following probabilities have been obtained:

$$p_1 = 0.4, \quad p_2 = 0.2, \quad p_3 = 0.2$$

Table 9.3   The algorithm of *Adaboost*

Input: the training set, T, consisting of m examples; and the user's choice of the induction technique

1. Let $i = 1$. For each $\mathbf{x}_j \in T$, set $p_1(\mathbf{x}_j) = 1/m$.
2. Create subset T_i consisting of m examples randomly selected according to the given probabilities. From T_i, induce C_i.
3. Evaluate C_i on each example, $\mathbf{x}_j \in T$.
   Let $e_i(\mathbf{x}_j) = 1$ if C_i misclassified $\mathbf{x}_j$ and $e_i(\mathbf{x}_j) = 0$ otherwise.

   i)   Calculate $\epsilon_i = \sum_{j=1}^{m} p_i(\mathbf{x}_j)e_i(\mathbf{x}_j)$;
   ii) Calculate $\beta_i = \epsilon_i/(1 - \epsilon_i)$

4. Modify the probabilities of correctly classified examples by $p_{i+1}(\mathbf{x}_j) = p_i(\mathbf{x}_j) \cdot \beta_i$
5. Normalize the probabilities to make sure that $\sum_{j=1}^{m} p_{i+1}(\mathbf{x}_j) = 1$.
6. Unless a termination criterion has been met, set $i = i + 1$ and go to 2.

The sum of these values is $0.4 + 0.2 + 0.2 = 0.8$. Dividing each of the three probabilities by 0.8 will give the following normalized values:

$$p_1 = \frac{0.4}{0.8} = 0.5, \quad p_2 = \frac{0.2}{0.8} = 0.25, \quad p_3 = \frac{0.2}{0.8} = 0.25$$

It is easy to verify that the new probabilities now sum up to 1:

$$p_1 + p_2 + p_3 = 0.5 + 0.25 + 0.25 = 1.0$$

An Illustration Table 9.4 shows how *Adaboost* modifies the probabilities of the individual examples. At the beginning, each example has the same chance of being selected for inclusion of the first training set, T_1. Once the first classifier, C_1, has been induced from T_1, the classifier is applied to each single example from the original set, T.

The probability of those examples that have been correctly classified by C_1 (examples x_1 through x_7) is then reduced according to $p_{i+1}(x_j) = p_i(x_j) \cdot \beta_i$ where $\beta_i = \epsilon_i/(1 - \epsilon_i)$ and $\epsilon_i = \sum_{j=1}^{m} p_i(x_j)e_i(x_j)$. The resulting probabilities are then normalized in order to make sure that they sum up to 1 as required by the theory of probability.

After this, the second training set, T_2, is created, and then classifier C_2 induced from it. Based on the observed behavior of C_2 when applied to T, we calculate the values of ϵ_2 and β_2; these are then used when modifying the correctly classified examples' probabilities, $p(x_1) \ldots p(x_{10})$. The process thus indicated continues until a predefined user-specified termination criterion is reached. This criterion can be based on the maximum permitted number of classifiers having been reached, or on the classification accuracy of the whole "assembly" having achieved a certain threshold, or on some other requirement.

Weighted Majority Voting Once the classifiers have been induced, the example to be classified is presented to them in parallel, and the final classification decision is reached by a *weighted majority voting* whereby each classifier has been assigned a certain weight according to its classification record (see below). The final decision is not obtained by the plain vote used in the previous sections. Rather, the voting mechanism used in *Adaboost* is somewhat less-then-democratic in that each classifier has a different strength defined by its *weight*, denoted as w_i.

When presented with an example, each classifier returns a class label. The final decision is then obtained by comparing the sum, W_{pos}, of the weights of the classifiers voting for the positive class with the sum, W_{neg}, of the weights of the classifiers voting for the negative class.

For instance, suppose there are seven classifiers with the following weights: $[0.2, 0.1, 0.3, 0.7, 0.2, 0.9, 0.8]$. Suppose, further, that the first four of these return pos, and last three return neg. Plain voting will return the pos label, seeing that this class is supported by the greater number of votes.

Table 9.4 An example illustrating how *Adaboost* modifies the probabilities of the individual examples

Suppose that the training set, T, consists of ten examples denoted as $x_1 \ldots x_{10}$. The total number of examples being $m = 10$, all initial probabilities are set to $p(x_i) = 1/m = 0.1$:

$p_1(x_1)$	$p_1(x_2)$	$p_1(x_3)$	$p_1(x_4)$	$p_1(x_5)$	$p_1(x_6)$	$p_1(x_7)$	$p_1(x_8)$	$p_1(x_9)$	$p_1(x_{10})$
0.1	0.1	0.1	0.1	0.1	0.1	0.1	0.1	0.1	0.1

According to this probability distribution, the examples for inclusion in T_1 are selected. From T_1, classifier C_1 is induced.

Suppose that, when applied to T, classifier C_1 classified correctly examples $x_1 \ldots x_7$ (for these, $e_1(x_j) = 0$) and misclassified examples $x_8 \ldots x_{10}$ (for these, $e_1(x_j) = 1$).

The weighted error is then obtained as follows:

$$\epsilon_1 = \Sigma_{j=1}^{10} p_1(x_j) \cdot e_1(x_j) = 0.3$$

From here, the multiplicative term for probability modifications is calculated:

$$\beta_1 = \epsilon_1/(1 - \epsilon_1) = 0.43$$

The probabilities are modified by $p(x_j) = p(x_j) \cdot \beta_1$
Here are the new (not yet normalized) values:

$p_2(x_1)$	$p_2(x_2)$	$p_2(x_3)$	$p_2(x_4)$	$p_2(x_5)$	$p_2(x_6)$	$p_2(x_7)$	$p_2(x_8)$	$p_2(x_9)$	$p_2(x_{10})$
0.043	0.043	0.043	0.043	0.043	0.043	0.043	0.1	0.1	0.1

After normalization, we obtain the following values:

$p_2(x_1)$	$p_2(x_2)$	$p_2(x_3)$	$p_2(x_4)$	$p_2(x_5)$	$p_2(x_6)$	$p_2(x_7)$	$p_2(x_8)$	$p_2(x_9)$	$p_2(x_{10})$
0.07	0.07	0.07	0.07	0.07	0.07	0.07	0.17	0.17	0.17

Note that these ten probabilities sum up to 1.0. The next classifier, C_2, is then induced from a training set T_2 whose examples have been selected from T according to these last probabilities.

By contrast, *weighted majority voting* will separately sum up the weights supporting the positive class, obtaining $W_{pos} = 0.2 + 0.1 + 0.3 + 0.7 = 1.3$, and then sum up the weights supporting the negative class: $W_{neg} = 0.2 + 0.9 + 0.8 = 1.9$. The master classifier will then label the example as negative because, as we can see, $W_{neg} > W_{pos}$.

How to Obtain the Weights of the Individual Classifiers Each classifier is assigned a weight according to its performance: the higher the classifier's reliability, the higher its weight. On the other hand, the weight can in principle even be negative—if it is believed that the classifier is more often wrong than right. It is important to be aware of methods to find the concrete weights.

Many alternative possibilities exist. The inventors of *Adaboost* suggested specific formulas which facilitated mathematical analysis of the technique's behavior.

Practically speaking, though, one can just as well use the perceptron-learning algorithm from Chap. 4. The idea is to begin with equal weights for all classifiers, and then present the system, one by one, with the training examples. Each time the master classifier misclassifies an example, we increase or decrease the weights of the individual classifiers according to the relation between the master classifier's hypothesis, $h(\mathbf{x})$, and the training example's true class, $c(\mathbf{x})$.

Of course, other methods can be used. One might consider the WINNOW that we know is good at weeding out those classifiers that do not contribute much to the overall system's performance (as if these classifiers were irrelevant attributes describing an example).

What Have You Learned?

To make sure you understand this topic, try to answer the following questions. If you have problems, return to the corresponding place in the preceding text.

- Describe the mechanism that *Adaboost* uses when selecting the training examples to be included in the training set, T_i, from which the i-th classifier is to be induced.
- Explain how, after the induction of the i-th classifier, *Adaboost* modifies for each example its probability that it will be chosen for inclusion in T_{i+1}.
- Explain the principle of the *weighted majority voting* that *Adaboost* uses when deciding about a concrete example's class label.
- How are the weights of the individual classifiers obtained?

9.4 Variations on the Boosting Theme

The essence of boosting is to combine several imperfect classifiers that tend to complement each other. Schapire was the first to suggest a concrete way of inducing the classifiers, with *bagging* and *Adaboost* being the most popular alternatives.

The number of variations on this theme is virtually inexhaustible. It is good to be aware of them, and this is why this section presents at least some of the most important ones.

Randomizing the Attribute Set Rather than inducing each classifier from a different training subset (as in the previous versions of boosting), one may consider using the same training examples, but always described by a different subset of attributes.

Again, the input is the set, T, of the training examples—and the set, A, of the attributes used to describe them. Instead of the random subsets of examples (as in *bagging*), we choose N random subsets of attributes, $A_i \subset A$. The i-th classifier ($i \in [1, N]$) is then induced from the examples from T described by attributes from

A_i. As before, the classifiers' outputs are combined by the *weighted majority voting*, the weights being obtained (for instance) by perceptron learning.

The approach is useful in domains marked by a great many attributes of which most are suspected to be either irrelevant or redundant. Typically, classifiers induced using less valuable attribute sets will exhibit poor classification performance; and as such, they will receive low (or even negative) weights.

The approach can easily be combined with classical *bagging*: the idea is that, for each classifier, a different set of examples, and also a different set of attributes, should be used.

Non-homogeneous Boosting So far, all boosting approaches presented here assumed that the individual classifiers are induced from somewhat different data, but always using the same induction technique. But there is no reason for this always having to be the case. The so-called *non-homogeneous boosting* does the exact opposite: each classifier is induced from the same data, but with a different machine-learning technique. The classifiers' outputs are then, again, combined by *weighted majority voting*.

The main advantage of this approach is the way in which it reduces error. As we will see in Chap. 10, the errors committed by any classifier fall into two fundamental categories. The first is caused by the *variance* in the available data: from a different training set, a somewhat different classifier will be induced, and this will lead to different errors. The second source of error is the *bias* inherent in the classifier. For instance, a linear classifier cannot avoid misclassifying some examples if the decision surface separating the positive examples from the negative is highly non-linear.

Importantly, *non-homogeneous boosting* is known to reduce both kinds of error: *variance*-related errors (this reduction happens in all boosting algorithms) as well as *bias*-related errors (which is an advantage specific to *non-homogeneous boosting*).

Stacking The idea of *non-homogeneous boosting* is to take the outputs of a group of classifiers, and then reach the final classification decision by the *weighted majority voting*. If you think of it, two layers are involved here: at the lower level are the base-level classifiers; and at the upper level is the master classifier combining their outputs. Note that the master classifier itself has to be induced; for instance, with the help of perceptron learning—because this is essentially a linear classifier.

The principle can be generalized to the so-called *stacking* approach. As before, a set of diverse classifiers is used at the lower lever. The upper level then goes beyond the bounds of a linear classifier. Indeed, any paradigm can be used for the master classifier: Bayesian, nearest-neighbor based, a decision tree, or a neural network. The linear classifier used in *non-homogeneous boosting* is just one out of many possibilities.

Likewise, the base-level classifiers may come from the most diverse paradigms. Sometimes, however, the individual classifiers differ only in the concrete choice of parameter values. For instance, there can be a few decision trees differing in the extent of pruning. At any rate, the base-level classifiers should all differ in the way they classify the examples.

Table 9.5 The class labels suggested by the six base-level classifiers are used as attributes to redescribe the examples

	x_1	x_2	x_3	...	x_m
Classifier 1	1	1	0	...	0
Classifier 2	0	0	1	...	1
Classifier 3	1	1	0	...	1
Classifier 4	1	1	0	...	1
Classifier 5	0	1	0	...	1
Classifier 6	0	0	0	...	1
Real class	1	1	0	...	1

Each column then represents a training example to be used for the induction of the master classifier

The method is illustrated in Table 9.5. The rows represent the individual classifiers. Specifically, six classifiers have been induced here, each by a different machine-learning technique. The columns represent the training examples. The field in the i-th row and the j-th column contains a one if the i-th classifier labels the j-th example as positive; otherwise, it contains a zero. Each column can thus be interpreted as a binary vector that redescribes the corresponding example. This new training set is then presented to a machine-learning program that induces from it the master classifier.

What Have You Learned?

To make sure you understand this topic, try to answer the following questions. If you have problems, return to the corresponding place in the preceding text.

- Explain what this section meant by "randomizing the attribute set." What is the main advantage of the approach? How can it be combined with the classical *bagging*?
- Explain the principle of *non-homogeneous boosting*. What is its main advantage from the perspective of the errors committed by the classifier?
- Explain the two-layered principle of "stacking." In what sense do we say that stacking is a generalization of *non-homogeneous boosting*?

9.5 Cost-Saving Benefits of the Approach

In some machine-learning techniques, computational costs grow very fast with the growing size of the training set. In this event, an experimenter will observe that induction from half of the examples incurs only a small fraction of the costs incurred by induction from all examples. Likewise, the induction technique's costs may grow very fast with the growing number of attributes.

Fig. 9.4 In some
machine-learning techniques,
computational costs grow
quickly with the growing size
of the training set. Quite
often, induction from half of
the examples incurs only a
small fraction of the costs
incurred by induction from all
examples

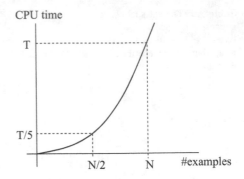

An Illustration The situation is visualized in Fig. 9.4. Here, the time needed for
a hypothetical machine-learning technique to induce a classifier from N examples
is T; however, the time needed to induce a classifier from 50% of the examples is
only one fifth of it, $0.2T$. From this follows that to induce two classifiers, one from
the first half of the training set, the other from the second half, we will need only
$2 \times 0.2T = 0.4T$ of the time that would have been needed for the induction of the
classifier from the whole set, T. The computational savings thus amount to 60% of
the original costs.

Generalization to K Classifiers Following the same logic, we may consider
induction of K classifiers, each from a different subset, T_i, such that the size of each
of them is m (a user-set constant) which is supposed to be much smaller than N.
The classifiers induced from these sets will then vote, just as they do in *bagging*. In
many cases, the induction of the individual classifiers will take only a tiny fraction of
the original time—and yet the classification performance of the "voting assembly"
thus obtained may compare favorably with that of a classifier induced from the
entire training set. Similar observations can be made also in the case of *Schapire's
boosting* and *Adaboost*.

To conclude, the boosting paradigm may not only improve classification performance. Quite often, it will lead to significant savings in the computational costs
associated with the induction. This observation is particularly important in domains
marked by many examples described by many attributes. Here, induction can be
very expensive, and any idea that helps reduce computational costs is more than
welcome.

Comments on Practical Implementation In *bagging*, the number of examples
to be included in T_i is the same as in the original training set; no computational
savings in the sense illustrated in Fig. 9.4 are thus possible. On the other hand, we
have to remember that the assumption about the sizes of T_i was used only to make it
possible to present *bagging* as a bootstrapping technique. In practical applications,
the assumption is unnecessary, and can thus be relaxed. In other words, the size, m,
of the sets T_i can be a user-set constant, just as it is in *Adaboost*.

In the case of *Schapire's Boosting* and *non-homogeneous boosting*, the reader will recall that the sizes of T_i, are user-specified parameters. The same applies to *stacking*.

The Costs of Example Selection When considering the savings in computational costs, however, we must not forget that the price for them is a certain unavoidable overhead. To be more specific, additional costs are associated with the need to select the examples to be included in the next T_i.

In the case of *bagging*, these costs are so small as to be easily neglected. They are higher in *Adaboost*, but even here they are still affordable. The situation is different in the case of *Schapire's boosting*. Here, the search for the examples that satisfy the conditions for T_2 and T_3 can be quite expensive—especially when a great number of training examples have been made available, but only a small percentage of them satisfy the required conditions.

What Have You Learned?

To make sure you understand this topic, try to answer the following questions. If you have problems, return to the corresponding place in the preceding text.

- Elaborate on the statement that "computational costs of induction can be greatly reduced by exploiting the idea of the voting assemblies induced from small subsets of examples."
- Discuss the circumstances under which one can expect that the boosting idea will reduce computational costs.
- How expensive (computationally) is it to create the training subsets, T_i, in each of the boosting techniques?

9.6 Summary and Historical Remarks

- A popular machine learning approach induces a set of classifiers, each from a different subset of the training examples. When presented with an example, each classifier offers a class. Based on these suggestions, a master classifier makes the final decision.
- The simplest application of this idea is the so-called *bagging*. Here, the subsets used for the induction of the individual classifiers are obtained from the original training set, T, by a bootstrapping mechanism. Suppose that T contains m examples. Then, when creating T_i, we choose m examples "with replacement." Some examples may then appear in T_i more than once, while others will not appear there at all.
- *Schapire's boosting* induces three classifiers, C_1, C_2, and C_3, making sure that they complement each other as much as possible. This complementarity is

achieved by the way the training subsets are created: the composition of T_1 is random; the composition of T_2 is such that C_1 experiences 50% error rate on this set; and T_3 consists of examples on which C_1 and C_2 disagree. Each of the three subsets contains the same number of examples.

- By contrast, *Adaboost* chooses the examples probabilistically in a way that makes each T_i consist primarily of examples on which the previous classifiers failed. Another difference is that the final class label is obtained by *weighted majority voting*.

- A few variations on the theme exist. One of them, *randomization of attribute sets*, induces the classifiers from the same training examples which, however, are each time described by a different subset of the attributes. Another one, *non-homogeneous boosting*, uses always the same training set (and the same attributes), but induces each classifier by a different induction technique. The final class is then obtained by *weighted majority voting*. Finally, *stacking* resembles non-homogeneous boosting, but the output is decided by a master classifier that has been trained on the outputs of the base-level classifiers.

Historical Remarks The idea of boosting was invented by Schapire [85] who pointed out that, in this way, even the performance of very weak induction paradigms can thus be "boosted"—hence the name. The somewhat more practical idea underlying *Adaboost* was published by Freund and Schapire [28], whereas the *bagging* approach was explored by Breiman [9]; its application to decision trees, known under the term *random forests*, was published by Breiman [10]. The principle of *stacking* (under the name *Stacking Generalization*) was introduced by Wolpert [99].

9.7 Solidify Your Knowledge

The exercises are to solidify the acquired knowledge. The suggested thought experiments will help the reader see this chapter's ideas in a different light and provoke independent thinking. Computer assignments will force the readers to pay attention to seemingly insignificant details they might otherwise overlook.

Exercises

1. Suppose the probabilities of the training examples to be used by *Adaboost* are those listed in Table 9.6. From these, a training subset, T_i, has been created, and from T_i, classifier C_i was induced. Suppose that C_i then misclassifies examples x_2 and x_9. Show how the probabilities of all training examples are re-calculated, and then normalize the probabilities.

2. Suppose that eight classifiers have labeled an example. Let the weights of the classifiers returning the pos label be $[0.1, 0.8, 0.2]$, and let the weights of the

Table 9.6 The probabilities of ten training examples

$p(\mathbf{x_1})$	$p(\mathbf{x_2})$	$p(\mathbf{x_3})$	$p(\mathbf{x_4})$	$p(\mathbf{x_5})$	$p(\mathbf{x_6})$	$p(\mathbf{x_7})$	$p(\mathbf{x_8})$	$p(\mathbf{x_9})$	$p(\mathbf{x_{10}})$
0.07	0.07	0.07	0.07	0.07	0.07	0.07	0.17	0.17	0.17

classifiers returning the neg label be $[-0.1, 0.3, 0.3, 0.4, 0.9]$. What label is going to be returned by a master classifier that relies on *weighted majority voting*?

3. Return to Table 9.5 that summarizes the class labels returned for some of the m examples by six different classifiers. Suppose a 3-NN-based master classifier is asked to label an example, and that the three nearest neighbors (the columns in Table 9.5) are $\mathbf{x_1}$, $\mathbf{x_2}$, and $\mathbf{x_3}$. What final label is returned?

Give It Some Thought

1. Recall how *Schapire's boosting* chooses the examples for inclusion in the training sets T_2, and T_3. Discuss possible situations under which it is impossible to find enough examples for these subsets. Identify a situation where enough examples *can* be found, but the search for them is impractically expensive. Conversely, under what circumstances can it be affordable to identify all the necessary examples even when recursion is used?

2. Give some thought to the *stacking* approach. Think of a situation under which it will disappoint. Conversely, suggest a situation where *stacking* will outperform the less general *non-homogeneous boosting*.

Computer Assignments

1. Implement the basic algorithm of *Adaboost*. The number of voting classifiers is determined by a user-set constant. Another user-set constant specifies the number of examples in each training set, T_i. The weights of the individual classifiers are obtained with the help of *perceptron learning*.

2. Apply the program implemented in the previous task to some of the benchmark domains from the UCI repository.[2] Make observations about this program's performance on different data. For each domain, plot a graph showing how the overall accuracy of the resulting classifier depends on the number of subclassifiers. Also, observe how the error rate on the training set and the error rate on the testing set tend to converge with the growing number of classifiers.

3. Implement the *stacking* algorithm for different base-level learning algorithms and for different types of the master classifier. Apply the implemented program to a few benchmark domains, and observe its behavior.

---

[2] www.ics.uci.edu/~mlearn/MLRepository.html.

Chapter 10
Some Practical Aspects to Know About

The engineer who wants to avoid disappointment has to be aware of certain machine-learning aspects that, for the sake of clarity, our introduction to the basic techniques had to neglect. To present some of the most important ones is the task for this chapter.

The first thing to consider is bias: to be able to learn, the learner has to build on certain assumptions about the problem at hand, thus reducing the size of the search space. The next important point has to do with the observation that an increase in the size of the training set can actually hurt the learner's chances if most of the training examples belong only to one class. After this, we will discuss the question how to deal with classes whose definitions tend to change with context or in time. The last part focusses on some more mundane aspects such as unknown attribute values, the selection of the most useful sets of attributes, and the problem of multi-label examples.

10.1 A Learner's Bias

Chapter 7 boldly declared that there is "no learning without bias." The point was that an unconstrained (and hence extremely large) hypothesis space is bound to contain many hypotheses that only by mere chance correctly classify the entire training set while still erring a lot on future examples. There is another, more practical side to it. To be able to find something, you need to know where to look; and the smaller the place where that something is hidden, the higher the chances of finding it.

A Simple Example Suppose we are to identify the property shared by the following set of integers: $\{2, 3, 10, 20, 12, 21, 22, 28\}$. In the language of machine learning, these constitute positive examples. Alongside these, also negative examples are provided: $\{1, 4, 5, 11\}$. In these, the property is *not* present.

© Springer International Publishing AG 2017
M. Kubat, *An Introduction to Machine Learning*,
DOI 10.1007/978-3-319-63913-0_10

A student trying to find the answer typically explores various notions offered by the number theory, such as primes, odd numbers, integers whose values exceed a certain threshold, results of arithmetic operations, and so on. After a lot of effort, some property satisfying the training examples is found. Usually, however, the discovered rule is ridiculously complicated and awkward to say the least.

And yet, there is a simple solution which the students almost never hit upon. Thing is, the underlying property does not come from the realm of arithmetics. What the positive examples have in common (and the negative examples lack) is that they all begin with the letter t: two, three, ..., all the way to twenty-eight. Conversely, none of the integers in the set of negative examples begins with a t.

The reason this simple solution is so difficult to find is that most people search for it in the wrong place: arithmetics. Expressed in the language of machine learning, they rely on the wrong *bias*. Once they are given the correct answer, their mindset will be willing to take this experience into account in the future. If you give them another puzzle of a similar nature, they will subconsciously think not only about arithmetics, but also about the English vocabulary—they will incorporate into their thinking also this new bias.

Representational Bias Versus Procedural Bias As far as machine learning is concerned, biases come in different forms. A so-called *representational bias* is determined by the language in which we want the classifier to be formulated. For instance, in domains with continuous-valued attributes, one possible representational bias can consist in the choice of a linear classifier; another can be the preference for polynomials, and yet another the preference for neural networks. If all attributes are discrete-valued, the engineer may prefer conjunctions of attribute values, or perhaps even decision trees. Of course, all of these biases then have its advantages as well as shortcomings.

Apart from this, the engineer usually also has to opt for a certain *procedural bias*. By this we mean preference to a certain method of searching for the solution, the selection of a specific machine-learning procedure. For instance, one such bias relies on the assumption that pruning will improve the classification performance of a decision tree on future data. Another procedural bias is the choice of a concrete set of parameters in a neural-network's training. And yet another, in the field of linear classifiers, can be the engineer's decision to employ the *perceptron learning* instead of WINNOW—or the other way round.

The Strength of a Bias Versus the Correctness of a Bias Suppose the engineer wants to decide whether to approach the given machine-learning problem with a linear classifier or with a neural network. If the positive and negative examples are linearly separable, then the linear classifier is clearly the better choice. While both paradigms are capable of finding the solution, neural networks tend to overfit the training set, thus poorly generalizing to future data. On the other hand, if the tentative boundary separating the two classes is highly non-linear, then the linear classifier will lack the necessary flexibility, whereas neural networks will probably manage quite easily. The reader now begins to understand that each bias has two critical aspects: strength, and correctness.

A bias is *strong* if it defines only a narrow class of classifiers. In this sense, the bias of linear classifiers is much stronger than that of neural networks: the former only allow for linear decision surfaces, while the latter can model virtually *any* decision surface.

A bias is *correct* if it is the right one for the task at hand. For instance, the linear classifier's bias is correct only in a domain where the positive examples are linearly separable from the negative ones. A conjunction of boolean attributes is correct only if the underlying class can indeed be described by a conjunction of attributes. Of course, the opposite term, *incorrect* bias, is not a crisp concept. Some gradation is involved; some biases are only slightly incorrect, others significantly so.

A Useful Rule of Thumb: Occam's Razor Ideally, the engineer wants to use a bias (representational or procedural) that is correct. And if there is a possibility to choose between two or more biases that are all correct, the stronger bias is to be preferred because it has a higher chance of success—this is what we learned in Chap. 7 where the advice to choose the simpler solution was presented under the name of the *Occam's Razor*.

Unfortunately, we rarely know in advance the correctness/incorrectness of all possible biases. An educated guess is the best we can hope for. In some paradigms, say, high-order polynomials, the bias is so weak that there is a high probability that a classifier from this class will classify the entire training set with zero error rate; and yet its performance on future data is uncertain on account of its problems with PAC-learnability. Strengthening the bias (say, by reducing a polynomial's order) will reduce the VC-dimension, increasing the chances on future data—but only as long as the bias remains correct. At a certain moment, further strengthening of the bias will do more harm than good because the bias becomes incorrect, perhaps very much so.

What we need to remember is the existence of an almost inescapable trade-off: a mildly incorrect but strong bias can be better than a correct but very weak bias. But what the term, "mildly incorrect bias," means in a concrete application can usually be decided only based on the engineer's experience or by additional experimentation (see Chap. 11).

"Lifelong Learning" In some applications, the machine-learning software is to learn a series of concepts or classes, all of which are expected to have a solution within the realm of the same specific bias. In this event, it makes sense to organize the learning procedure in two tiers. At the lower level, the task is to identify the most appropriate bias; at the higher level, the software induces the classifier using this bias. The term used for this strategy, "lifelong learning," reminds us of something typical of our own human difficulties in learning: the need to "learn how to learn" in a given field.

Two Sources of the Classifier's Errors The observations made so far will help us get a better grasp of the two main sources of a classifier's errors.

The first is the *variance* in the training examples. Thing is, the data used for the induction of the concrete classifier almost never capture all aspects of the underlying class. This is partly due to the way the training set has been created. In some applications, the training set has been created at random. In other domains, it consists of examples available at the given moment, which involves a great deal of randomness, too. And in yet others, the training set has been created by an expert who has chosen the examples which he believes best represent the given class. The last case is inevitably subjective, and thus no less unreliable than the previous two. In view of all this, one can easily imagine that a different training set might be created for the same domain. And here is the point. From a different training set, a somewhat different classifier will be induced, and this different classifier will make somewhat different errors on future data. This is what we mean by saying that variance in the training data is an important source of errors. Its negative effect can often be reduced if we use really large training sets.

The second source of error is *bias-related*. If the two classes, pos and neg, are not linearly separable, then any linear classifier is bound to misclassify certain minimum percentage of future examples. Bias-related errors cannot be reduced below a certain limit because they are inherent in the very nature of the selected type of classifier.

It is instructive to give some thought to the *trade-off* between the two sources. For one thing, the bias-related error can be reduced if we resort to a machine-learning paradigm known to have a weaker bias. Unfortunately, one of the unintended consequences of such a decision is higher variance. Conversely, variance can in principle be reduced by strengthening the bias—which, if incorrect, will increase the frequency of bias-related errors.

What Have You Learned?

To make sure you understand this topic, try to answer the following questions. If you have problems, return to the corresponding place in the preceding text.

* Explain the difference between the representational bias and the procedural bias. Illustrate each type by examples.
* Explain the difference between the strong and weak bias. Explain the difference between the correct and incorrect bias. Discuss the interrelation of the two dichotomies.
* What has this section taught us about the two typical causes of a classifier's underperformance on future data?

10.2 Imbalanced Training Sets

When discussing the oil-spill domain, Sect. 8.2 pointed out that well-documented images of oil spills are relatively rare. Indeed, the project could rely only on a few dozen positive examples while the negative examples were abundant. Such imbalanced representation of the two classes is not without serious consequences for machine learning. This section will explain the cause of the difficulties, then proceed to some very simple ways of reducing their negative impact.

A Simple Experiment Suppose we have at our disposal a training set that is so small that it consists of only 50 positive examples and 50 negative examples. Let us subject this set to a fivefold crossvalidation[1]: we divide it into five equally sized parts; then, in five different experimental runs, we always remove one of the parts, induce a classifier from the union of the remaining four, and test the classifier on the removed part. In this manner, we eliminate, or at least reduce, the effect of randomness in the choice of the concrete training set. At the end, we write down the average results of the testing: classification accuracy on the positive examples, classification accuracy on the negative examples, and the geometric mean of the two classification accuracies.

Suppose now that we realize we have many more negative examples at our disposal than we originally thought. In order to find out how this newly discovered bounty is going affect the learning, we add to the previous training set another 50 negative examples (the positive examples remain the same), repeat the experimental procedure, and then write down the new result. We then continue in the same spirit, always adding to the training set another batch of 50 negative examples while keeping the same original 50 positive examples.

Observation If we plot the results of the above series of experiments, we will obtain a graph that, in all likelihood, will look very much like the one shown in Fig. 10.1 where the 1-NN classifier was used. The reader can see that as the number of the majority-class examples increases, the induced classifiers become biased toward this class, gradually converging to a situation where the classification accuracy on the negative examples (the majority class) approaches 100%, while the classification accuracy on the positive examples (the minority class) drops to well below 20%. The geometric mean of the two values keeps dropping, too.

The observation may appear somewhat counterintuitive. Surely the induced classifiers should become more powerful if more training examples are made available, even if these added examples all happen to belong to the same class? In turns out, however, that the unexpected behavior described above is typical of many machine-learning techniques. Engineers usually call it the problem of *imbalanced class representation*.

Majority-Class Undersampling (The Mechanical Approach) The experiment has convinced us that adding more examples from the majority class may cause degradation of the induced classifier's performance on the minority class. This may be a serious shortcoming. Thus in the oil-spill domain, the minority class represents the oil spills, the primary target of the machine-learning undertaking. In medical diagnosis, any disorder we want to recognize is typically a minority class, too. And

---

[1] An evaluation methodology introduced in Sect. 11.5.

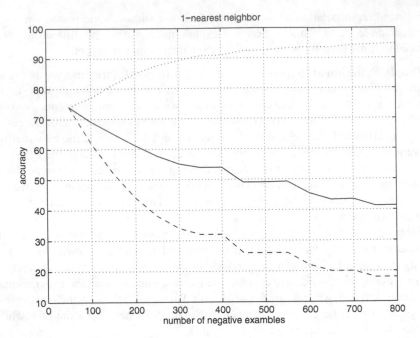

Fig. 10.1 *Dotted curve*: classification accuracy on the neg class; *dashed curve*: classification accuracy on the pos class; *solid curve*: geometric means of the two classification accuracies

the same applies to software whose task is to alert a company to misuse of its product (e.g., a wrongful use of calling cards, or credit-card fraud). In domains of this kind, it is the minority class that interests us. We now know that blindly adding more and more majority-class examples to the training set is likely to do more harm than good.

Suppose we are provided with a heavily imbalanced training set where, say, nine out of ten examples are negative. In this event, we will often benefit from the removal of many negative examples. In the simplest possible approach, this removal can be made at random: for instance, each negative example will face a 50% chance of being deleted from the training set. As we noticed above, the classifier induced from this reduced set is likely to outperform a classifier induced from the entire training set.

Identifying the Cause The mechanical solution indicated in the previous paragraph will hardly satisfy the thoughtful engineer who wants to understand *why* the data-removing trick worked—or, conversely, why increasing the number of majority-class examples may have such detrimental consequences.

Suppose the 1-NN classifier uses a training set where the vast majority of the examples are negative, and only a few are positive. Moreover, the data are known to suffer from a considerable amount of class-label noise. Limiting itself to an easy-to-visualize two-class domain, the left part of Fig. 10.2 shows one such training set.

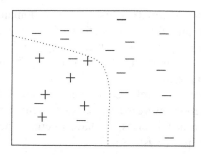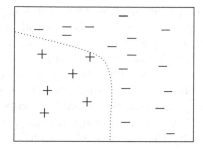

Fig. 10.2 In noisy domains where negative examples heavily outnumber positive examples, the removal of negative examples that participate in Tomek links may increase classification performance

The reader is sure to have noticed the point: in consequence of the noise, the nearest neighbor of almost every positive example is negative. In reality, these neighbors are probably positive, their negative labels being explained by errors made in the course of the creation of the training set. Be it as it may, the 1-NN classifier misclassifies these positive examples, and this is why there are so many false negatives, and only a few (if any) false positives.

Of course, not all machine-learning paradigms will suffer from this situation as dramatically as the 1-NN classifier. But most of them *will* suffer to some degree, and we now understand the reason why.

An Informed Solution: One-Sided Selection Knowing the source of our troubles, we are ready to suggest a remedy. To wit, the cause of our woes is the presence of many class-noisy examples in the positive region; the situation should therefore improve if we remove primarily *these* examples (rather than resorting to a random selection as in the mechanical approach suggested above).

In Chap. 3, we encountered a simple algorithm capable of identifying "suspicious" examples: the technique of Tomek links. The reader will recall that two examples, $(\mathbf{x}, \mathbf{y})$, are said to participate in a Tomek link if three conditions are satisfied: (1) each of the two examples has a different class label; (2) the nearest neighbor of $\mathbf{x}$ is $\mathbf{y}$; and (3) the nearest neighbor of $\mathbf{y}$ is $\mathbf{x}$. In the situation depicted in Fig. 10.2, many of the noisy examples on the left indeed do participate in Tomek links. This indicates that we may do improve the classifier's behavior if we delete from the training set the negative participants of each Tomek-link pair. The principle is known as *one-sided* selection because only one side of the Tomek link is selected for inclusion in the training set.

Applying the technique to the training set shown in the left part of Fig. 10.2, we will obtain something like the smaller training set shown in the right part. It is easy to see that the frequency of false negatives is now going to be lower. The efficiency of this methods is usually higher than just mechanical removal of randomly picked negative examples.

The Opposite Solution: Oversampling the Minority Class In some domains, however, the training set is so small that any further reduction of its size by undersampling is impractical. Even the majority-class examples are sparse, here; the deletion of any one of them may remove some critical aspect of the learning task, and thus jeopardize the performance of the induced classifier.

In this event, the opposite approach is sometimes preferred. Rather than removing majority-class examples, we *add* examples representing the minority class. Since we do not have at our disposal real examples from this class, we have to create them artificially. This can be done in two fundamental ways:

1. For each example from the minority class, create one copy and add this copy into the training set. Alternatively, two or more copies for each example can thus be created and added.
2. For each example from the minority class, create its slightly modified version and add it into the training set. The modification is made by minor (random) changes in continuous attributes; much less useful, though still possible, are changes in discrete attribute values.

The left part of Fig. 10.2 helps us understand why this works. To the neighborhood of some of the "afflicted" positive examples (those whose nearest neighbors have been turned negative by noise), minority-class oversampling inserts additional positive examples; as a result, the 1-NN classifier is no longer misled. The principle helps improve the behavior of other types of classifiers, too.

What Have You Learned?

To make sure you understand this topic, try to answer the following questions. If you have problems, return to the corresponding place in the preceding text.

- What does the term *imbalanced training set* refer to? Explain the main reason why induction from imbalanced training sets so often leads to disappointing results.
- What is the essence of majority-class undersampling? Explain the mechanical approach, and then proceed to the motivation and principle of the *one-sided selection* that uses Tomek links.
- Explain the principle of minority-class oversampling. Describe and discuss the two alternative ways of creating new examples that are to be added to the training set.

10.3   Context-Dependent Domains

Up till now, we have tacitly assumed that the underlying "meaning" of a given class is fixed and immutable, that a single classifier, once induced, will under all circumstances exhibit the same (or at least similar) behavior. This, however, is not always the case.

Context-Dependent Classes   Some classes change their essence with circumstances. If you think of that, this is the case of many concepts used in daily life. Thus the meaning of "fashionable dress" changes in time, and different cultures have a different idea of what they want to wear. "State-of-the-art technology" was something else a 100 years ago that it is today. Even the intended meaning of such notorious terms as "democracy" or "justice" depends on political background and historical circumstances. And if you want a more technical example, consider the problems encountered by speech-recognition software: everybody knows that the same word is often pronounced differently in England than in North America; but the software should "understand" speakers from both backgrounds.

Context-Dependent Features   For the needs of this book, *context* is understood as a "a feature that has no bearing on the class if taken in isolation, but still affects the class when combined with other features."

For instance, suppose you want to induce a classifier capable of suggesting medical diagnosis, of recognizing X based on a set of symptoms. Some attributes, say, `gender`, do not have any predictive power; the patient being `male` is no proof of `prostate-cancer`; but the attribute value `gender=female` is a clear indication that the class is *not* `prostate-cancer`. This, of course, was an extreme sample. In other diagnoses, the impact of `gender` will be limited to influencing the critical values of certain laboratory tests, say, $p = 0.5$ being a critical threshold for male patients and $p = 0.7$ for female patients. Alternatively, the prior probabilities will be affected, `breast-cancer` being more typical of females, although men can suffer from it, too.

Induction in Context-Dependent Domains   Suppose you want to induce a speech-recognition system, and you have a set of training examples coming both from British and American speakers. Suppose the attribute vector describing each example contains the "context" attribute, the speaker's origin. The other attributes capture the properties of the concrete digital signal. Each class label represents a different phoneme.

For the induction of a classifier that for each attribute vector decides which phoneme it represents, the engineer can essentially follow two different strategies. The first takes advantage of the contextual attribute, and divides the training examples into two subsets, one for British speakers and one for American speakers; then it induces a separate classifier from each of these training subsets. The second strategy mixes all examples in one big training set, and induces one "universal" classifier.

Practical experience indicates that, in applications of this kind, the first strategy performs better, provided that the real-time system in which the classifiers are embedded knows which of them to use. This decision can be assisted by an additional two-valued classifier that is trained to distinguish British speakers from American speakers.

Concept Drift Sometimes, the context changes in time. The "fashionable dress" example mentioned earlier belongs to this category, as do the political terms. In this event, machine-learning specialists talk about a so-called *concept drift*. What they have in mind is that, in the course of time, the essence or meaning of a class drifts from one context to another.

The drift has many aspects. One of them is the extent to which the context has changed the meaning of the class. In some rare domains, this change is so substantial that the induced classifier becomes virtually useless, and a new one has to be induced. Much more typical, however, is a less severe change that results only in a minor reduction of the classification performance. The old classifier can then still be used, perhaps after some fine-tuning.

Another feature worth consideration is the "speed" of the drift. At one extreme is an abrupt change. At a certain moment, one context is simply replaced, at it were, by another. More typically, however, the change is gradual in the sense that there is a certain "transition" period during which one context is, step by step, replaced by the other. In this event, the engineer may ask how fast the transition is, and whether (or when) the concept drift will necessitate special actions.

Induction of Time-Varying Classes Perhaps the simplest scenario in which concept drift is encountered is the one shown in Fig. 10.3. Here, a classifier is faced with a stream of examples that arrive one at a time, either in regular or irregular intervals. Each time an example arrives, the classifier labels it with a class. There may or may not be a feedback loop that tells the system (immediately or after some delay) whether the classification was correct, and if not, what the correct class label should have been.

If there is a reason to suspect the possibility of an occasional concept drift, it may be a good idea to take advantage of a sliding window such as the one shown in the picture. The classifier is then induced only from the examples "seen through the window." Each time a new example arrives, it is added to the window. Whenever deemed appropriate, older examples are removed, either one at a time, or in groups,

Fig. 10.3 A window passes over a stream of examples; "current classifier" is periodically updated to reflect changes in the underlying class. Occasionally, the system can retrieve some of the previous classifiers if the underlying context recurs

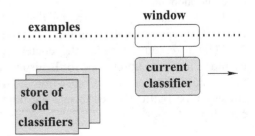

such as "the oldest 25% examples." The motivation for the deletion is simple: the engineer wants the window to contain only recent examples, suspecting that older ones may belong to an outdated context.

As already mentioned, the classifier is supposed to reflect only the examples contained in the window. In the simplest implementation, the classifier is re-induced each time the window contents change. Alternatively, the change in the window contents may only trigger a modification/adaptation of the classifier.

Figure 10.3 shows yet another aspect of this learning paradigm: sometimes, an older context may reappear (e.g., due to a certain "seasonality"). For this reason, it may be a good idea to store previously induced versions of the classifier, just in case they might be re-used in the future.

Engineering Issues in the Sliding-Window Approach There are certain essential issues that the engineer wishing to implement the sliding-window approach needs to consider. The first is the question of the size of the window. If it is too small, then the examples it contains may not be sufficient for successful learning. If it is too big, then it may contain examples that come from outdated contexts. Ideally, then, the window should grow (no old examples deleted) as long as it can be assumed that the context has not changed. When a change is detected, a certain number of the oldest examples should be deleted because they are no longer trusted.

This leads us to the next important question: how to recognize that a context has changed? One simple solution relies on the feedback about the classifier's behavior. The change of context is then identified by a sudden drop in the classification performance.

Finally, there is the question of how many of the oldest examples to delete from the window. The answer will depend on how gradual the context change is, and also on the extent of this change. At one extreme, an abrupt and considerable change will call for the deletion of *all* examples. At the other extreme, a very slow transition between two very similar contexts will necessitate the deletion of only a few of the oldest examples.

What Have You Learned?

To make sure you understand this topic, try to answer the following questions. If you have problems, return to the corresponding place in the preceding text.

- Give examples of domains where the meaning of a given class varies in time and/or geographical location. Give examples of domains where previous meanings recur in time.
- Describe the basic scenario that is based on a stream of time-ordered examples. Explain the principle of the sliding-window approach to induction in time-varying domains.
- Discuss briefly the basic engineering issues encountered in the sliding-window approach.

10.4   Unknown Attribute Values

In many domains, certain attribute values are not known. A patient refused to give his age, a measurement device failed, and some information got lost—or is unavailable for any other reason. As a result, we get an imperfect training set such as the one shown in Table 10.1 where some of the values are replaced with question marks. In some domains, the question marks represent a considerable portion of all attribute-value fields, and this may complicate the learning task. The engineer needs to understand what kind of damage the unknown values may cause, and what solutions exist.

Adverse Effects   In the case of the plain version of the k-NN classifier, the distance between two vectors can only be calculated if all values in the vectors are known. True, the distance metric can be modified so that it quantifies also the distance between, say, `red` and `unknown`; but distances calculated in this manner tend to be rather ad hoc.

The situation is not any better in the case of linear and polynomial classifiers. Without the knowledge of all attribute values, it is impossible to calculate the weighted sum, $\Sigma w_i x_i$, whose sign tells the classifier which class label to choose. Likewise, unknown attribute values complicate the use of Bayesian classifiers and neural networks.

Decision trees are more flexible, in this sense. When classifying an example, it is quite possible that the attribute whose value is unknown will not have to be tested (will not find itself on the path from the root node to the terminal node).

Trivial Approaches to Filling-In Missing Values   In a domain with a sufficiently large training set that contains only a few question marks, there is usually no harm in removing all examples that have unknown attribute values. This, however, will become impractical in domains where the number of question marks is so high

Table 10.1 Training examples with missing attribute values

Example	Shape	Crust		Filling		Weight	Class
		Size	Shade	Size	Shade		
ex1	Circle	Thick	Gray	Thick	Dark	7	pos
ex2	Circle	Thick	White	Thick	Dark	2	pos
ex3	Triangle	Thick	Dark	Thick	Gray	2	pos
ex4	Circle	Thin	White	?	Dark	3	pos
ex5	Square	Thick	Dark	?	White	4	pos
ex6	Circle	Thick	White	Thin	Dark	?	pos
ex7	Circle	Thick	Gray	Thick	White	6	neg
ex8	Square	Thick	?	Thick	Gray	5	neg
ex9	Triangle	Thin	Gray	Thin	Dark	5	neg
ex10	Circle	Thick	Dark	Thick	?	?	neg
ex11	Square	Thick	White	Thick	Dark	9	neg
ex12	Triangle	Thick	White	Thick	Gray	8	neg

that the removal of all affected examples would destroy most of the training set, disposing in the process of valuable information.

In this event, we may try to replace the question marks with *some* values, even though these may be incorrect. This is easily done. When the attribute is discrete, then we may simply replace the question mark with the attribute's most frequent value. Thus in Table 10.1, example ex_8, the unknown value of `crust-shade` will be replaced with `white` because this is the most frequent value of this attribute in this particular training set. In the case of a continuous-valued attribute, the average value can be used. In ex_6 and ex_{10}, the value of `weight` is unknown. Among the 10 examples where it *is* known, the average value is `weight=5.1`, and this is the value we will use in ex_6 and ex_{10}.

When doing so, caution is called for. The reader has to keep in mind that using the most frequent or average values will render the examples' description unreliable, perhaps even dubious. The technique should therefore be used sparingly. When many values are missing, more sophisticated methods (such as the one below) should be used.

Learning to Fill-in Missing Values Sometimes, using the most common or average values can mislead the learning program. A better idea of how to fill the empty slots is built around the observation that attributes are rarely independent from each other. For instance, the taller the man, the greater his body weight. If the `weight` of someone with `height=6.5` is unknown, it would be foolish to use the average weight calculated over the whole population; after all, our rather tall individual is certainly heavier than the average person. Seeking a way out, we will probably do better calculating the average weight among those with `height > 6`.

So much for a pair of mutually dependent attributes. Quite often, however, the interrelations are more complicated than that, easily involving three or more attributes. One simple mechanism to predict unknown values in situations of this kind will rely on the idea of decision-tree induction. A pseudocode of the technique is provided in Table 10.2.

The idea is quite simple. Suppose that *at* is an attribute that has, in the training set, many question marks. We want to replace the question marks with the most likely values. We decide to so by means of a decision tree. To this end, we convert

Table 10.2 An algorithm to determine unknown attribute values

---

Let T be the original training set.

Let *at* be the attribute with unknown values.

1. Create a new training set, T', in which *at* becomes the class label; the examples are described by all the remaining attributes, the former class label (e.g., `pos` versus `neg`) being treated like just another attribute.
2. Remove from T' all examples in which the value of *at* is unknown.
3. From this final version of T', induce a decision tree.
4. Use the decision tree thus induced to determine the values of *at* in those examples in which its values were unknown.

---

the original training set, T, into a new training set, T', where the original class label (e.g., pos or neg) becomes one of the attributes, whereas at will be treated as a class label. From this training set, we remove all examples whose values of at are unknown. From the rest, we induce the decision tree, and then use the decision tree to fill in the missing values.

What Have You Learned?

To make sure you understand this topic, try to answer the following questions. If you have problems, return to the corresponding place in the preceding text.

- What are the main difficulties posed by unknown attribute values? What are their typical consequences for classifier induction?
- Describe the trivial ways of dealing with examples with some unknown attribute values. Discuss their limitations.
- Explain how the idea of decision-tree induction can be used when we want to determine the unknown attribute values.

10.5   Attribute Selection

In many domains, the training examples are described by great many attributes: tens of thousands, or even more. Learning from data sources of this kind can be prohibitively expensive. Besides, we have to face problems with learnability, and also with issues related to irrelevant or redundant attributes. To avoid unnecessary disappointments, the engineer needs to be acquainted with methods to select the most appropriate attributes.

Irrelevant and Redundant Attributes   Not all attributes are created equal. Some of them are irrelevant in the sense that their values do not have any effect on an example's class. Others are redundant in the sense that their values can be obtained from values of other attributes: for instance, age can be obtained from the value of date-of-birth. Attributes of these kinds can mislead certain induction techniques. For instance, irrelevant attributes (and, to a lesser degree, also redundant attributes) distort the vector-to-vector distances calculated by the k-NN classifier. Other paradigms, such as decision trees, are less vulnerable, but even they may suffer from excessive computational costs.

Extremely Long Attribute Vectors   Some domains, such as automated text categorization, are marked by tens of thousands of attributes, and this often causes problems. One of the difficulties is the reduced learnability: the induced classifiers are prone to overfit the training data, disappointing the user when tested on future examples. Also the computational costs can be impractical, especially when a multilayer neural network is used: each additional attribute increases the number of weights to be trained, thus adding to the calculations.

Moreover, examples described by thousands of attributes are inevitably sparse, which is known to mislead many machine-learning approaches. For instance, the problem of sparsity in k-NN classifiers was explained in Chap. 3.

And yet it is known that, for all intents and purposes, most of the attributes are useless, and as such, should be disposed of.

Filter Approaches to Attribute Selection Perhaps the simplest approach to attribute selection is based on what machine learning calls *filtering*. The idea is to calculate for each attribute its "utility" for the classification task at hand, and then order them according to this criterion. The intention to select the top N percent. The choice of the "cut-off" point, N, is usually made by trial and error.

When ordering the attributes, the *information gain* from Sect. 6.3 can be used if the attributes are discrete. Thanks to the existence of mechanisms for binarization (see Sect. 6.4), information gain can actually be employed even in the case of continuous-valued attributes; for this, however, statistical approaches to correlation measurement are usually preferred.

One reason to criticize attribute filtering is that this approach ignores the relations that exist among the attributes. This makes it difficult, almost impossible, to identify redundant attributes. As we know, a redundant attribute does not bring any additional information beyond that provided by the other attributes; and yet its information gain can be high.

There is a relatively simple way to overcome this weakness. We induce a decision tree, and then use only those attributes that are encountered in the tests in the tree's internal nodes. The careful reader will recall that this approach was used in some of the simple applications discussed in Chap. 8.

Wrapper Approaches to Attribute Selection More powerful, but also more computationally expensive, is the so-called *wrapper* approach to attribute selection. Here is the underlying principle. Suppose we want to compare the quality of two attribute sets, A_1 and A_2. From the original training set, T, we create two training sets, T_1 and T_2. In both, all examples have the same class labels as in T. However, T_1 describes the examples by A_1 and T_2 uses A_2. From the two newly created training subsets, two classifiers are induced and evaluated on some independent evaluation set, T_E. The attribute set that results in the higher performance is better.

This is the principle used in the search-based algorithm whose pseudocode is provided in Table 10.3. The input consists of a training set, T, and of a set of attributes, A. The output is a subset, $S \in A$, of the most useful attributes. At the beginning, S is empty. At each step, the approach chooses the best attribute from A, and adds it to S. What is "best" is determined by the classification performance (on an independent testing set) of the classifier induced from the examples described by the attributes from S. The algorithm is terminated if no addition to S leads to an improvement of the classification performance—or if there are no more attributes to be added to S.

Table 10.3 An wrapper approach to sequential attribute selection

Divide the available set of pre-classified examples into two parts, T_T and T_E. Let A be the set of attributes. Create an empty set, S.

1. For every attribute, $at_i \in A$:

    (i)   add at_i to S; let all examples in T_T and T_E be described by attributes from S;
    (ii) induce a classifier from T_T, then evaluate its performance on T_E; denote this performance by p_i;
    (iii) remove at_i from S.

2. Identify the attribute that resulted in the highest value of p_i. Remove this attribute from A and add it to S.
3. If $A = \emptyset$, stop; if the latest addition did not increase performance, remove this last attribute from S and stop, too. In both cases, S is the final set of attributes.
4. Return to step 1.

What Have You Learned?

To make sure you understand this topic, try to answer the following questions. If you have problems, return to the corresponding place in the preceding text.

- In what sense do we say that some attributes are less useful than others? Why is the engineer often incapable of choosing the right attributes for example description?
- Explain and discuss the principle of filter-based approaches to attribute selection.
- Describe the principle of the *wrapper* approaches to attribute selection. Explain how the principle can be employed in a simple search-based approach to attribute selection.

10.6   Miscellaneous

Some issues worth knowing about do not merit a separate section in an introductory text, and yet they cannot be ignored. Let us briefly summarize them here.

Lack of Regularity in the Data Suppose you are asked to induce a classifier from training examples that have been created by a random-number generator; all attribute values are random, and so are the class labels. Obviously, there is no regularity in such data—and yet—machine learning techniques are often capable of inducing from them a classifier with zero error rate on the training set. Of course, this perfect behavior on the training set will not translate into similar behavior on future examples.

This observation suggests a simple mechanism to be used when measuring the amount of regularity in data. The idea is simply to divide the data in two subsets— one for training and one for testing. The classifier is induced from the training set, and then applied to the testing set. In the case of random data, we will observe only a small (if any) error rate on the training examples but the results on the testing examples will be dismal. Conversely, the more regularity there is, in the data, the better the results on the testing set.

Classes That Can Be Linearly Ordered In some domains, each example is labeled with one out of several (perhaps even many) classes. In this context, we have to mention the special case where the different class labels can be ordered. For instance, suppose that the output class is month, with values january through december. In domains of this kind, it would be a great (and misleading) simplification to assume that misclassifying june for may is the same as misclassifying june for december.

Not only in performance evaluation, but also during the induction of such classes, some attention should therefore be devoted to the ordering of the individual classes. One possibility is to begin by grouping neighboring class labels; only after inducing a classifier for each group should we consider the possibility of fine-tuning within the group.

In the case of class month, we may perhaps first induce classifier for seasons, each comprising 3 months, and only after this, a separate classifier for each month.

Regression In some applications, the expected output is not a discrete-valued class, but rather a number from a continuous range. For instance, this can be the case when the software is to predict a value of a stock-market index. Problems of this kind are called *regression*. In this book, we do not address them. The simplest way of dealing with them within the framework of machine learning is to replace the continuum with subintervals, and then treat each subinterval as a separate class. Note that the task would then belong to the category of "classes that can be linearly ordered" mentioned in the previous paragraph.

What Have You Learned?

To make sure you understand this topic, try to answer the following questions. If you have problems, return to the corresponding place in the preceding text.

- Explain how to use machine learning when measuring the amount of regularity in data. Give examples of domains where this regularity can be expected to be low.
- What are *multi-label* domains? What is the simplest approach to induction in these domains? What typical problems does the machine-learning engineer encounter in them?

10.7   Summary and Historical Remarks

- Chapter 7 offered mathematical arguments supporting the claim that "there is no learning without bias." Certain practical considerations have convinced us that this is indeed the case.
- Sometimes, the meaning of the underlying class depends on a concrete context; this context can change in time, in which case we are facing the problem of time-varying classes.
- The classical machine-learning techniques from the earlier chapters of this book assume that both (or all) classes are adequately represented in the training set. Quite often, however, this requirement is not satisfied, and the engineer has to deal with the difficulties caused by the problem of *imbalanced training sets*.
- The most typical approaches to the problem of *imbalanced training sets* are majority-class undersampling and minority-class oversampling.
- In many training sets, some attribute values are unknown, and this complicates the use of certain induction techniques. One possible solution is to use (in place of the unknown values) the most frequent or the average values of the given attributes.
- Quite often, the engineer is faced with the necessity to select the most appropriate set of attributes. Two fundamental approaches can be used here: the *filtering* techniques and the *wrapper* techniques.
- In domains with more than two classes, it sometimes happens that the individual classes can be ordered. This circumstance can affect performance evaluation. For instance, if the task is to recognize a concrete month, then it is not the same thing if the classifier's output missed the target by 1 month or by 5. Even the learning procedure should then perhaps be modified accordingly.
- Sometimes, the output is not a discrete-valued class, but rather a value from a continuous range. This type of problem is called *regression*. This book does not address regression explicitly.

Historical Remarks   The idea to distinguish different biases in machine learning was pioneered by Gordon and desJardin [34]. The principle of lifelong learning was first mentioned by Thrun and Mitchell [88]. The early influential papers addressing the issue of context were published by Turney [90] and Katz et al. [40]. Induction of time-varying concepts was introduced by Kubat [49] and some early algorithms were described by Widmer and Kubat [97]. The oldest paper on multi-label classification known to the author of this book was published by McCallum [57]. The Wrapper approach to attribute selection is introduced by Kohavi [44].

10.8   Solidify Your Knowledge

The exercises are to solidify the acquired knowledge. The suggested thought experiments will help the reader see this chapter's ideas in a different light and provoke independent thinking. Computer assignments will force the readers to pay attention to seemingly insignificant details they might otherwise overlook.

Table 10.4 A simple
exercise in "unknown values"

Example	Shape	Crust		Filling		Class
		Size	Shade	Size	Shade	
ex_1	Circle	Thick	Gray	Thick	Dark	pos
ex_2	Circle	Thick	White	Thick	Dark	pos
ex_3	Triangle	Thick	Dark	?	Gray	pos
ex_4	Circle	Thin	White	Thin	?	pos
ex_5	Square	Thick	Dark	Thin	White	pos
ex_6	Circle	Thick	White	Thin	Dark	pos
ex_7	Circle	Thick	Gray	Thick	White	neg
ex_8	Square	Thick	White	Thick	Gray	neg
ex_9	Triangle	Thin	Gray	Thin	Dark	neg
ex_{10}	Circle	Thick	Dark	Thick	White	neg
ex_{11}	Square	Thick	White	Thick	Dark	neg
ex_{12}	Triangle	?	White	Thick	Gray	neg

Exercises

1. Consider the training set shown in Table 10.4. How will you replace the missing values (question marks) with the most frequent values? How will you use a decision tree to this end?
2. Once you have replaced the question marks in Table 10.4 with concrete values, identify the two attributes that offer the highest information gain.

Give It Some Thought

1. The text emphasized the difference between two basic types of error: those caused by the wrong bias(representational or procedural), and those that are due to variance in the training data. Suggest an experimental procedure that would give the engineer an idea about how much of the overall error rate can in a given domain be explained by either of these two sources.
2. Boosting algorithms are known to be relatively robust with respect to variance-based errors. Explain why this is the case. Further on, *non-homogeneous boosting* presented in Sect. 9.4 is known to reduce bias-based errors. Again, offer some explanation.
3. Suppose that a Bayesian classifier is to be employed in an imbalanced two-class domain where examples from one class heavily outnumber examples from the other class. Will this classifier be as sensitive to this situation as the nearest-neighbor approach? Support your answer by concrete arguments. Suggest an experimental verification.

4. In this section, the problem of imbalanced training sets was explored only within the framework of two-class domains where each example is either positive or negative. How does the same problem generalize to domains that have more than two classes? Suggest some concrete situations where imbalanced classes in such multi-class domains are or are not a problem.

5. Consider the case of linearly ordered classes mentioned in Sect. 10.6. Using the hint provided in the text, suggest a machine-learning scenario addressing this issue.

Computer Assignments

1. Write a computer program that accepts as input a training set where many attribute-values are missing, and outputs an improved training set where the missing values of discrete attributes have been replaced with the most frequent values, and the missing values of continuous attributes with the average values. Implement a computer program that will experimentally ascertain whether the missing-values replacement helps or harms the performance of a decision tree induced from such data.

2. Choose some public-domain data, for instance from the UCI repository.[2] Make sure this domain has at least one binary attribute. The exercise suggested here will assume that this binary attribute represents a context. Divide the training data into two subsets, each for a different context (a different value of the binary attribute). Then induce from each subset the corresponding context-dependent classifier. Assuming that it is at each time clear which of the two classifiers to use, how much will the average performance of these two classifiers be better than that of a "universal" classifier that has been induced from the original training set?

---

[2]www.ics.uci.edu/~mlearn/MLRepository.html.

Chapter 11
Performance Evaluation

The previous chapters pretended that performance evaluation in machine learning is a fairly straightforward matter. All it takes is to apply the induced classifier to a set of examples whose classes are known, and then count the number of errors the classifier has made. In reality, things are not as simple. Error rate rarely paints the whole picture, and there are situations in which it can even be misleading. This is why the conscientious engineer wants to be acquainted with other criteria to assess the classifiers' performance. This knowledge will enable her to choose the one that is best in capturing the behavioral aspects of interest.

So much for the evaluation of classifiers. Somewhat different is the question is how to compare the suitability of alternative machine-learning techniques for induction in a given domain. Dividing the set of pre-classified examples randomly into two subsets (one for induction, the other for testing) may not be the best thing to do, especially if the training set is small; random division may then result in subsets that do not represent the given domain properly. To obtain more reliable results, repeated random runs are necessary.

This chapter addresses both issues, explaining alternative criteria to quantify the performance of classifiers, and then discussing some strategies commonly used in experimental evaluation of machine-learning algorithms. The question of statistical evaluation of the results is relegated to the next chapter.

11.1 Basic Performance Criteria

Let us begin with formal definitions of error rate and classification accuracy. After this, we will take a look at the consequences of the decision to refuse to classify an example if the evidence favoring the winning class is weak.

© Springer International Publishing AG 2017
M. Kubat, *An Introduction to Machine Learning*,
DOI 10.1007/978-3-319-63913-0_11

Table 11.1 The basic quantities used in the definitions of performance criteria

| | | Labels returned by the classifier | |
		pos	neg
True labels:	pos	N_{TP}	N_{FN}
	neg	N_{FP}	N_{TN}

For instance, N_{FP} is the number of *false positives*: negative examples misclassified by the classifier as positive

Correct and Incorrect Classification Let us first define four fundamental quantities that will be used throughout this chapter. When testing a classifier on an example whose real class is known, we can encounter only the following four different outcomes: (1) the example is positive and the classifier correctly recognizes it as such (*true positive*); (2) the example is negative and the classifier correctly recognizes it as such (*true negative*); (3) the example is positive, but the classifier labels it as negative (*false negative*); and (4) the example is negative, but the classifier labels it as positive (*false positive*).

When applying the classifier to an entire set of examples (whose real classes are known), each of these four outcomes will occur a different number of times—and these numbers are then employed in the performance criteria defined below. The symbols representing the four outcomes are summarized in Table 11.1. Specifically, N_{TP} is the number of *true positives*, N_{TN} is the number of *true negatives*, N_{FP} is the number of *false positives*, and N_{FN} is the number of *false negatives*. In the entire example set, T, only these four categories are possible; therefore, the size of the set, $|T|$, equals the sum, $|T| = N_{FP} + N_{FN} + N_{TP} + N_{TN}$.

Note that the number of correct classifications is the number of *true positives* plus the number of *true negatives*, $N_{TP} + N_{TN}$; and the number of errors is the number of *false positives* plus the number of *false negatives*, $N_{FP} + N_{FN}$.

Error Rate and Classification Accuracy A classifier's *error rate*, E, is the frequency of errors made by the classifier over a given set of examples. It is calculated by dividing the number of errors, $N_{FP} + N_{FN}$, by the total number of examples, $N_{TP} + N_{TN} + N_{FP} + N_{FN}$.

$$E = \frac{N_{FP} + N_{FN}}{N_{FP} + N_{FN} + N_{TP} + N_{TN}} \tag{11.1}$$

Sometimes, the engineer prefers to work with the opposite quantity, *classification accuracy*, *Acc*: the frequency of correct classifications made by the classifier over a given set of examples. Classification accuracy is calculated by dividing the number of correct classifications, $N_{TP} + N_{TN}$, by the total number of examples. Note that $Acc = 1 - E$.

$$Acc = \frac{N_{TP} + N_{TN}}{N_{FP} + N_{FN} + N_{TP} + N_{TN}} \tag{11.2}$$

Rejecting an Example When discussing the problem of optical character recognition, Sect. 8.1, suggested that the classifier should sometimes be allowed to refuse to classify an example if the evidence supporting the winning class is not strong enough. The motivation is quite simple: in some domains, the penalty for misclassification can be much higher than the penalty for not making any classification at all.

An illustrative example is not difficult to find. Thus the consequence of a classifier's refusal to return the precise value of the ZIP code is that the decision where the letter should be sent will have to be made by a human operator. To be sure, this manual processing is more expensive than automatic processing, but not excessively so. On the other hand, an incorrect value returned by the classifier results in the letter being sent to a wrong destination, which can cause a serious delay in delivery. This latter cost is often much higher than the cost of "manual" reading. Similarly, an incorrect medical diagnosis is often more expensive than no diagnosis at all; lack of knowledge can be remedied by additional tests, but a wrong diagnosis may result in choosing a treatment that does more harm than good.

This is why the classifier should sometimes refuse to classify an example if the evidence favoring either class is insufficient. In some machine-learning paradigms, the term *insufficient evidence* is easy to define. Suppose, for instance, that, in a 7-NN classifier, four neighbors favor the pos class, and the remaining three favor the neg class. The final count being four versus three, the situation seems "too close to call." More generally, the engineer may define a threshold for the minimum difference between the number of votes favoring the winning class and the number of votes favoring the runner-up class.

In bayesian classifiers, the technique is easily implemented, too. If the difference between the probabilities of the two most strongly supported classes falls short of a user-specified minimum, the example is rejected as too ambiguous to classify. Something similar can be done also in neural networks: compare the signals returned by the corresponding output neurons—and refuse to classify if there is no clear-cut winner.

In other classifiers, such as decision trees, implementation of the rejection mechanism is not so straightforward, and is only made possible by the implementation of "additional tricks."

Advantages and Disadvantages of a Rejection to Classify The classifier that occasionally refuses to make a decision about an example's class is of course less likely to go wrong. No wonder that its error rate will be lower. Indeed, the more examples are rejected, the lower the error rate. But the caution should not be exaggerated. It may look like a good thing that the error rate is reduced almost to zero. But if this low rate is achieved only thanks to the refusal to classify almost all examples, the classifier becomes impractical. Which of these two aspects (low error rate versus rare classifications) plays a more important role will depend on the concrete circumstances of the given application.

Figure 11.1 graphically illustrates the essence of the trade-off involved in decisions of this kind. The horizontal axis represents a parameter capable of

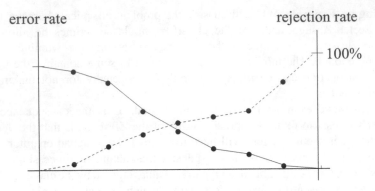

Fig. 11.1 Error rate can be increased by allowing the classifier to refuse to classify an example if available evidence fails to "convince" it. At the extreme, the error rate drops to $E = 0$ because all examples are rejected

adjusting the rejection rate. As we move from left to right, the rejection rate increases, whereas the error rate goes down until, at the extreme, all examples are rejected. At this point, the zero error rate is a poor consolation for having a classifier that never does anything. The lesson is clear. Occasional rejection of unclear examples makes a lot of sense, but the principle should be handled with care.

What Have You Learned?

To make sure you understand this topic, try to answer the following questions. If you have problems, return to the corresponding place in the preceding text.

- Define the terms *false negative, false positive, true negative*, and *true positive*.
- Specify the formulas for classification accuracy and error rate. How are the two criteria mutually interrelated?
- When should a classifier reject an example (i.e., when should it refuse to classify)? How would you implement this behavior in diverse types of classifiers?
- How does rejection rate relate to error rate? What is the trade-off between the two? Under what circumstances can refusal to classify be useful, and under what circumstances does it hurt?

11.2 Precision and Recall

In some applications, negative examples outnumber positive ones by a wide margin. When this happens, error rate offers a misleading picture of classification performance. To see why, just consider the case where only 2% of all examples are positive, and all the remaining 98% are negative. A "classifier" that returns the negative class for any example in the set will be correct 98% of the time—which may look like a remarkable feat. And yet, the reader will agree, a classifier that never recognizes a positive example is useless.

Imbalanced Classes Revisited This observation is worth keeping in mind because domains with imbalanced classes are quite common. We encountered some of them in Chaps. 8 and 10, and other applications can be found. Thus in automated information retrieval, the user may want to find a scientific document dealing with, say, "performance evaluation of classifiers." No matter how attractive the topic may appear to this particular person, papers dealing with it will represent only a small fraction of the millions of documents available in the digital library. Likewise, patients suffering from a specific medical disorder are in the entire population relatively rare. And the same goes for any undertaking that seeks to recognize a rare event such as a default on mortgage payments or a fraudulent use of a credit card. A seasoned engineer will go so far as to say that the majority of realistic applications are in some degree marked by the phenomenon of imbalanced classes.

In domains of this kind, error rate and classification accuracy will hardly tell us anything reasonable about the classifier's practical utility. Rather than averaging the performance over both (or all) classes, we need criteria that focus on a class which, while important, is represented by only a few examples. Let us take a quick look at some of them.

Precision By this we mean the percentage of true positives, N_{TP}, among all examples that the classifier has labeled as positive: $N_{TP} + N_{FP}$. The value is thus obtained by the following formula:

$$Pr = \frac{N_{TP}}{N_{TP} + N_{FP}} \tag{11.3}$$

Put another way, *precision* is the probability that the classifier is right when labeling an example as positive.

Recall By this we mean the probability that a positive example will be correctly recognized as such (by the classifier). The value is therefore obtained by dividing the number of true positives, N_{TP}, by the number of positives in the given set: $N_{TP} + N_{FN}$. Here is the formula:

$$Re = \frac{N_{TP}}{N_{TP} + N_{FN}} \tag{11.4}$$

Note that the last two formulas differ only in the denominator. This makes sense. Whereas *precision* is the frequency of true positives among all examples deemed positive by the classifier, *recall* is the frequency of the same true positives among all positive examples in the set.

Table 11.2 Illustration of the two criteria: *precision* and *recall*

Suppose a classifier has been induced. Evaluation on a testing set gave the results summarized in this table:

	Labels returned by the classifier	
	pos	neg
True labels: pos	20	50
neg	30	900

From here, the following values of precision, recall, and accuracy are obtained:

$$\text{precision} = \frac{20}{50} = 0.40; \quad \text{recall} = \frac{20}{70} = 0.29; \quad \text{accuracy} = \frac{920}{1000} = 0.92$$

Suppose the classifier's parameters were modified with the intention to improve its behavior on positive examples. After the modification, evaluation on a testing set gave the results summarized in the table below.

	Labels returned by the classifier	
	pos	neg
True labels: pos	30	70
neg	20	880

From here, the following values of precision, recall, and accuracy are obtained.

$$\text{precision} = \frac{30}{50} = 0.60; \quad \text{recall} = \frac{30}{100} = 0.30; \quad \text{accuracy} = \frac{910}{1000} = 0.91$$

The reader can see that precision has considerably improved, while recall remained virtually unaffected. Note that classification accuracy has not improved, either.

Illustration of the Two Criteria Table 11.2 illustrates the behavior of the two criteria in a simple domain with an imbalanced representation of two classes, pos and neg. The induced classifier, while exhibiting an impressive classification accuracy, suffers from poor *precision* and *recall*. Specifically, *precision* of $Pr = 0.40$ means that of the 50 examples labeled as positive by the classifier, only 20 are indeed positive, the remaining 30 being nothing but false positives. With *recall*, things are even worse: out of the 70 positive examples in the testing set, only 20 are correctly identified as such by the classifier.

Suppose that the engineer decides to improve the situation by modifying some of the classifier's internal parameters, and suppose that this modification results in an increased number of true positives (from $N_{TP} = 20$ to $N_{TP} = 30$) and also a drop in the number of false positives (from $N_{FP} = 30$ to $N_{FP} = 20$). On the other hand, the number of false negatives has gone up, too: from $N_{FN} = 50$ to $N_{FN} = 70$. The calculations in Table 11.2 indicate that *recall* was thus barely affected, but *precision* has improved, from $Pr = 0.40$ to $Pr = 0.60$. Classification accuracy remained virtually unchanged—actually, it has even gone down a bit, in spite of the improved *precision*.

When High Precision Matters In some domains, *precision* is more important than *recall*. For instance, when you purchase something from an e-commerce web site, their recommender system often reacts with a message to the effect that, "Customers who have bought X buy also Y." The obvious intention is to cajole you into buying Y as well.

Recommender systems are sometimes created with the help of machine learning techniques applied to the company's historical data.[1] When evaluating their performance, the engineer wants to achieve high *precision*. The customers better be happy about the recommended merchandise, or else they will ignore the recommendations in the future.

The value of *recall* is here unimportant. The list offered on the web site has to be of limited size, and so it does not matter much that the system identifies only a small percentage of all items that the customers may like.

When High Recall Matters In other domains, by contrast, *recall* is more important. This is often the case in medical diagnosis. A patient suffering from X, and properly diagnosed as such, represents a true positive. A patient suffering from X but *not* diagnosed as such represents a false negative, something the doctor wants to avoid—which means that N_{FN} should be small. In the definition of *recall*, $Re = \frac{N_{TP}}{N_{TP}+N_{FN}}$, the number of false negatives appears in the denominator; consequently, a small value of N_{FN} implies a high value of *recall*.

ROC Curves In many classifiers, tweaking certain parameters can modify the values of N_{FP} and N_{FN}, thus affecting (at least to some extent) the classifier's behavior, for instance, by improving *recall* at the cost of worsened *precision* or vice versa. This possibility can be useful in domains where the user has an idea as to which of these two quantities is more important.

The reader will find it easy to suggest various ways to do the "tweaking." Thus in the k-NN classifier, the engineer may request that an example be labeled as negative unless some very strong evidence supports the positive label. For instance, if four out of seven nearest neighbors are positive, the classifier can be instructed still to return the negative label (in spite of the small majority recommending the positive class). In this way, the number of false positives can be reduced, though this often means to the price in terms of an increased number of false negatives. Even stronger reduction in N_{FP} (and an increase in N_{FN}) can be achieved by requesting that any example be deemed negative unless at least five (or six) of the seven nearest neighbors are positive.

---

[1] The concrete techniques employed to this end are somewhat more advanced than those discussed in this textbook, and are thus not treated here.

Fig. 11.2 Example of ROC curves for two classifiers, $c1$ and $c2$. The parameters of the classifiers can be used to modify the numbers of false positives and false negatives

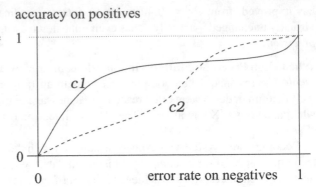

Something similar is easy to implement in Bayesian classifiers and in neural networks. The idea is the same: label the example with the preferred class unless strong evidence indicates that this decision is incorrect.

The behavior of the classifier under different values of its parameters can be visualized by the so-called ROC curve, a graph where the horizontal axis represents error rate on the negative examples, and the vertical axis represents classification accuracy on the positive examples. Example ROC curves for two classifiers, $c1$ and $c2$, are shown in Fig. 11.2. Ideally, we would like to reach the upper-left corner that represents zero error rate on the negatives and 100% accuracy on the positives. For various reasons, this is rarely possible.

An important question to ask is which of the two, $c1$ or $c2$ is better. This can only be answered based on the specific needs of the concrete application. All we can say by just looking at the graph is that $c1$ outperforms $c2$ on the positives in the region with low error rate on negatives. As the error rate on the negative examples increases, $c2$ outperforms $c1$ on the positive examples. Again, whether this is good or bad can only be decided by the user. See the comments concerning the question when to prefer *precision* and when *recall*.

What Have You Learned?

To make sure you understand this topic, try to answer the following questions. If you have problems, return to the corresponding place in the preceding text.

- In what kind of data would we rather evaluate the classifier's performance by the quantities known as *precision* and *recall* (instead of, say, error rate)?
- What formulas help us calculate the concrete values of these criteria? How do the two formulas differ?
- Under what circumstances do we prefer high *precision*, and under what circumstances do we place more emphasis on high *recall*?

- Explain the nature of an ROC curve. What extra information does the curve convey about a classifier's behavior? How does the ROC curve help the user in the choice between two alternative classifiers?

11.3 Other Ways to Measure Performance

Apart from error rate, classification accuracy, *precision*, and *recall*, other criteria are sometimes used, each reflecting a somewhat different aspect of the classifier's behavior. Let us take at least a cursory look at some of the most important ones.

Combining *Precision* and *Recall* in F_β Using two different (and sometimes contradictory) performance criteria can be awkward; it is in the human nature to want to quantify things by a single number. Especially in the case of *precision* and *recall*, attempts have been made somehow to combine the two in one quantity. The best-known such formula, F_β, is defined as follows:

$$F_\beta = \frac{(\beta^2 + 1) \times Pr \times Re}{\beta^2 \times Pr + Re} \tag{11.5}$$

The parameter, $\beta \in [0, \infty)$, enables the user to weigh the relative importance of the two criteria. If $\beta > 1$, then more weight is given to *recall*. If $\beta < 1$, then more weight is apportioned to *precision*. It would be easy to show that F_β converges to *recall* when $\beta \to \infty$, and to *precision* when $\beta = 0$.

Quite often, the engineer does not really know which of the two, *precision* or *recall*, is more important, and by how much. In that event, she prefers to work with the neutral value of the parameter, $\beta = 1$:

$$F_1 = \frac{2 \times Pr \times Re}{Pr + Re} \tag{11.6}$$

A Numeric Example Suppose that an evaluation of a classifier on a testing set resulted in the values summarized in the upper part of Table 11.2. For these, the table established the values of *precision* and *recall*, respectively, as $Pr = 0.40$ and $Re = 0.29$. Using these numbers, we will calculate F_β for the following concrete settings of the parameter: $\beta = 0.2, \beta = 1$, and $\beta = 5$.

$$F_{0.2} = \frac{(0.2^2 + 1) \times 0.4 \times 0.29}{0.2^2 \times 0.4 + 0.29} = \frac{0.121}{0.306} = 0.39$$

$$F_1 = \frac{(1^2 + 1) \times 0.4 \times 0.29}{0.4 + 0.29} = \frac{0.232}{0.330} = 0.70$$

$$F_5 = \frac{(5^2 + 1) \times 0.4 \times 0.29}{5^2 \times 0.4 + 0.29} = \frac{3.02}{10.29} = 0.29$$

Sensitivity and Specificity The choice of the concrete criterion is often influenced by the given application field—with its specific needs and deep-rooted traditions that should not be ignored. Thus the medical domain has become accustomed to assessing performance of their "classifiers" (not necessarily developed by machine learning) by *sensitivity* and *specificity*. In essence, these quantities are nothing but *recall* measured on the positive and negative examples, respectively. Let us be more concrete:

Sensitivity is *recall* measured on positive examples:

$$Se = \frac{N_{TP}}{N_{TP} + N_{FN}} \tag{11.7}$$

Specificity is *recall* measured on negative examples:

$$Sp = \frac{N_{TN}}{N_{TN} + N_{FP}} \tag{11.8}$$

In the machine-learning literature, evaluation in terms of *sensitivity* and *specificity* is rare, but it is still good to be aware of these two criteria. After all, we may be asked to use them when applying machine learning to medical data—and quite often it is the customer who is the ultimate boss.

Gmean When inducing a classifier in a domain with imbalanced class representation, the engineer sometimes wants to achieve similar performance on both classes, pos and neg. In this event, the geometric mean, gmean, of the two classification accuracies (on the positive examples and on the negative examples) is used:

$$\text{gmean} = \sqrt{acc_{\text{pos}} \times acc_{\text{neg}}} = \sqrt{\frac{N_{TP}}{N_{TP} + N_{FN}} \times \frac{N_{TN}}{N_{TN} + N_{FP}}} \tag{11.9}$$

Note that *gmean* is actually the square root of the product of two numbers: *recall* on positive examples and *recall* on negative examples—or, in other words, the product of *sensitivity* and *specificity*.

Perhaps the most important aspect of *gmean* is that it depends not only on the concrete values of the two terms under the square root symbol, acc_{pos} and acc_{neg}, but also on how close the two values are to each other. A simple numeric example will convince us that this is indeed the case.

Thus the arithmetic average of 0.75 and 0.75 is $(0.75 + 0.75)/2 = 0/75$; also the arithmetic average of 0.55 and 0.95 is $(0.55 + 0.95)/2 = 0.75$. However, the geometric means of the first pair is $\sqrt{0.75 \times 0.75} = 0.75$ whereas the geometric means of the second pair is $\sqrt{0.55 \times 0.95} = 0.72$, a smaller number. We can see that the *gmean* is indeed smaller when the two numbers are different; and the more different they are, the lower the value of *gmean*.

Another Numeric Example Again, suppose that the evaluation of a classifier on a testing set resulted in the values summarized in the upper part of Table 11.2. The values of *sensitivity*, *specificity*, and *gmean* are calculated as follows:

$$Se = \frac{20}{50 + 20} = 0.29$$

$$Sp = \frac{900}{900 + 30} = 0.97$$

$$\text{gmean} = \sqrt{\frac{20}{50 + 20} \times \frac{900}{900 + 30}} = \sqrt{0.29 \times 0.97} = 0.53$$

Cost Functions In the world of machine-learning applications, not all errors carry the same penalty. False positives may be more costly than false negatives or vice versa. Comparisons of alternative classifiers are further complicated by the fact that, in some domains, the costs of the two types of error cannot even be expressed in the same (or at least comparable) units.

Consider the oil spill-recognition problem discussed in Chap. 8. A false positive here means that a "lookalike" is incorrectly taken for an oil spill (false positive). When this happens, a plane is unnecessarily sent to the given location to verify the case. The costs incurred by this error are those associated with the flight. By contrast, a false negative means that a potential environmental hazard has gone undetected, something difficult to express in monetary terms.

What Have You Learned?

To make sure you understand this topic, try to answer the following questions. If you have problems, return to the corresponding place in the preceding text.

- Explain how F_β combines *precision* and *recall*. Write down the formula and discuss how different values of β give more weight either to *precision* or *recall*. Which value of β gives equal weight to both?
- Give the formulas for the calculation of *sensitivity* and *specificity*. Explain their nature, and show their relation to *recall*. When are the two criteria used?
- Give the formula for the calculation of *gmean*. Explain the nature of this quantity, and show its relation to *recall*. When is this criterion used?
- Under what circumstances can the costs associated with false positives be different from the costs associated with false negatives? Suggest a situation where the comparison of the costs is almost impossible.

11.4   Learning Curves and Computational Costs

The first four sections of this chapter dealt with the problem of performance evaluation of the induced classifiers. Let us now turn our attention to the evaluation of the learning algorithm itself. How efficient is the given induction technique computationally? And how good are the classifiers it induces? Will better results be achieved if we choose some other machine-learning framework?

In this section, we will give some thought to the costs of learning—in terms of the number of examples needed for successful induction, as well as in terms of computational time consumed. The other aspect, the ability to induce a tool with high classification performance will be addressed in the following section.

The Learning Curve   When evaluating a human subject's ability to learn how to solve a certain problem, psychologists rely on a *learning curve*, a notion that machine learning has borrowed for its own purposes.

From our perspective, the learning curve simply shows how the classification performance of the induced classifier depends on the size of the training set. Two such curves are shown in Fig. 11.3. The horizontal axis represents the number of training examples; and the vertical axis represents the classification accuracy of the classifier induced from these examples. Usually, though not always, this classification accuracy is evaluated on independent testing examples.

Most of the time, a larger training set means higher classification performance—at least until the moment when no further improvement is possible. Ideally, we would like to achieve maximum performance from the smallest possible training set. For one thing, training examples can be expensive to obtain, and their source can be limited no matter how much we are willing to spend on them. For another, the more examples we use, the higher the computational costs on induction.

Comparing Learners with Different Learning Curves   Figure 11.3 shows the learning curves of two learners, l_1 and l_2. The reader can see that the learning curve of the former, l_1, rises very quickly, only to level off at a point beyond which virtually no improvement is possible—the limitation may be imposed by an

Fig. 11.3 A learning curve shows how the achieved classification accuracy depends on the number of examples used during learning

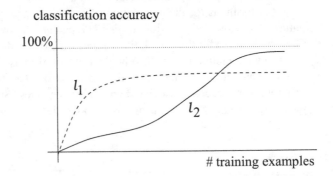

incorrect bias (see Sect. 10.1). By contrast, the learning curve of the second learner, l_2, does not grow so fast, but in the end achieves higher levels of accuracy than l_1.

Which of the two curves indicates a preferable learner depends on the circumstances of the given application. When the source of the training examples is limited, the first learner is clearly more appropriate. If the examples are abundant, we will prefer the other learner, assuming of course that the computational costs are not prohibitive.

Computational Costs There are two aspects to computational costs. First is the time needed for the induction of the classifier from the available data. The second is the time it takes to classify a set of examples with the classifier thus induced.

Along these lines, the techniques described in this book cover a fairly broad spectrum. As for induction costs, the cheapest is the basic version of the k-NN classifier: the only "computation" involved is to store the training examples.[2] On the other hand, the k-NN classifier is expensive in terms of the classification costs. For instance, if we have a million training examples, each described by ten thousand attributes, then tens of billions of arithmetic operations will have to be carried out to classify a single example. When asked to classify millions of examples, even a very fast computer will take quite some time.

The situation is different in the case of decision trees. These are cheap when used to classify examples: usually only a moderate number of single-attribute tests are needed. However, induction of decision trees can take a lot of time if many training examples described by many attributes are available.

Induction costs and classification costs of the other classifiers vary, and the engineer is well advised to consider these costs when choosing the most appropriate machine-learning paradigm for the given application. Just as important is a solid understanding of how these costs depend on the number of the training examples, on the number of attributes describing them, and sometimes also on the required accuracy (e.g., in the case of neural networks).

What Have You Learned?

To make sure you understand this topic, try to answer the following questions. If you have problems, return to the corresponding place in the preceding text.

- What are the two main aspects of costs associated with a given machine-learning paradigm?
- What does the learning curve tell us about a machine learning algorithm's behavior? What shape of the learning curve do we want in the ideal case? What should we expect in reality?
- Under what circumstances is a steeper learning curve with lower maximum better than a flatter curve with a higher maximum?

---

[2]Some computation will be necessary if we decide to remove noisy or redundant examples.

11.5   Methodologies of Experimental Evaluation

The reader understands that different domains will benefit from different induction techniques. The choice is usually not difficult to make. Some knowledge of the available training data often helps us choose the most appropriate paradigm; for instance, if a high percentage of the attributes are suspected to be irrelevant, then decision trees are likely to be more successful than a nearest-neighbor classifier.

However, the success of a given technique also depends on the values of various parameters. Although even here certain time-tested rules of thumb can help, the best parameter setting is usually found only by experimentation.

Baseline Approach and Its Limitations   The basic scenario is simple. The set of pre-classified examples is divided into two subsets, one used for training, the other for testing. The training-testing session is repeated for different parameter settings, and the one that results in the highest performance is then chosen.

This, however, can only be done if a great many pre-classified examples are available. In domains where examples are scarce, or expensive to obtain, a random division into a pair of the training and testing sets will lack objectivity. Either of the two sets can, by mere chance, misrepresent the given domain adequately. Statisticians are telling us that both the training set and the testing set should have more or less the same distribution of examples. In small sets, of course, this cannot be guaranteed.

Random Subsampling   When the set of pre-classified examples is small, the engineer usually prefers to repeat the training-testing procedure several times. In each run, the set of examples is randomly divided into two parts, one for training, the other for testing. The measured performances are recorded, and then averaged. Care has to be taken that the individual data-splits are mutually independent.

Once the procedure has been repeated N times (typically, $N = 10$ or $N = 5$), the results are reported in terms of the average classification accuracy and the standard deviation, say, 84.2 ± 0.6. For the calculation of averages and standard deviations, the formulas from Chap. 2 are used: the average, μ, is obtained by Eq. (2.12); the standard deviation, σ, is obtained as the square root of variance, σ^2, which is calculated using Eq. (2.13).

N-Fold Crossvalidation   For more advanced statistical evaluation, experienced experimenters often prefer the so-called N-fold crossvalidation. The principle is shown in Fig. 11.4. To begin with, the set of pre-classified examples is divided into N equally sized (or almost equally-sized) subsets which the machine-learning jargon sometimes (not quite correctly) refers to as "folds."

N-fold crossvalidation then runs N experiments. In each, one of the N subsets is removed so as to be used only for testing (this guarantees that, in each run, a different testing set is used). The training is then carried out on the union of the remaining $N - 1$ subsets. Again, the results are averaged, and the standard deviation calculated.

Fig. 11.4 *N*-fold cross-validation divides the training set into *N* equally sized subsets. In each of the *N* experimental runs, it withholds a different subset for testing, inducing the classifier from the union of the remaining *N* − 1 subsets

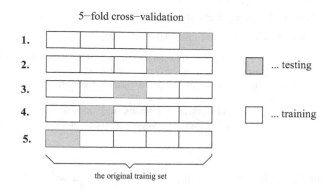

5−fold cross−validation

□ ... testing

□ ... training

the original trainig set

The advantage of *N*-fold crossvalidation as compared to random subsampling is that the testing sets are disjoint (non-overlapping), which is deemed advantageous for certain types of statistical evaluations (see Chap. 12).

Stratified Approaches Suppose you are dealing with a domain with 60 positive and 940 negative examples. If you rely on *N*-fold crossvalidation with $N = 10$, each of the "folds" is likely to contain a different number of positive examples. On average, there will be six positives in each fold, but the concrete numbers can vary significantly. In a domain where one of the classes is relatively rare, it can even happen that some of the "folds" will not contain any single positive example at all. The situation is typically encountered in domains with heavily imbalanced class representation.

In this event, the experienced experimenter prefers a so-called *stratified* approach to *N*-fold crossvalidation. The idea is to make sure that each of the *N* "folds" contains (approximately) the same representation of the examples from the individual classes. For instance, when using the fivefold crossvalidation in a domain with 60 and 940 positive and negative examples, respectively, each fold should consist of 200 examples of which 12 are positive.

The same principle is often used in the random-subsampling approach—which, admittedly, is in its stratified version no longer totally random. Again, the idea is to make sure that each training set, and each testing set, has about the same representation of each class.

5 × 2 Crossvalidation (5 × 2cv) There is yet another approach to experimental evaluation of machine-learning techniques, the so-called 5 × 2 crossvalidation, sometimes abbreviated as 5 × 2cv. This may actually be the most popular methodology in machine learning. The principle is built around a combination of random subsampling and twofold crossvalidation.

To be more specific, 5 × 2cv divides the set of pre-classified examples into two equally sized parts, T_1 and T_2. Next, it uses T_1 for training and T_2 for testing, and then the other way round: T_2 for training and T_1 for testing. The procedure is then repeated five times, each time with a different random division into two subsets. All

Table 11.3 The algorithm for 5×2 cross-validation (5×2 CV)

Let T be the original set of pre-classified examples.

1. Divide T randomly into two equally-sized subsets. Repeat the division five times. The result is five pairs of subsets denoted as T_{i1} and T_{i2} (for $i = 1, \ldots 5$).
2. For each of these pairs, use T_{i1} for training and T_{i2} for testing, and then the other way round.
3. For the ten training/testing sessions thus obtained, calculate the mean value and the standard deviation of the chosen performance criterion.

in all, ten learning/testing sessions are thus created. The principle is summarized by the pseudocode in Table 11.3.

Again, many experimenters prefer to work with the stratified version of this methodology, making sure that the representation of the individual classes is about the same in each of the ten parts used in the experiments.

The No-Free-Lunch Theorem It would be foolish to expect some machine-learning technique to be a holy grail, a mechanism to be preferred under all circumstances. Nothing like this exists. The reader by now understands that each paradigm has its advantages that make it succeed in some domains—and shortcomings that make it fail miserably in others. Only systematic experiments can tell the engineer which type of classifier, and which induction algorithm, to select for the task at hand. The truth of the matter is that no machine-learning approach will outperform all other machine-learning approaches under all circumstances.

Mathematicians have been able to prove the validity of this statement by a rigorous proof. The result is known under the (somewhat fancy) name of "no-free-lunch theorem."

What Have You Learned?

To make sure you understand this topic, try to answer the following questions. If you have problems, return to the corresponding place in the preceding text.

- What is the difference between N-fold cross-validation and random subsampling? Why do we sometimes prefer to employ the stratified versions of these methodologies?
- Explain the principle of the 5×2 cross-validation (5×2 cv), including its stratified version.
- What does the so-called no-free-lunch theorem tell us?

11.6 Summary and Historical Remarks

- The basic criterion to measure classification performance is error rate, E, defined as the percentage of misclassified examples in the given set. The complementary quantity is classification accuracy, $Acc = 1 - E$.
- When the evidence for any class is not sufficiently strong, the classifier should better reject the example to avoid the danger of a costly misclassification. Rejection rate then becomes yet another important criterion for the evaluation of classification performance. Higher rejection rate usually means lower error rate; beyond a certain point, however, the classifier's utility will degrade.
- Criteria for measuring classification performance can be defined by the counts (denoted as $N_{TP}, N_{TN}, N_{FP}, N_{FN},$) of true positives, true negatives, false positives, and false negatives, respectively.
- In domains with imbalanced class representation, error rate can be a misleading criterion. A better picture is offered by the use of *precision* ($Pr = \frac{N_{TP}}{N_{TP}+N_{FP}}$) and *recall* ($Re = \frac{N_{TP}}{N_{TP}+N_{FN}}$).
- Sometimes, *precision* and *recall* are combined in a single criterion, F_β, that is defined by the following formula:

$$F_\beta = \frac{(\beta^2 + 1) \times Pr \times Re}{\beta^2 \times Pr + Re}$$

The value of the user-set parameter β determines the relative importance of *precision* ($\beta < 1$) or *recall* ($\beta > 1$). When the two are deemed equally important, we use $\beta = 1$, obtaining the following:

$$F_1 = \frac{2 \times Pr \times Re}{Pr + Re}$$

- Less common criteria for classification performance include *sensitivity*, *specificity*, and *gmean*.
- In domains where an example can belong to more than one class at the same time, the performance is often evaluated by an average taken over the performances measured along the individual classes. Two alternative methods of averaging are used: *micro-averaging* and *macro-averaging*.
- Another important aspect of a machine-learning technique is how many training examples are needed if a certain classification performance is to be reached. The situation is sometimes visualized by means of a *learning curve*. Also worth the engineer's attention are the computational costs associated with induction and with classification.
- When comparing alternative machine-learning techniques in domains with limited numbers of pre-classified examples, engineers rely on methodologies known as random subsampling, N-fold cross-validation, and the 5×2 cross-validation. The stratified versions of these techniques make sure that each training set (and testing set) has the same proportion of examples for each class.

Historical Remarks Most of the performance criteria discussed in this chapter are well established in the statistical literature, and have been used for such a long time that it is difficult to trace their origin. The exception is the relatively recent *gmean* that was proposed to this end by Kubat et al. [51].

The idea to refuse to classify examples where the *k*-NN classifier cannot rely on a significant majority was put forward by Hellman [36] and later analyzed by Louizou and Maybank [56]. The principle of 5×2 cross-validation was suggested, and experimentally explored, by Dietterich [22]. The no-free-lunch theorem was published by Wolpert [100].

11.7 Solidify Your Knowledge

The exercises are to solidify the acquired knowledge. The suggested thought experiments will help the reader see this chapter's ideas in a different light and provoke independent thinking. Computer assignments will force the readers to pay attention to seemingly insignificant details they might otherwise overlook.

Exercises

1. Suppose that the evaluation of a classifier on a testing set resulted in the counts summarized in the following table:

		Labels returned by the classifier	
		pos	neg
True labels:	pos	50	50
	neg	40	850

   Calculate the values of *precision, recall, sensitivity, specificity*, and *gmean*.
2. Using the data from the previous question, calculate F_β for different values of the parameter: $\beta = 0.5, \beta = 1$, and $\beta = 2$.
3. Suppose that an evaluation of a machine-learning technique using fivefold cross-validation resulted in the following error rates measured in the testing sets:
   $E_{11} = 0.14, E_{12} = 0.16, E_{13} = 0.10, E_{14} = 0.15, E_{15} = 0.18$
   $E_{21} = 0.17, E_{22} = 0.15, E_{23} = 0.12, E_{24} = 0.13, E_{25} = 0.20$
   Calculate the mean value of the error rate as well as the standard deviation, σ, using the formulas from Chap. 2 (do not forget that standard deviation is the square root of variance, σ^2).

Give It Some Thought

1. Suggest a domain where *precision* is much more important than *recall*; conversely, suggest a domain where it is the other way round, *recall* being more important than *precision*.

   (Of course, use different examples than those mentioned in this chapter.)
2. What aspects of the given domain is reflected in the pair, *sensitivity* and *specificity*? Suggest circumstances under which these two give a better picture of the classifier's performance than *precision* and *recall*.
3. Suppose that, for a given domain, you have induced two classifiers: one with very high *precision*, the other with high *recall*. What can be gained from the combination of the two classifiers? How would you implement this combination? Under what circumstances will the idea fail?
4. Try to think about the potential advantages and shortcomings of random subsampling in comparison with *N*-fold crossvalidation.

Computer Assignments

1. Assume that some machine-learning experiment resulted in a table where each row represents a testing example. The first column contains the examples' class labels ("1" or "0" for the positive and negative examples, respectively), and the second column contains the labels suggested by the induced classifier.

   Write a program that calculates *precision*, *recall*, as well as F_β for a user-specified β.

   Write a program that calculates the values of the other performance criteria.
2. Suppose that the training set has the form of a matrix where each row represents an example, each column represents an attribute, and the rightmost column contains the class labels.

   Write a program that divides this set randomly into five pairs of equally sized subsets, as required by the 5×2 cross-validation technique. Then write another program that creates the subsets in the *stratified* manner where each subset has approximately the same representation of each class.
3. Write a program that accepts two inputs: (1) a set of class labels of multi-label testing examples, and (2) the labels assigned to these examples by a multi-label classifier. The output consists of *micro*-averaged and *macro*-averaged values of *precision* and *recall*.
4. Write a computer program that accepts as input a training set, and outputs *N* subsets to be used in *N*-fold crossvalidation. Make sure the approach is *stratified*. How will your program have to be modified if you later decide to use the 5×2 crossvalidation instead of the plain *N*-fold crossvalidation?

Chapter 12
Statistical Significance

Suppose you have evaluated a classifier's performance on an independent testing set. To what extent can you trust your findings? When a flipped coin comes up *heads* eight times out of ten, any reasonable experimenter will suspect this to be nothing but a fluke, expecting that another set of ten tosses will give a result closer to reality. Similar caution is in place when measuring classification performance. To evaluate classification accuracy on a testing set is not enough; just as important is to develop some notion of the chances that the measured value is a reliable estimate of the classifier's true behavior.

This is the kind of information that an informed application of mathematical statistics can provide. To acquaint the student with the requisite techniques and procedures, this chapter introduces such fundamental concepts as standard error, confidence intervals, and hypothesis testing, explaining and discussing them from the perspective of the machine-learning task at hand.

12.1 Sampling a Population

If we test a classifier on several different testing sets, the error rate on each of them will be different—but not totally arbitrary: the distribution of the measured values cannot escape the laws of statistics. A good understanding of these laws can help us estimate how representative the results of our measurements really are.

An Observation Table 12.1 contains one hundred zeros and ones, generated by a random-number generator whose parameters have been set to make it return a zero 20% of a time, and a one 80% of the time. The real percentages in the generator's output are of course slightly different than what the setting required. In this particular case, the table contains 82 ones and 18 zeros.

© Springer International Publishing AG 2017
M. Kubat, *An Introduction to Machine Learning*,
DOI 10.1007/978-3-319-63913-0_12

Table 12.1 A set of binary
values returned by a
random-number generator set
to return a one 80% of the
time

0	0	1	0	1	1	1	0	1	1	6
1	1	0	1	1	1	1	1	1	1	9
1	1	1	0	1	1	1	1	1	1	9
1	1	1	1	1	1	0	0	1	1	8
1	1	1	0	1	0	1	0	1	1	7
1	1	1	1	1	1	1	1	1	1	10
1	1	1	1	1	1	1	1	0	1	9
1	1	1	0	1	1	1	0	1	1	8
1	1	1	0	1	0	1	1	1	1	8
1	0	1	1	1	1	1	0	1	1	8
9	8	9	5	10	8	9	5	9	10	82

In reality, there are 82 ones and 18 zeros. At
the ends of the rows and columns are the corresponding sums

The numbers on the side and at the bottom of the table tell us how many ones are
found in each row and column. Based on these, we can say that the proportions of
ones in the first two rows are 0.6 and 0.9, respectively, because each row contains
10 numbers. Likewise, the proportions of ones in the first two columns are 0.9 and
0.8. The average of these four proportions is $(0.6 + 0.9 + 0.9 + 0.8)/4 = 0.80$, and
the standard deviation is 0.08.[1]

For a statistician, each row or column represents a *sample* of the population. All
samples have the same size: $n = 10$. Now, suppose we increase this value to, say,
$n = 30$. How will the proportions be distributed then?

Returning to the table, we can see that the first three rows combined contain
$6 + 9 + 9 = 24$ ones, the next three rows contain $8 + 7 + 10 = 25$ of them,
the first three columns contain $9 + 8 + 9 = 26$, and the next three columns contain
$5 + 10 + 8 = 23$. Dividing each of these numbers by $n = 30$, we obtain the following
proportions: $\frac{24}{30} = 0.80$, $\frac{25}{30} = 0.83$, $\frac{26}{30} = 0.87$, and $\frac{23}{30} = 0.77$. Calculating the
average and the standard deviation of these four values, we get 0.82 ± 0.02.

If we compare the results observed in the case of $n = 10$ with those for
$n = 30$, we notice two things. First, there is a minor difference between the
average calculated for the bigger samples (0.82) versus the average calculated for
the smaller samples (0.80). Second, the bigger samples exhibit a clearly smaller
standard deviation: 0.02 for $n = 30$ versus 0.08 for $n = 10$. Are these observations
explained by mere coincidence, or are they the consequence of some underlying
law?

---

[1]Recall that *standard deviation* is the square root of *variation*; this, in turn, is calculated by
Eq. (2.13) from Chap. 2.

Estimates Based on Random Samples The answer is provided by a theorem that says that estimates based on samples become more accurate with the growing sample size, n. Further on, the larger the samples, the smaller the variation of the estimates from one sample to another.

Another theorem, the so-called *central limit theorem*, states that the distribution of the individual estimates can be approximated by the Gaussian normal distribution which we already know from Chap. 2—the reader will recall its signature bell-like shape. However, this approximation is known to be reasonably accurate only if the proportion, p, and the sample size, n, satisfy the following two conditions:

$$np \geq 10 \tag{12.1}$$

$$n(1-p) \geq 10 \tag{12.2}$$

If the conditions are *not* satisfied (if at least one of the products is less than 10), the distribution of estimates obtained from the samples cannot be approximated by the normal distribution without certain loss in accuracy.

Sections 12.2 and 12.3 will elaborate on how the normal-distribution approximation can help us establish our confidence in the measured performance of the induced classifiers.

An Illustration Let us check how these conditions are satisfied in the case of the samples of Table 12.1. We know that the proportion of ones in the original population was determined by the user-set parameter of the random-number generator: $p = 0.8$. Let us begin with samples of size $n = 10$. It turns out that none of the two conditions is satisfied because $np = 10 \cdot 0.8 = 8 < 10$ and $n(1-p) = 10 \cdot 0.2 = 2 < 10$. Therefore, the distribution of the proportions observed in these small samples cannot be approximated by the normal distribution.

In the second attempt, the sample size was increased to $n = 30$. As a result, we obtain $np = 30 \cdot 0.8 = 24 > 10$, and this means that Condition 12.1 is satisfied. At the same time, however, Condition (12.2) is *not* satisfied because $n(1-p) = 30 \cdot 0.2 = 6 < 10$. Even here, therefore, the normal distribution does not offer sufficiently accurate approximation.

The situation will change if we increase the sample size to $n = 60$. Doing the math, we easily establish that $np = 60 \cdot 0.8 = 48 \geq 10$ and also $n(1-p) = 60 \cdot 0.2 = 12 \geq 10$. We can therefore conclude that the distribution of the proportions of ones in samples of size $n = 60$ can be approximated with the normal distribution without any perceptible loss in accuracy.

The Impact of p Note how the applicability of the normal distribution is affected by p, the proportion of ones in the entire population. It is easy to see that, for different values of p, different sample sizes are called for if the two conditions are to be satisfied. Relatively small size is sufficient if $p = 0.5$; but the more the proportion differs from $p = 0.5$ to either side, the bigger the samples that we need.

To get a better idea of what this means in practice, recall that we found the sample size of $n = 60$ to be sufficient in a situation where $p = 0.8$. What if, however, we decide to base our estimates on samples of the same size, $n = 60$, but in a domain where the proportion is higher, say, $p = 0.95$? In this event, we will realize that $n(1 - p) = 60 \cdot 0.05 = 3 < 10$, which means that Condition (12.2) is not met, and the distribution of the proportions in samples of this size cannot be approximated by the normal distribution. For this condition to be satisfied in *this* domain, we would need a sample size of at least $n = 200$. Since $200 \cdot 0.05 = 10$, we have just barely made it. By the way, note that, on account of the symmetry of the two conditions, (12.1) and (12.2), the same minimum size, $n = 200$, will be called for in a domain where $p = 0.05$ instead of $p = 0.95$.

Parameters of the Distribution Let us return to our attempt to estimate the proportion of ones based on sampling. We now know that if the samples are large enough, the distribution of estimates made in different samples can be approximated by the normal distribution whose mean equals the (theoretical) proportion of ones that would have been observed in the entire population if such an experiment were possible.

The other parameter of a distribution is the standard deviation. In our context, statisticians prefer the term *standard error*, a terminological subtlety essentially meant to indicate the following: whereas "standard deviation" refers to a distribution of *any* variable (such as `weight`, `age`, or `temperature`), the term "standard error" is used when we refer to variations of estimates from one sample to another. And this is what interests us in the case of our proportions.

Let us denote the standard error by s_E. Mathematicians have established that its value can be calculated from the sample size, n, and the theoretical proportion, p, using the following formula:

$$s_E = \sqrt{\frac{p(1 - p)}{n}} \tag{12.3}$$

For instance, if $n = 50$ and $p = 0.80$, then the standard error is as follows:

$$s_E = \sqrt{\frac{0.80 \cdot 0.20}{50}} = 0.06$$

When expressing this result in plain English, some engineers prefer to say that the standard error is 6%.

The Impact of n; Diminishing Returns Note how the value of the standard error goes the other way than the sample size, n. To be more specific, the larger the samples, the lower the standard error and vice versa. Thus in the case of $n = 50$ and $p = 0.80$, we obtained $s_E = 0.06$. If we use larger samples, say, $n = 100$, the standard error will drop to $s_E = \sqrt{\frac{0.8 \cdot 0.2}{100}} = 0.04$. The curve defined by the normal distribution thus becomes narrower, and the proportions in different samples will tend to be closer to p.

This said, we should also be aware of the fact that increasing the sample size brings *diminishing returns*. Let us illustrate this statement using a simple example. The calculations carried out in the previous paragraph convinced us that, when proceeding from $n = 50$ to $n = 100$ (doubling the sample size), we managed to reduce s_E by two percentage points, from 6 to 4%. If, however, we do the same calculation for $n = 1000$, we get $s_E = 0.013$, whereas $n = 2000$ results in $s_E = 0.009$. In other words, doubling the sample size from 1000 to 2000, we only succeeded in reducing the standard error from 1.3 to 0.9%, which means that the only reward for doubling the sample size was the paltry 0.4%.

This last observation is worth remembering—for very practical reasons. In many domains, pre-classified examples are difficult or expensive to obtain; the reader will recall that this was the case of the oil-spill domain discussed in Sect. 8.2. If acceptable estimates of proportions can be made using a relatively small testing set, the engineer will not want to go into the trouble of trying to procure additional examples; the miniscule benefits may not justify the extra costs.

What Have You Learned?

To make sure you understand the topic, try to answer the following questions. If needed, return to the appropriate place in the text.

- Write down the formulas defining the conditions to be satisfied if the distribution of the proportions obtained from random samples are to follow the normal distribution.
- Explain how the entire-population proportion, p, affects the sample size, n, that is necessary for the proportions measured on different samples to follow the normal distribution.
- What is the mean value of a set of estimates that have been made based on different samples? Also, write down the formula that calculates the standard error.
- Elaborate on the statement that "increasing the sample size brings only diminishing returns."

12.2 Benefiting from the Normal Distribution

The previous section investigated the proportions of ones in samples taken from a certain population. The sample size was denoted by n, and the theoretical proportion of ones in the whole population was denoted by p. This theoretical value we do not know; the best we can do is estimate it based on our observation of a sample. Also, we have learned that, while the proportion in each individual sample is different, the

Fig. 12.1 Gaussian (normal)
distribution whose mean
value is p

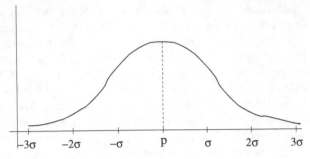

distribution of these values can often be approximated by the normal distribution—
the approximation being reasonably accurate if Conditions (12.1) and (12.2) are
satisfied.

The normal distribution can help us decide how much to trust the classification
accuracy (or, for that matter, any other performance criterion) that has been
measured on one concrete testing set. To be able to do so, let us take a brief look at
how to calculate so-called *confidence values*.

Re-formulation in Terms of a Classifier's Performance Suppose the ones and
zeros in Table 12.1 represent correct and incorrect classifications, respectively, as
they have been made by a classifier being evaluated on a testing set that consists of
one hundred examples (one hundred being the number of entries in the table). In this
event, the proportion of ones gives the classifier's accuracy, whereas the proportion
of zeros defines its error rate.

Evaluation of the classifier on a different testing set will of course result in
different values of the classification accuracy or error rate. But when measured on
great many testing sets, the individual accuracies will be distributed in a manner
that, as we have seen, roughly follows the normal distribution.

Properties of the Normal Distribution Figure 12.1 shows the fundamental shape
of the normal distribution. The vertical axis represents the *probability density
function* as we know it from Chap. 2. The horizontal axis represents classification
accuracy. The mean value, denoted here as p, is the *theoretical* classification
accuracy which we would obtain if we had a chance to evaluate the classifier
on all possible examples from the given domain. This theoretical value is of course
unknown, which is why our intention is to estimate it on the basis of a concrete
sample—the available set of testing examples.

The bell-like shape of the density function reminds us that most testing sets
will yield classification accuracies relatively close to the mean, p. The greater
the distance from p, the smaller the chance that this particular performance will
be obtained from a random testing set. Note also that, along the horizontal axis,
the graph highlights certain specific distances from p: the multiples of σ, the
distribution's standard deviation—or, when we deal with sample-based estimates,
the standard error of these estimates.

Table 12.2 For the normal
distribution, with mean p and
standard deviation σ, the left
column gives the percentage
of values found in the interval
$[p - z^*\sigma, \ p + z^*\sigma]$

Confidence level (%)	z^*
68	1.00
90	1.65
95	1.96
98	2.33
99	2.58

The formula defining the normal distribution was introduced in Sect. 2.5 where it was called the Gaussian "bell" function. Knowing the formula, we can establish the percentage of values found within a specific interval, $[a, b]$. The size of the entire area under the curve (from minus infinity to plus infinity) is 1. Therefore, if the area under the curve within the range of $[a, b]$ is 0.80, we can say that 80% of the performance estimates are found in this interval.

Identifying Intervals of Interest Not all intervals are equally important. For the needs of classifier evaluation, we are interested in those that are centered at the mean value, p. For instance, the engineer may want to know what percentage of values will be found in $[p - \sigma, \ p + \sigma]$. Conversely, she may want to know the size of the interval (again, centered at p) that contains 95% of all values.

Strictly speaking, questions of this kind can be answered with the help of mathematical analysis. Fortunately, we do not need to do the math ourselves because others have done it before, and we can take advantage of their findings. Some of the most useful results are shown in Table 12.2. Here, the left column lists percentages called *confidence levels*; for each of these, the right column specifies the interval that comprises the given percentage of values. Note that the length of the interval is characterized by z^*, the number of standard deviations to either side of p. More formally, therefore, the interval is defined as $[p - z^*\sigma, \ p + z^*\sigma]$.

Here is how the table is used for practical purposes. Suppose we want to know the size of the interval that contains 95% of the values. This percentage is found in the third row. We can see that the number on the right is 1.96, and this is interpreted as telling us that 95% of the values are in the interval $[p - 1.96 \cdot \sigma, \ p + 1.96 \cdot \sigma]$. Similarly, 68% of the values are found in the interval $[p - \sigma, \ p + \sigma]$—this is what we learn from the first row in the table.

Standard Error of Sample-Based Estimates Let us take a look at how to employ this knowledge when evaluating classification accuracies. Suppose that the testing sets are all of the same size, n, and suppose that this size satisfies Conditions (12.1) and (12.1) that allow us to use the normal distribution. We already know that the average of the classification accuracies measured on great many independent testing sets will converge to the *theoretical accuracy*, the one that would have been obtained by testing the classifier on all possible examples.

The standard error[2] is calculated using Eq. (12.3). For instance, if the theoretical classification accuracy is $p = 0.70$, and the size of each testing set is $n = 100$, then the standard error of the classification accuracies obtained from great many different testing sets is calculated as follows:

$$s_{acc} = \sqrt{\frac{p(1-p)}{n}} = \sqrt{\frac{0.7(1-0.7)}{100}} = 0.046 \qquad (12.4)$$

After due rounding, we will say that the classification accuracy is 70% plus or minus 5%. Note, again, that the standard error will be lower if we use a larger testing set. This makes sense: the larger the testing set, the more thorough the evaluation, and thus the higher our confidence in the value thus obtained.

Let us now ask what value we are going to obtain if we evaluate the classifier on some other testing sets of the same size. Once again, we answer the question with the help of Table 12.2. First of all, we find the row representing 95%. In this row, the right column gives the value $z^* = 1.96$; and this is interpreted as telling us that 95% of all results will be in the interval $[p - 1.96 \cdot s_{acc}, p + 1.96 \cdot s_{acc}] = [0.80 - 1.96 \cdot 0.46, 0.80 + 1.96 \cdot 0.46] = [0.61, 0.79]$.

Do not forget, however, that this will only be the case if the testing set has the same size, $n = 100$. For a different n, Eq. (12.4) will give us a different standard error, s_{acc}, and thus a different interval.

Two Important Reminders It may be an idea to remind ourselves of what exactly the normal-distribution assumption is good for. Specifically, if the distribution is normal, then we can use Table 12.2 from which we learn the size of the interval (centered a p) that contains the given percentage of values.

On the other hand, the formula for standard error (Eq. (12.3)) is valid generally, even if the distribution is not normal. For the calculation of standard error, the two conditions, (12.1) and (12.2), do *not* have to be satisfied.

What Have You Learned?

To make sure you understand the topic, try to answer the following questions. If needed, return to the appropriate place in the text.

- How do the considerations from the previous section apply to the evaluation of an induced classifier's performance?
- What kind of information can we glean from Table 12.2? How can this table be used when quantifying the confidence in the classification-accuracy value obtained from a testing set of size n?

---

[2] As explained in Sect. 12.1 in connection with the distribution of results obtained from different samples, we prefer the term *standard error* to the more general *standard deviation*.

- How will you calculate the standard error of estimates based on a given testing set? How does this standard error depend on the size of the testing set?

12.3 Confidence Intervals

Let us now focus on how the knowledge gained in the previous two sections can help us specify the experimenter's confidence in the classifier's performance as measured on the given testing data.

Confidence Interval: An Example Now that we understand how the classification accuracies obtained from different testing sets are distributed, we are ready to draw conclusions about how confident we can be in our expectation that the value measured on one concrete testing set is close to the true theoretical value.

Suppose the size of the testing set is $n = 100$, and let the classification accuracy measured on this testing set be $acc = 0.85$. For the training set of this size, the standard error is as follows:

$$s_{acc} = \sqrt{\frac{0.85 \cdot 0.15}{100}} = 0.036 \qquad (12.5)$$

Checking the normal-distribution conditions, we realize that they are both satisfied here because $100 \cdot 0.85 = 85 \geq 10$ and $100 \cdot 0.15 = 15 \geq 10$. This means that we can take advantage of the z^*-values listed in Table 12.2. Using this table, we easily establish that 95% of all values are found in the interval $[acc - 1.96 \cdot s_{acc},\ acc + 1.96 \cdot s_{acc}]$. For $acc = 0.85$ and $s_{acc} = 0.036$, we realize that the corresponding interval is $[0.85 - 0.07,\ 0.85 + 0.07] = [0.78, 0.92]$.

What this result is telling us is that, based on the evaluation on the given testing set, we can say that, with 95% confidence, the real classification accuracy finds itself somewhere in the interval $[0.78, 0.92]$. This interval is usually called the *confidence interval*.

Two New Terms: Confidence Level and Margin of Error Confidence intervals reflect specific *confidence levels*—those defined by the percentages listed in the left column of Table 12.2. In our specific case, the confidence level was 95%.

Each confidence level defines a different confidence interval. This interval can be re-written as $p \pm M$ where p is the mean and M is the so-called *margin of error*. For instance, in the case of the interval $[0.78, 0.92]$, the mean was $p = 0.85$ and the margin of error was $M = z^* s_{acc} = 1.96 \cdot 0.036 = 0.07$.

Choosing the Confidence Level In the example discussed above, the requested confidence level was 95%, a fairly common choice. For another confidence level, a different confidence interval would have been obtained. Thus for 99%, Table 12.2 gives $z^* = 2.8$, and the confidence interval is $[0.85 - 2.58 \cdot s_{acc},\ 0.85 + 2.58 \cdot s_{acc}] = [0.76, 0.94]$. Note that this interval is longer than the one for confidence level 95%.

This was to be expected: the chance that the real, theoretical, classification accuracy finds itself in a longer interval is higher. Conversely, it is less likely that the theoretical value will fall into some narrower interval. Thus for the confidence level of 68% (and the standard error rounded to $s_{acc} = 0.04$), the confidence interval is $[0.85 - 0.04, \ 0.85 + 0.04] = [0.81; \ 0.89]$.

Importantly, we must not forget that, even in the case of confidence level 99%, one cannot be absolutely sure that the theoretical value will fall into the corresponding interval. There is still that 1% probability that the measured value will be outside this interval.

Another Parameter: Sample Size   The reader now understands that the length of the confidence interval depends on the standard error, and that the standard error, in turn, depends on the size, n, of the testing set (see Eq. (12.3)). Essentially, the larger the testing set, the stronger the evidence in favor of the measured value, and thus the narrower the confidence interval. This is why we say that the margin of error and the training-set size are in *inverse relation*: as the training-set size increases, the margin of error decreases.

Previously, we mentioned that a higher confidence level results in a longer confidence interval. If we think this interval to be too big, we can make it shorter by using a bigger testing set, and thus a higher value of n (which decreases the value of the standard error of the measured value).

There is a way of deciding how large the testing set should be if we want to limit the margin of error to a certain maximum value. Here is the formula calculating the margin of error:

$$M = z^* s_{acc} = z^* \sqrt{\frac{p(1-p)}{n}} \qquad (12.6)$$

Solving this equation for n (for specific values of M. p, and z^*) will give us the required testing-set size.

A Concluding Remark   The method of establishing the confidence interval for the given confidence level was explained using the simplest performance criterion, classification accuracy. Yet the scope of the method's applicability is much broader: the uncertainty of *any* variable that represents a proportion can thus be quantified. In the context of machine learning, we can use the same approach to establish our confidence in any of the performance criteria from Chap. 11, be it *precision, recall,* or some other quantity.

But we have to be careful we do it right. For one thing, we must not forget that the distribution of the values of the given quantity can only be approximated by the normal distribution if the conditions (12.1) and (12.2) are satisfied. Second, we have to make sure we understand the meaning of n when calculating the standard error using Eq. (12.3). For instance, the reader remembers that *precision* is calculated with the formula, $\frac{N_{TP}}{N_{TP}+N_{FP}}$: the percentage of true positives among all examples labeled by the classifier as positive. This means that we are dealing with a proportion of true positives in a sample of the size $n = N_{TP} + N_{FP}$. Similar considerations have to be made in the case of *recall* and some other performance criteria.

What Have You Learned?

To make sure you understand the topic, try to answer the following questions. If needed, return to the appropriate place in the text.

- Explain the meaning of the term, *confidence interval*. What is meant by the *margin of error*?
- How does the size of the confidence interval (and the margin of error) depend on the user-specified confidence level? How does it depend on the size of the testing set?
- Discuss the calculations of confidence intervals for some other performance criteria such as *precision* and *recall*.

12.4 Statistical Evaluation of a Classifier

A claim about a classifier's performance can be confirmed or refuted experimentally, by testing the classifier on a set of pre-classified examples. One possibility for the statistical evaluation of the results thus obtained is to follow the algorithm from Table 12.3. Let us illustrate the procedure on a simple example.

A Simple Example Suppose a machine-learning specialist tells you that the classifier he has induced has classification accuracy $acc = 0.78$. Faithful to the dictum, "trust but verify," you decide to find out whether this statement is correct. To this end, you prepare $n = 100$ examples whose class labels are known, and then set about measuring the classifier's performance on this testing set.

Table 12.3 The algorithm for statistical evaluation of a classifier's performance

---

1. For the given size, n, of the testing set, and for the claimed classification accuracy, acc, check whether the conditions for normal distribution are satisfied:

$$n \cdot acc \geq 10 \text{ and } n \cdot (1 - acc) \geq 10$$

2. Calculate the standard error by the usual formula:

$$s_{acc} = \sqrt{\frac{acc(1 - acc)}{n}}$$

3. Assuming that the normal-distribution assumption is correct, find in Table 12.2 the z^*-value for the requested level of confidence. The corresponding confidence interval is $[acc - z^* \cdot s_{acc}, \ acc + z^* \cdot s_{acc}]$.

4. If the value measured on the testing set finds itself outside this interval, reject the claim that the accuracy equals acc. Otherwise, assume that the available evidence is insufficient for the rejection.

Let the experiment result in giving us classification accuracy 0.75. Well, this is less than the promised 0.78, but then: is this observed difference still within reasonable bounds? To put it another way, is there a chance that the specialist's claim was correct, and that the lower performance measured on the testing set can be explained by the variations implied by the random nature of the employed testing data? After all, a different testing set is likely to result in a different classification accuracy.

Checking the Conditions for Normal Distribution The first question to ask is whether the distribution of the performances thus obtained can be approximated by the normal distribution. A positive answer will allow us to base our statistical evaluation on the values from Table 12.2.

Verification of Conditions (12.1) and (12.2) is quite easy. Seeing that $np = 100 \cdot 0.75 = 75 \geq 10$, and that $n(1 - p) = 100 \cdot 0.25) = 25 \geq 10$, we realize that the conditions are satisfied and the normal-distribution assumption can be used.

Finding the Confidence Interval for the 95%-Confidence Level Suppose that you are prepared to accept the specialist's claim ($acc = 0.78$) if there is at least a 95% chance that such performance will make it possible to observe that the classification accuracy on a random testing set is $acc = 0.75$. This will be possible if 0.75 finds itself within the corresponding confidence interval, centered at 0.78. Let us find out whether this is the case.

The corresponding row in the table informs us that $z^* = 1.96$; this means that 95% of accuracies obtained on random testing set will find themselves in the interval $[acc - 1.96 \cdot s_{acc}, \; acc + 1.96 \cdot s_{acc}]$, where $acc = 0.78$ is the original claim and s_{acc} is the standard error to be statistically expected for testing sets of the given size, n.

In our concrete training set, the size is $n = 100$. The standard error is calculated as follows:

$$s_{acc} = \sqrt{\frac{acc(1 - acc)}{n}} = \sqrt{\frac{0.75 \cdot 0.25}{100}} = 0.043 \tag{12.7}$$

We conclude that the confidence interval is $[0.78 - 1.96 \cdot 0.043, \; 0.78 + 1.96 \cdot 0.043]$ which, after evaluation and due rounding, is $[0.70, 0.86]$.

A Conclusion Regarding the Specialist's Claim Evaluation on our own training set resulted in classification accuracy $acc = 0.75$, a value that finds itself within the confidence interval corresponding to the chosen confidence level of 95%.

This is encouraging. For the given claim, $acc = 0.78$, there is a 95% probability that our evaluation on a random testing set will give us a classification accuracy somewhere within the interval $[0.70, 0.86]$. This, indeed, is what happened in this particular case. And so, although our result, $acc = 0.75$, is somewhat lower than the specialist's claim, we have to admit that our experimental evaluation failed to provide convincing evidence against the claim. In the absence of such evidence, we accept the claim as valid.

Type-I Error in Statistical Evaluation: False Alarm The reader now understands the fundamental principle of statistical evaluation. Someone makes a statement about performance. Based on the size of our testing set (and assuming normal distribution), we calculate the size of the interval that is supposed to contain the given 95% of all values. There is only a 5% chance that, if the original claim is correct, the result of testing will be outside this interval. This is why we reject any hypothesis whose testing results landed in this less-than-5% region. We simply assume that it is rather unlikely that such difference would be observed.

This said, such difference should still be expected in 5% cases. We have to admit that there exists some small danger that the evaluation of the classifier on a random testing set will result in a value outside the given confidence interval. In this case, rejecting the specialist's claim would be unfair. Statisticians call this the *type-I* error: the false rejection of an otherwise correct claim; a rejection that is based on the fact that certain results are untypical.

If we do not like to face this danger, we can reduce it by increasing the required confidence level. If we choose 99% instead of the 95%, false alarms will be less frequent. But this reduction is not gained for free—as will be explained in the next paragraph.

Type-II Error in Statistical Evaluation: Failing to Detect an Incorrect Claim Also the opposite case is possible. To wit, the initial claim is false, and yet the classification accuracy obtained from our testing falls within the given confidence interval. When this happens, we are forced to conclude that our experiment failed to provide sufficient evidence against the claim; the claim thus has to be accepted.

The reader may find this unfortunate, but this is indeed what sometimes happens. An incorrect claim is not refuted. Statisticians call this the *type-II* error. It is typical of those cases where very high confidence level is required: so broad is the corresponding interval that the results of testing will almost never fall outside; the experimental results then hardly ever lead to the rejection of the initial claim.

The thing to remember is the inevitable trade-off between the two types of error. By increasing the confidence level, we reduce the danger of the *type-I* error, but only at the cost of increasing the danger of the *type-II* error; and vice versa.

What Have You Learned?

To make sure you understand the topic, try to answer the following questions. If needed, return to the appropriate place in the text.

- Explain how to evaluate statistically the results of an experimental measurement of a classifier's performance on a testing set.
- What is meant by the term, *type-I error* (false alarm)? What can be done to reduce the danger of making this error?
- What is meant by the term, *type-II error* (missed detection)? What can be done to reduce the danger of making this error?

12.5   Another Kind of Statistical Evaluation

At this moment, the reader understands the essence of statistical processing of experimental results, and knows how to use it when evaluating the claims about a given classifier's performance. However, much more can be accomplished with the help of statistics.

Do Two Testing Sets Represent Two Different Contexts?   Chapter 10 mentioned the circumstance that, sometimes, a different classifier should perhaps be induced for a different context—such as the British accent as compared to the American accent. Here is how statistics can help us identify such situations in the data.

Suppose we have tested two classifiers on two different testing sets. The classification accuracy in the first test is $\hat{p}_1$ and the classification accuracy in the second test is $\hat{p}_2$ (the letter "p" alluding to the *proportion* of correct answers). The sizes of the two sets are denoted by n_1 and n_2. Finally, let the average proportion of correctly classified examples in the two sets combined be denoted by $\hat{p}$.

The statistics of interest is defined by the following formula:

$$z = \frac{\hat{p}_1 - \hat{p}_2}{\sqrt{\hat{p}(1 - \hat{p})(\frac{1}{n_1} + \frac{1}{n_2})}} \tag{12.8}$$

The result is compared to the critical value for the given confidence level—the value can be found in Table 12.2.

A Concrete Example   Suppose the classifier was evaluated on two testings sets whose sizes are $n_1 = 100$ and $n_2 = 200$. Let the classification accuracies measured on the two be $\hat{p}_1 = 0.82$ and $\hat{p}_2 = 0.74$, respectively, so that the average classification accuracy on the two sets combined is $\hat{p} = 0.77$. The reader will easily verify that the conditions for normal distribution are satisfied.

Plugging these values into Eq. (12.8), we obtain the following:

$$z = \frac{0.82 - 0.74}{\sqrt{0.77(1 - 0.77)(\frac{1}{100} + \frac{1}{200})}} = 1.6. \tag{12.9}$$

Since this value is lower than the one given for the 95% confidence level in Table 12.2, we conclude that the result is within the corresponding confidence interval, and therefore accept that the two results are statistically indistinguishable.

What Have You Learned?

To make sure you understand the topic, try to answer the following questions. If needed, return to the appropriate place in the text.

- Why should we be concerned that a classifier is being applied to a wrong kind of data?
- What formula is used here? How do you carry our the evaluation?

12.6 Comparing Machine-Learning Techniques

Sometimes, we need to know which of two alternative machine-learning techniques is more appropriate for the given class-recognition problem. The usual methodology is here somewhat different than in the previous sections, though it is built around similar principles. Instead of lengthy theorizing, let us illustrate this kind of evaluation on a simple example with concrete data.

Experimental Methodology As discussed in Chap. 11, one way to compare, experimentally, two machine-learning algorithms is to use 5×2 crossvalidation. The reader will recall that, in this method, the set of available pre-classified data is divided into two equally sized subsets, T_{11} and T_{12}. First, the two machine-learning techniques are used to induce their classifiers from T_{11}, and these classifiers are then tested on T_{12}; then, the two classifiers are both induced from T_{12} and tested on T_{11}. The process is repeated five times, each time for a different random division of the set of data into two subsets, T_{i1} and T_{i2}.

As a result, we obtain ten pairs of testing-set classification accuracies (or error rates, precisions, recalls, or any other performance criterion of choice). The question to ask is then formulated as follows: "Are the differences between the ten pairs of results statistically significant?"

Example of Experimental Results: Paired Comparisons Let us denote the i-th pair of the sets by T_{i1} and T_{i2}, respectively. Suppose we are comparing two machine-learning algorithms, ML1 and ML2, evaluating them in the ten experimental runs as explained in the previous paragraph. Suppose that the results are those listed in Table 12.4. In this table, each column is headed with the name of the test set used in the corresponding experimental run. The fields in the table give the classification accuracies (in percentages) that have been achieved on the corresponding test sets by classifiers induced by the two alternative induction techniques. The last row specifies the differences between the two classification accuracies. Note that the differences are either positive or negative.

Evaluating these results, we realize that the average difference is $\overline{d} = 2.0$, and that the standard deviation of these differences is $s_d = 4.63$.

The Principle of Statistical Evaluation of Paired Differences Observing the mean value of differences, $\overline{d}$ (with standard deviation, s_d), we have to ask: is this difference statistically significant? In other words, is this difference outside what we previously called a confidence interval for a given confidence level, say, 95%? Note that the midpoint of this confidence interval is $\hat{d} = 0$.

Table 12.4 Example experimental results of a comparison of two alternative machine-learning techniques, ML1 and ML2

	T_{11}	T_{12}	T_{21}	T_{22}	T_{31}	T_{32}	T_{41}	T_{42}	T_{51}	T_{52}
ML1	78	82	99	85	80	95	87	57	69	73
ML2	72	79	95	80	80	88	90	50	73	78
d	6	3	4	5	0	7	−3	7	−4	−5

The numbers in the first two rows give classification accuracies (in percentages), the last row gives the differences, d

Table 12.5 Some probabilities of the t-values for nine degrees of freedom

Degrees of freedom	Confidence level			
	0.10%	0.05%	0.02%	0.0%1
9	1.83	2.26	2.81	3.35

As compared to the methods discussed in the previous sections, we have to point out two major differences. First of them is the fact that, instead of proportions, we are now dealing with mean values, d. The second is the circumstance that we can no longer rely on the normal distribution because the number, n, of the values on which the statistical evaluation rests is small, and the standard deviation has only been estimated on the basis of the given ten observations (it is not known for the entire population).

In this situation, we have to resort to another theoretical distribution, the so-called t-distribution. Its shape is similar to the normal distribution (the "bell" shape), but it is flatter. Also, its "flatness" or "steepness" depends on what is called the number of *degrees of freedom*. In the case of 10 testing sets, there are $10 - 1 = 9$ degrees of freedom. Some typical t values for the case of 9 degrees of freedom are shown in Table 12.5.[3]

Calculating the t-Values in Paired Tests Statistical evaluation using the t-tests is essentially the same as in the case of normal distribution. For the mean difference, $\overline{d}$, and the standard deviation s_d, the t_9-value (the subscript refers to the number of degrees of freedom) is calculated by the following formula, where n is the number of tests:

$$t_9 = \frac{\overline{d} - 0}{s_{\overline{d}}/\sqrt{n}} \tag{12.10}$$

The result is then compared with the critical thresholds associated with concrete levels of confidence. These are listed in Table 12.5. Specifically, in the case of the results from Table 12.4, we obtain the following:

---

[3]With more degrees of freedom, the curve would get closer to the normal distribution, becoming almost indistinguishable from it for 30 or more degrees of freedom.

$$t_9 = \frac{2-0}{4.63/\sqrt{10}} = 1.35 \tag{12.11}$$

Seeing that the obtained value is less than the 2.26 listed for the 95% confidence level in the table, we conclude that the experiment has failed to refute (for the given confidence level) the hypothesis that the two techniques lead to comparable classification accuracies. We therefore accept the claim.

What Have You Learned?

To make sure you understand the topic, try to answer the following questions. If needed, return to the appropriate place in the text.

- Explain the principle of the 5×2-crossvalidation technique which results in a set of 10 paired results.
- Why cannot we use the normal distribution as in the previous sections? What other distribution is used here?
- Write down the formula calculating the t-value. Explain how the value for each individual variable in this formula is obtained.

12.7 Summary and Historical Remarks

- The essence of statistical evaluation is to draw conclusions about the behavior of an entire population based on our observations made on a relatively small sample.
- Different samples will yield different results, but these different values are bound to be distributed according to statistical laws. Knowledge of these laws helps a machine-learning engineer to calculate his or her confidence in the classification performance as measured on concrete testing set.
- The most typical distribution of the "sampling results" is the Gaussian normal distribution. It can only be used if two essential conditions are satisfied. For training-set size n and for the mean value p, the conditions are as follows: $np \geq 10$ and $n(1-p) \geq 10$.
- Regardless whether the distribution is normal or not, the standard error of the classification accuracies, acc, measured on different testing sets is given by the following formula:

$$s_{acc} = \sqrt{\frac{acc(1-acc)}{n}} = \sqrt{\frac{0.75 \cdot 0.25}{100}} = 0.043$$

- For each confidence level, the normal-distribution assumption leads to a specific z^* value (see Table 12.2). Having calculated the standard error, and having chosen a confidence level, we establish the confidence interval by the following formula:

$$[acc - z^* s_{acc}, \ acc + z^* s_{acc}]$$

The term $z^* s_{acc}$ is referred to as the *margin of error.* For different performance metrics, similar formulas are used.

- Suppose we are testing a claim about certain classification accuracy, *acc.* If the result of experimental evaluation falls into the confidence interval defined by the chosen confidence level, we assume we do not have enough evidence against the claim regarding the value of *acc.* If the result is outside the value, we reject the claim.
- When comparing two machine-learning techniques, we often rely on the 5×2-crossvalidation technique, subjecting the results to the so-called *t*-test. This test relies on the *t*-distribution instead of the normal distribution. The *t*-distribution has a slightly different shape for different numbers of *degrees of freedom.*

Historical Remarks The statistical methods discussed in this chapter are so old and well established that textbooks of statistics no longer care to give credit to those who developed them. From the perspective of machine learning, however, we must mention that the idea of applying *t*-tests to experimental results obtained from 5×2 crossvalidation was advocated by Dietterich [22].

12.8 Solidify Your Knowledge

The exercises are to solidify the acquired knowledge. The suggested thought experiments will help the reader see this chapter's ideas in a different light and provoke independent thinking. Computer assignments will force the readers to pay attention to seemingly insignificant details they might otherwise overlook.

Exercises

1. Suppose we intend to evaluate the statement that a certain classifier's accuracy is $p = 0.80$. What size n, of the testing set is needed if we want to be able to rely on the normal distribution and Table 12.2?
2. Suppose that a certain classifier accuracy has been specified as $p = 0.9$. What is the value of the standard error, s_E, if the classifier is to be evaluated on testing sets of size $n = 400$? Determine the size of the confidence interval for the 95% confidence level. Do not forget to check the validity of Conditions (12.1) and (12.2).
3. Suppose your company is offered a classifier with stated accuracy $p = 0.85$. When the company decides to test the validity of this statement on a testing set of 200 examples, the measured value is 0.81, which of course is less than what was promised. Is there at least 95% chance that the original claim was correct? What about 99%?

Give It Some Thought

1. Suppose you test a classifier's performance, using the 95%-confidence interval. What if you change your mind and decide to use the 99%-confidence instead? You will increase tolerance, but what is the price for this?

Computer Assignments

1. This assignment assumes that the reader has already implemented a program dividing the data into the five "folds" needed for the evaluation of the performance using the 5×2-CV methodology. Another assumption is that the reader has implemented at least two class-induction programs.

   Write a program comparing the two induction techniques using the 5×2-CV methodology, evaluating the results using t-tests.

Chapter 13
Induction in Multi-Label Domains

All the techniques discussed in the previous chapters assumed that each example is labeled with one and only one class. In realistic applications, however, this is not always the case. Quite often, an example is known to belong to two or more classes at the same time, sometimes to *many* classes. For machine learning, this poses certain new problems. After a brief discussion of how to deal with this issue within the framework of classical paradigms, this chapter describes the currently most popular approach: *binary relevance.*

The idea is to induce a binary classifier separately for each class, and then to use all these classifiers in parallel. More advanced versions of this technique seek to improve classification performance by exploiting mutual interrelations between classes. As yet another alternative, the chapter discusses also the simple mechanism of *class aggregation.*

13.1 Classical Machine Learning in Multi-Label Domains

Let us begin with an informal definition of a multi-label domain. After this, we will take a look at how to address the problem within the classical paradigms introduced in earlier chapters.

What is a Multi-Label Domain? In many applications, the traditional requirement that an example should be labeled with one and only one class is hard to satisfy. Thus a text document may represent nutrition, diet, athletics, popular science, and perhaps quite a few other categories. Alternatively, an image may at the same time represent summer, cloudy weather, beach, sea, seagulls, and so on. Something similar is the case in many other domains.

The number of classes with which an average example is labeled differs from one application to another. In some of them, almost every example has a great many labels selected from perhaps thousands of different classes. At the other end of the

© Springer International Publishing AG 2017
M. Kubat, *An Introduction to Machine Learning*,
DOI 10.1007/978-3-319-63913-0_13

spectrum, we find domains where only some examples belong to more than one class, the majority being labeled with only a single one.

Whatever the characteristics of the concrete data, the task for machine learning is to induce a classifier (or a set of classifiers) satisfying two basic requirements. First, the tool should for a given example return as many of its true classes as possible; missing any one of them would constitute a *false negative*. At the same time, the classifier should not label the example with a class to which the example does not belong—each such "wrong" class would constitute a *false positive*.

Neural Networks Chapter 5 explained the essence of a *multilayer perception*, MLP, a popular architecture of artificial neural networks. The reader will recall that the output layer consists of one neuron for each class, the number of inputs equals the number of attributes, and the ideal size of the hidden layer reflects the complexity of the classification problem at hand.

On the face of it, using an MLP in multi-label domains should not pose any major problems. For instance, suppose the network has been presented with a training example that is labeled with classes C_3, C_6, and C_7. In this event, the target values for training will be set to, say, $t_i = 0.8$, in the case of output neurons with indices $i \in \{3, 6, 7\}$, and to $t_i = 0.2$ for all the other output neurons.[1] The backpropagation-of-error technique can then be used in the same manner as in single-label domains.

A Word of Caution Multilayer perceptions may not necessarily be the best choice here. Indeed, multi-label domains have been less intensively studied, in the neural-networks literature, than other approaches, and not without reason. For one thing, the training of plain MLPs is known to be vulnerable to local minima, and there is always the architecture-related question: what is the best number of hidden neurons if we want to strike a reasonable compromise between overfitting the data if the network is too large, and suffering from insufficient flexibility if the network is too small?

Also the notoriously high computational costs can be a reason for concern. The fact that each training example can belong to more than one class certainly complicates the learning process. Sensing the difficulty of the task, the engineer is sometimes tempted to increase the number of hidden neurons. This, however, not only adds to the already high computational costs, but also increases the danger of overfitting.

It is always good to keep in mind that training neural networks is more art than science. While a lot can be achieved through ingenuity and experience, beginners are often disappointed. And in the case of a failure, the machine-learning expert should be prepared to resort to some alternative, less dangerous technique.

---

[1]The reader will recall that the target values 0.8 and 0.2 are more appropriate for the backpropagation-of-error algorithm than 1 and 0. See Chap. 5.

Nearest-Neighbor Classifiers Another possibility is the use of nearest-neighbor classifiers with which we got acquainted in Chap. 3. When example $\mathbf{x}$ is presented, the k-NN classifier first identifies the example's k nearest neighbors. Each of these may have been labeled with a set of classes, and the simplest classification attempt in a multi-label domain will label $\mathbf{x}$ with the union of these sets. For instance, suppose that $k = 3$, and suppose that the sets of class labels encountered in the three nearest neighbors are $\{C_1, C_2\}$, $\{C_2\}$, and $\{C_1, C_3\}$, respectively. In this event, the classifier will classify $\mathbf{x}$ as belonging to C_1, C_2, and C_3.

A Word of Caution This approach is practical only in domains where the average number of classes per example is moderate, say, less than three. Also the number of voting neighbors, k, should be small. Unless these two requirements are satisfied, too many class labels may be returned for $\mathbf{x}$, and this can give rise to too many false positives, which, in turn, leads to poor *precision*. At the same time, however, the multitude of returned labels also reduces the number of false negatives, which improves *recall*. In some domains, this is what we want. In others, *precision* is critical, and its low value may not be acceptable.

As so often in this paradigm, the engineer must resist the temptation to increase the number of the nearest neighbors in the hope that spreading the vote over more "participants" will give a chance to less frequent classes. The thing is, some of these "nearest neighbors" might then be too distant from $\mathbf{x}$, and thus inappropriate for classification purposes.

A Note on Other Approaches Machine learning scientists have developed quite a few other ways of modifying traditional machine-learning paradigms for the needs of multi-label domains. Among these, very interesting appear to be attempts to induce multi-label decision trees. But since they are somewhat too advanced for an introductory text, we will not present them here. After all, comparable classification performance can be achieved by simpler means—and these will be the subject of the rest of this chapter.

What Have You Learned?

To make sure you understand the topic, try to answer the following questions. If needed, return to the appropriate place in the text.

- Suggest an example of a multi-label domain. What is the essence of the underlying machine-learning task? In what way will one multi-label domain differ from another?
- Explain the simple method of multi-label training in multilayer perceptions. What practical difficulties might discourage you from using this paradigm?
- Describe the simple way of addressing a multi-label domain by a k-NN classifier. Discuss its potential pitfalls.

13.2   Treating Each Class Separately: Binary Relevance

Let us now proceed to the main topic of this section, the technique of *binary relevance*. We will begin by explaining the principle, and then discuss some of its shortcomings and limitations.

The Principle of Binary Relevance The most common approach to multi-label domains induces a separate binary classifier for each class: in a domain with N classes, N classifiers are induced. When classifying a future example, all these classifiers are used in parallel, and the example receives all classes for which the classifiers returned the positive label.

For the induction of these classifiers, the training data have to be modified accordingly. Here is how. For the i-th class ($i \in [1, N]$), we create a training set, T_i, that consists of the same examples as the original training set, T, the only difference being in labeling: in T_i, an example's class label is 1 if the list of class labels for this example in T contains C_i; otherwise, the label in T_i is 0.

Once the new training sets have been created, we apply to each of them a *baseline learner* that is responsible for the induction of the individual classifiers. Common practice applies the same baseline learner to each T_i. Typically, we use to this end some of the previously discussed machine-learning techniques such as perception learning, decision-tree induction, and so on.

Illustration of the Learning Principle Table 13.1 illustrates the mechanism with which the new training data are created. In the original training set, T, five different class labels can be found: $C_1, \ldots, C_5$. The *binary relevance* technique creates the five new training sets, $T_1, \ldots, T_5$, shown in the five tables below the original one.

Table 13.1 The original multi-label training set is converted into five new training sets, one for each class

	Classes
ex_1	C_1, C_2
ex_2	C_2
ex_3	C_1, C_3, C_5
ex_4	C_2, C_3
ex_5	C_2, C_4

T_1			T_2			T_3			T_4			T_5	
ex_1	1		ex_1	1		ex_1	0		ex_1	0		ex_1	0
ex_2	0		ex_2	1		ex_2	0		ex_2	0		ex_2	0
ex_3	1		ex_3	0		ex_3	1		ex_3	0		ex_3	1
ex_4	0		ex_4	1		ex_4	1		ex_4	0		ex_4	0
ex_5	0		ex_5	1		ex_5	0		ex_5	1		ex_5	0

Thus in the very first of them, T_1, examples ex_1 and ex_3 are labeled with 1 because these (and only these) two examples contain the label C_1 in the original T. The remaining examples are labeled with 0.

The baseline learner is applied separately to each of the five new sets, inducing from each T_i the corresponding classifier C_i.

An Easy-to-Overlook Pitfall In each of the training sets thus obtained, every example is labeled as a positive or negative representative of the given class. When the induced binary classifiers are used in parallel (to classify some **x**), it may happen that none of them returns 1. This means that no label for **x** has been identified. When writing the machine-learning software, we must not forget to instruct the classifier what to do in this event. Usually, the programmer chooses from the following two alternatives: (1) return a default class, perhaps the one most frequently encountered in T, or (2) reject the example as too ambiguous to be classified.

Discussion The thing to remember is that the idea behind *binary relevance* is to transform the multi-label problem into a set of single-label tasks that are then addressed by classical machine learning. To avoid disappointment, however, the engineer needs to be aware of certain difficulties which, unless properly addressed, may lead to underperformance. Let us briefly address them.

Problem 1: Imbalanced Classes Some of the new training sets, T_i, are likely to suffer from the problem of *imbalanced class* representation which was discussed in Sect. 10.2. In Table 13.1, this occurs in the case of sets T_4 and T_5. In each of them, only one example out of five (20%) is labeled as positive, and all others are labeled as negative. In situations of this kind, we already know, machine-learning techniques tend to be biased toward the majority class—in this particular case, the class labeled as 0.

The solution is not difficult to find. The two most straightforward approaches are majority-class undersampling or minority-class oversampling. Which of them to choose will of course depend on the domain's concrete circumstances. As a rule of thumb, one can base the decision on the size of the training set. In very big domains, majority-class undersampling is better; but when the examples are scarce, the engineer cannot afford to "squander" them, and thus prefers minority-class oversampling.

Problem 2: Computational Costs Some multi-label domains are very large. Thus the training set in a text categorization domain may consist of hundreds of thousands of examples, each described by tens of thousands of attributes and labeled with a subset of thousands of different classes. It stands to reason that to induce thousands of decision trees from a training set of this size will be expensive, perhaps prohibitively so. We can see that, when considering candidates for the baseline learner, we may have to reject some of them because of computational costs.

Another possibility is to resort to the technique discussed in Sect. 9.5 in the context of boosting techniques: for each class, we create multiple subsets of the training examples, some of them perhaps described by different subsets of attributes. The idea is to induce for each class a group of subclassifiers that then vote. If (in a given paradigm) learning from 50% of the examples takes only 5% of the time, considerable savings can be achieved.

Problem 3: Performance Evaluation Another question is how to measure the success or failure of the induced classifiers. Usually, each of them will exhibit different performance, some better than average, some worse than average, and some dismal. To get an idea of the big picture, some averaging of the results is needed. We will return to this issue in Sect. 13.7.

Problem 4: Neglecting Mutual Interdependence of Classes The baseline version of *binary relevance* treats all classes as if they were independent of each other. Quite often, this assumption is justified. In other domains, the classes *are* to some degree interdependent, but not much harm is done when this fact is ignored. But in some applications, the overall performance of the induced classifiers considerably improves if we find a way to exploit the class interdependence. To introduce methods of doing so will be the task for the next three sections.

What Have You Learned?

To make sure you understand the topic, try to answer the following questions. If needed, return to the appropriate place in the text.

- Describe the principle of *binary relevance*. How does it organize the learning process, and how are the induced classifiers used for the classification of future examples?
- What can render the computational costs of this approach prohibitively high? How will the engineer handle the situation?
- Why does *binary relevance* often lead to the problem of imbalanced classes? What remedies would you recommend?

13.3   Classifier Chains

In many applications, the classes are interrelated. The fact that a text document has been labeled as `nutrition` is sure to increase its chances of belonging also to `diet`—and decrease the probability that has something to do with `quantum mechanics`. In the context of binary relevance, this means that methods of exploiting class interdependence are likely to improve classification performance. One possibility of doing so is known as a *classifier chain*.

The Idea A very simple approach relies on a chain of classifiers such as the one in Fig. 13.1. Here, the classifiers are created one at a time, starting from the left. To begin with, the leftmost classifier is induced from the original examples labeled as positive or negative instances of class C_1 (recall the training set T_1 from Table 13.1).

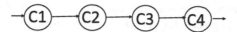

Fig. 13.1 With the exception of $C1$, the input of each classifier consists of the original attribute vector *plus* the label returned by the previous classifier

The second classifier is then induced from examples labeled as positive or negative instances of class C_2. To describe these latter examples, however, one extra attribute is added to the original attribute vector: the output of C_1. The same principle is then repeated in the course of the induction of all the remaining classifiers: for each, the training examples are described by the original attribute vector *plus* the class label returned by the previous classifier.

When using the classifier chain for the classification of some future example, **x**, the same pattern is followed. The leftmost classifier receives **x** described by the original attributes. To all other classifiers, the system presents **x** described by the original attribute vector plus the label delivered by the previous classifier. Ultimately, **x** is labeled with those classes whose classifiers returned 1.

An Important Assumption (Rarely Satisfied) In the classifier-chain technique, the ordering of the classes from left to right is the responsibility of the engineer. In some applications, this is easy because the classes form a logical sequence. Thus in document classification, `science` subsumes `physics`, which in turn subsumes `quantum mechanics`, and so on. If a document does not belong to `science`, it is unlikely to belong to `physics`, either; it thus makes sense to choose `science` as the leftmost node in the graph in Fig. 13.1, and to place `physics` next to it.

In other applications, class subsumption is not so obvious, but the sequence can still be used without impairing the overall performance. Even when the subsumptions are only intuitive, the engineer may always resort to a sequence backed by experiments: she can suggest a few alternative versions, test them, and then choose the one with the best results. Another possibility is to apply the classifier chain only to some of the classes (where the interrelations are known), treating the others as if only plain *binary relevance* was to be employed.

Hierarchically Ordered Classes Class interrelation does not have to be linear. It can acquire forms that can only be reflected by a more sophisticated data structure, perhaps a graph. In that case, we will need more advanced techniques such as the one described in Sect. 13.5.

One Shortcoming of the Classifier-Chain Approach More often than not, the engineer lacks any a priori knowledge about class interrelations. If she then still wants to employ classifier chains, the best she can do is to create the classifier sequence randomly. Of course, such ad hoc method cannot be guaranteed to work; to insist on an inappropriate classifier sequence may be harmful to the point where the classification performance of the induced system may fail to reach even that of plain *binary relevance*.

Sometimes, however, there is a way out. If the number of classes is manageable (say, five), the engineer may choose to experiment with several alternative sequences, and then choose the best one. But if the number of classes is greater, the necessity to try many alternatives will be impractical.

Error Propagation The fact that the classifiers are forced into a linear sequence makes them vulnerable to a phenomenon known as *error propagation*. Here is what it means. When a classifier misclassifies an example, the incorrect class label is passed on to the next classifier that uses this label as an additional attribute. An incorrect value of this additional attribute may then sway the next classifier to a wrong decision, which, too, is passed on, down the chain. In other words, an error of a single class may result in additional errors being made by subsequent classifiers. In this event, the *classifier chain* is likely to underperform. The thing to remember is that the overall error rate strongly depends on the quality of the earlier classifiers in the sequence.

One Last Comment The error-propagation phenomenon is less damaging if the classifiers do not strictly return either 0 or 1. Thus a Bayesian classifier calculates for each class its probability, a number from the interval [0, 1]. Propagating this probability through the chain is less harmful than the strict **pos** or **neg**. Similar considerations apply to some other classifiers such as those from the paradigm of neural networks (for instance, multilayer perceptions and radial-basis function networks).

What Have You Learned?

To make sure you understand the topic, try to answer the following questions. If needed, return to the appropriate place in the text.

- Discuss the typical impact that class interdependence can have on the performance of the *binary relevance* technique.
- Explain the principle of *classifier chain*. What can you say about the need to find a proper sequence of classifiers?
- Explain the problem of *error propagation* in classifier chains. Is there anything else to criticize about this approach?

13.4   Another Possibility: Stacking

In the light of the aforementioned drawbacks of classifier chains, some less risky alternative is needed. One possibility is to rely on *stacking*.

Architecture and Induction The essence is illustrated in Fig. 13.2. Here, the classifiers are arranged in two layers. The upper one represents plain *binary*

Fig. 13.2 The *stacking* technique. The upper-layer classifiers use as input the original attribute vector. For the lower-layer classifiers, this vector is extended by the vector of the class labels returned by the upper layer

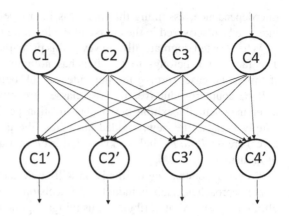

relevance (independently induced binary classifiers, one for each class). More interesting is the bottom layer. Here, the classifiers are induced from the training sets where the original attribute vectors are extended by the list of the class labels returned by the upper-layer classifiers. In the concrete case depicted in Fig. 13.2, each attribute vector is preceded by four new binary attributes (because there are four classes): the i-th attribute has value 1 if the i-th classifier in the upper layer has labeled the example as belonging to class i; otherwise, this attribute's value is 0.

Classification When the class labels of some future example $\mathbf{x}$ are needed, $\mathbf{x}$ is presented first to the upper-layer classifiers. After this, the obtained classes labels are added at the front of $\mathbf{x}$'s original attribute vector as N new binary attributes (assuming there are N classes), and the newly described example is presented in parallel to the lower-layer classifiers. Finally, $\mathbf{x}$ is labeled with the classes whose lower-layer classifiers have returned 1.

The underlying philosophy rests on the intuition that the performance of classifier C_i may improve if this classifier is informed about the "opinions" of the other classifiers—about the other classes to which $\mathbf{x}$ belongs.

An Example Consider an example that is described by a vector of four attributes with these values: $\mathbf{x} = \{a, f, r, z\}$. Suppose that the upper-layer classifiers return the following labels: $C_1 = 1, C_2 = 0, C_3 = 1, c_4 = 0$. In this event, the lower-layer classifiers are all presented with the following example description: $\mathbf{x} = \{1, 0, 1, 0, a, f, r, z\}$.

The classification behaviors of the lower-level classifiers can differ from those in the upper layer. For example, if the lower-layer classifiers return $1, 1, 1, 0$, the overall system will label $\mathbf{x}$ with C_1, C_2, and C_3, ignoring the original recommendations of the upper layer.

Some Comments Intuitively, this approach is more flexible than classifier chains because *stacking* makes it possible for any class to influence the recognition of any other class. The engineer does not provide any a priori information about class interdependence, assuming that the class interdependence (or the lack thereof) is likely to be discovered in the course of the learning process.

When treated dogmatically, however, this principle may do more harm than good. The fact that $\mathbf{x}$ belongs to C_i often has nothing to do with $\mathbf{x}$ belonging to C_j. If this is the case, forcing the dependence link between the two (as in Fig. 13.2) will be counterproductive. If most classes are mutually independent, the upper layer may actually exhibit better classification performance than the lower layer, simply because the newly added attributes (the classes obtained from the upper layer) are *irrelevant*—and we know that irrelevant attributes can impair the results of induction.

Proper understanding of this issue will guide the choice of the baseline learner. Some approaches, such as induction of decision trees, or WINNOW are capable of eliminating irrelevant attributes, thus mitigating the problem.

What Have You Learned?

To make sure you understand the topic, try to answer the following questions. If needed, return to the appropriate place in the text.

- How are interdependent classes addressed by the stacking approach? Discuss both the induction phase and the classification phase.
- In what situations will stacking outperform *binary relevance* and/or a *classifier chain*?
- Under what circumstances will you prefer *binary relevance* to stacking?

13.5 A Note on Hierarchically Ordered Classes

In some domains, the known class interdependence is more complicated than in the cases we have considered so far. Our machine-learning techniques then have to be modified accordingly.

An Example Figure 13.3 shows a small part of a class hierarchy that could have been suggested by a specialist preparing the data for machine learning. Each node in the graph represents one class.

The domain deals with the classification of text documents. The hierarchy is interpreted in a way reminiscent of decision trees. To begin with, some documents may belong to the class `machine learning`. A solid line emanating from the corresponding node represents "yes," and the dashed line represents "no." Among those documents that do belong to `machine-learning`, some deal with the topic `decision tree`, others with `k-NN classifiers`, and so on (for simplicity, most subclasses are omitted here).

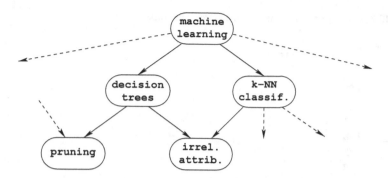

Fig. 13.3 Sometimes, the classes are hierarchically organized in a way known to the engineer in advance

In the picture, the relations are represented by arrows that point from parent nodes to child nodes. A node can have more than one parent, but a well-defined class hierarchy must avoid loops. The data structure defining class relations of this kind is known as a *directed acyclic graph*. In some applications, each node (except for the root node) has one and only one parent. This more constrained structure is known as a *generalization tree*.

Induction in Domains of This Kind   Induction of hierarchically ordered classes is organized in a way similar to *binary relevance*. For each node, the corresponding training set is constructed, and from this training set, the baseline learner induces a classifier. By doing so, the most common approach proceeds in a top-down manner where the output of the parent class instructs the choice of the examples for the induction of a child class.

Here is a way to carry this out in a domain where the classes are organized by a generalization tree. First, the entire original training set is used for the induction of the class located at the root of the tree. Next, the training set is divided into two parts, one containing training examples that belong to the root class, the other containing training examples that do *not* belong to this class. The lower-level classes are then induced only from the relevant training sets.

A Concrete Example   In the problem from Fig. 13.3, the first step is to induce a classifier for the class `machine learning`. Suppose that the original training set consists of the seven examples shown in Table 13.2. The labels of those examples are then used to decide which examples to include in the training sets for the induction of the child classes. For instance, note that only positive examples of `machine learning` are included in the training sets for `decision trees` and `k-NN classifiers`. Conversely, only negative examples of `machine learning` are included in the training set for the induction of `programming`.

Two Major Difficulties to Be Aware Of   The induction process is not as simple as it looks. The first problem complicating the task is, again, the phenomenon of *error propagation*. Suppose an example represents a text document from the field

Table 13.2 Illustration of a domain with hierarchically ordered classes

	Machine learning
ex_1	**pos**
ex_2	**pos**
ex_3	**pos**
ex_4	**pos**
ex_5	**neg**
ex_6	**neg**
ex_7	**neg**

Decision trees	
ex_1	1
ex_2	1
ex_3	0
ex_4	0

k-NN	
ex_1	0
ex_2	0
ex_3	1
ex_4	1

Programming	
ex_5	1
ex_6	0
ex_7	0

In some lower-level classes, the training sets contain only those training examples for which the parent classifier returned **pos**; in others, only those for which the parent classifier returned **neg**

circuit analysis. If this example is mistakenly classified as belonging to machine-learning, the classifier, misled by this information, will pass it on to the next classifiers, such as decision trees, thus potentially propagating the error down to lower levels.[2]

Another complication is that the training sets associated with the individual nodes in the hierarchy are almost always heavily *imbalanced*. Again, appropriate measures have to be taken—usually undersampling or oversampling.

Where Does the Class Hierarchy Come From? In some rare applications, the complete class hierarchy is available right from the start, having been created manually by the customer who has the requisite background knowledge about the concrete domain. This is the case of some well-known applications from the field of text categorization.

Caution is needed, though. Customers are not infallible, and the hierarchies they develop often miss important details. They may suffer from subjectivity—with consequences similar to those explained when we discussed classifier chains. In some domains, only parts of the hierarchy are known. In this event, the engineer has to find a way of incorporating this partial knowledge in the *binary relevance* framework discussed earlier.

---

[2]The reader has noticed that the issue is similar to the one we have encountered in the section dealing with classifier chains.

What Have You Learned?

To make sure you understand the topic, try to answer the following questions. If needed, return to the appropriate place in the text.

- Give an example of a domain where the individual classes are hierarchically ordered.
- Explain the training-set dividing principle for the induction of hierarchically ordered classifiers.
- What are the most commonly encountered difficulties in the induction of hierarchically ordered classes?

13.6 Aggregating the Classes

The situation is simpler if there are only a few classes. For instance, the number of all class-label combinations in a domain with three classes cannot exceed seven, assuming that each example is labeled with at least one class.

Creating New Training Sets In a domain of this kind, a sufficiently large training set is likely to contain a sufficient number of representatives for each class combination, and this makes it reasonable to treat each such combination as a separate class. The general approach is similar to those we have already seen: from the original training set, T, new training sets, T_i, are created, and from each, a classifier is induced by the baseline learner. However, do not forget that, in *class aggregation*, each T_i represents one combination of class labels, say, C_2 AND C_4.

When, in the future, some example $\mathbf{x}$ is to be classified, it is presented to all of these classifiers in parallel.

Illustration of the Principle The principle is illustrated in Table 13.3. Here, the total number of classes in the original training set, T, is three. Theoretically, the total number of class combinations should be seven. In reality, only five of these combinations are encountered in T because no example is labeled with C_3 alone, and no example is labeled with all three classes simultaneously. We therefore create five tables, each defining the training set for one class combination. Note that this approach deals only with those class combinations that have been found in the original training set. For instance, no future example will be labeled as belonging to C_3 alone. This may be seen as a limitation, and the engineer will have to find a way to address this issue in a concrete application.

Classification The programmer must not forget to specify what exactly is to be done in a situation where more than one of these "aggregated" classifiers returns 1. In some machine-learning paradigms, say, a Bayesian classifier, this is easy because the classifiers are capable of quantifying their confidence in the class

Table 13.3 In a domain with a manageable number of class-label combinations, it is often possible to treat each combination as a separate class

	Classes
ex_1	C_1, C_2
ex_2	C_2
ex_3	C_1, C_3
ex_4	C_2, C_3
ex_5	C_1

C_1			C_2			C_1 AND C_2			C_1 AND C_3			C_2 AND C_3	
ex_1	0		ex_1	0		ex_1	1		ex_1	0		ex_1	0
ex_2	0		ex_2	1		ex_2	0		ex_2	0		ex_2	0
ex_3	0		ex_3	0		ex_3	0		ex_3	1		ex_3	0
ex_4	0		ex_4	0		ex_4	0		ex_4	0		ex_4	1
ex_5	1		ex_5	0		ex_5	0		ex_5	0		ex_5	0

they recommend. If two or more classifiers return 1, the master classifier simply chooses the one with the highest confidence.

The choice is more complicated in the case of classifiers that only return 1 or 0 without offering any information about their confidence in the given decision. In principle, one may consider merging the sets of classes. For example, suppose that, for some example **x**, two classifiers return 1, and that one of the classifiers is associated with classes C_1, C_3, and C_4, and the other is associated with classes C_3 and C_5. In this event, **x** will be labeled with C_1, C_3, C_4, and C_5.

Note, however, that this may easily result in **x** being labeled with "too many" classes. The reader already knows that this may give rise to many false positives, and thus lead to low *precision*.

Alternative Ways of Aggregation In the approach illustrated in Table 13.3, the leftmost table (the one headed by C_1) contains only one positive label because there is only one training example in T labeled solely with this class. If we want to avoid having to deal with training sets that are so extremely imbalanced, we need a "trick" that would improve the class representations in T_i's.

Here is one possibility. In T_i, we will label with 1 each example whose set of class labels in the original T contains C_1. By doing so, we must not forget that ex_1 will thus be labeled as positive also in the table headed with (C_1 AND C_2).

Similarly, we will label with 1 all subsets of the set of classes found in a given training set. For instance, if an example is labeled with C_1, C_3, and C_4, we will label it with 1 in all training sets that represent nonempty subsets of $\{C_1, C_3, C_4\}$. This, of course, improves only training sets for relatively "small" combinations (combining, say, only one or two classes). For larger combinations, the problem persists.

A solution of the last resort will aggregate the classes only if the given combination is found in a sufficient percentage of the training examples. If the combination is rare, the corresponding T_i is not created. Although this means that the induced classifiers will not recognize a certain combination, this may not be such big loss if the combination is rare.

Some Criticism Class aggregation is not a good idea in domains where the number of class combinations is high, and the training set size is limited. If these two conditions are not satisfied, some of the newly created sets, T_i, are likely to contain no more than just a few positive examples, and as such will be ill-suited for machine learning: the training sets will be so *imbalanced* that all attempts to improve the situation by minority-class oversampling or majority-class undersampling are bound to fail—for instance, this will happen when a class combination is represented by just a single example.

As a rule of thumb, in domains with a great number of different class labels, where many combinations occur only rarely and some do not occur at all, the engineer will prefer plain *binary relevance* or some of its variations (chaining or stacking). Class aggregation is then to be avoided.

What Have You Learned?

To make sure you understand the topic, try to answer the following questions. If needed, return to the appropriate place in the text.

- Describe the principle of *class aggregation*. Explain separately the induction process and the way the induced classifiers are used to classify future examples.
- What possible variations on the class-aggregation theme do you know?
- What main shortcoming can render this approach impractical in many realistic applications?

13.7 Criteria for Performance Evaluation

We have mentioned earlier that performance evaluation in multi-label domains depends on averaging the results across classes. Let us introduce and briefly discuss some commonly used ways of doing so.

Macro-Averaging The simplest approach, *macro*-averaging, finds the values of the given criterion for each class separately, and then calculates their arithmetic average. Let L be the total number of classes. Here are the formulas that calculate macro-*precision*, macro-*recall*, and macro-F_1 from the values of these quantities for the individual classes:

$$Pr^M = \frac{1}{L} \sum_{i=1}^{L} pr_i$$

$$Re^M = \frac{1}{L} \sum_{i=1}^{L} re_i \qquad (13.1)$$

$$F_1^M = \frac{1}{L} \sum_{i=1}^{L} F_{1i}$$

Macro-averaging is suitable in domains where each class has approximately the same number of representatives. In some applications, this requirement is not satisfied, but the engineer may still prefer macro-averaging if he or she considers each class to be equally important, regardless of its proportional representation in the training set.

Micro-Averaging In the other approach, *micro*-averaging, each class is weighed according to its frequency in the given set of examples. In other words, the performance is averaged over all examples. Let L be the total number of classes. Here are the formulas for micro-*precision*, micro-*recall*, and micro-F_1:

$$Pr^\mu = \frac{\sum_{i=1}^{L} N_{TP_i}}{\sum_{i=1}^{L} (N_{TP_i} + N_{FP_i})}$$

$$Re^\mu = \frac{\sum_{i=1}^{L} N_{TP_i}}{\sum_{i=1}^{L} (N_{TP_i} + N_{FN_i})} \qquad (13.2)$$

$$F_1^\mu = \frac{2 \times Pr^\mu \times Re^\mu}{Pr^\mu + Re^\mu}$$

Note that F_1^μ is calculated from micro-*precision* and micro-*recall* and not from the observed classifications of the individual examples.

Micro-averaging is preferred in applications where the individual classes cannot be treated equally. For instance, the engineer may reason that good performance on dominant classes is not really compromised by poor performance on classes that are too rare to be of any importance.

A Numeric Example Let us illustrate these formulas using Table 13.4. Here, we can see five examples. For each of them, the middle column lists the correct class labels, and the rightmost column gives the labels returned by the classifier. The reader can see minor discrepancies in the sense that the classifier has missed some classes (causing false negatives). For instance, this is the case of class C_3 being missed in example ex_3. At the same time, the classifier labels some examples with incorrect class labels, which constitutes false positives. For instance, this is the case of example ex_1 being labeled with class C_3.

Table 13.4 Illustration of performance evaluation in multi-label domains

The following table gives, for five testing examples, the known class labels versus the class labels returned by the classifier.

	True classes	Classifier's classes
ex$_1$	C_1, C_2	$C_1, C_2, C_3,$
ex$_2$	C_2	$C_2, C_4,$
ex$_3$	C_1, C_3, C_5	$C_1, C_5,$
ex$_4$	C_2, C_3	$C_2, C_3,$
ex$_5$	C_2, C_4	$C_2, C_5,$

Separately for each class, here are the values of true positives, false positives, and false negatives. Next to them are the corresponding values for precision and recall, again separately for each class.

$$N_{TP_1} = 2 \quad N_{FP_1} = 0 \quad N_{FN_1} = 0 \quad Pr_1 = \frac{2}{2+0} = 1 \quad Re_1 = \frac{2}{2+0} = 1$$
$$N_{TP_2} = 4 \quad N_{FP_2} = 0 \quad N_{FN_2} = 0 \quad Pr_2 = \frac{4}{4+0} = 1 \quad Re_2 = \frac{4}{4+0} = 1$$
$$N_{TP_3} = 1 \quad N_{FP_3} = 1 \quad N_{FN_3} = 1 \quad Pr_3 = \frac{1}{1+1} = 0.5 \; Re_3 = \frac{1}{1+1} = 0.5$$
$$N_{TP_4} = 0 \quad N_{FP_4} = 1 \quad N_{FN_4} = 1 \quad Pr_4 = \frac{0}{0+1} = 0 \quad Re_4 = \frac{0}{0+1} = 0$$
$$N_{TP_5} = 1 \quad N_{FP_5} = 1 \quad N_{FN_5} = 0 \quad Pr_5 = \frac{1}{1+1} = 0.5 \; Re_5 = \frac{1}{1+0} = 1$$

This is how the macro-averages are calculated:

$$Pr^M = \frac{1+1+0.5+0+0.5}{5} = 0.6$$

$$Re^M = \frac{1+1+0.5+0+1}{5} = 0.7$$

Here is how the micro-averages are calculated.

$$Pr^\mu = \frac{2+4+1+0+1}{(2+0)+(4+0)+(1+1)+(0+1)+(1+1)} = 0.73$$

$$Re^\mu = \frac{2+4+1+0+1}{(2+0)+(4+0)+(1+1)+(0+1)+(1+0)} = 0.8$$

These discrepancies are then reflected in the numbers of true positives, false positives, and false negatives. These, in turn, make it possible to calculate for each class its *precision* and *recall*. After this, the table shows the calculations of the macro- and micro-averages of these two criteria.

Averaging the Performance over Examples So far, the true and false positive and negative examples were counted across individual classes. However, in domains where an average example belongs to a great many classes, it can make sense to average over the individual examples.

The procedure is in principle the same as before. When comparing the true class labels with those returned for each example by the classifier, we obtain the numbers of true positives, false positives, and false negatives. From these, we easily obtain the macro-averages and micro-averages. The only thing we must keep in mind is that the average is not taken over the classes, but over examples—thus in macro-averages, we divide the sum by the number of examples, not by the number of classes.

What Have You Learned?

To make sure you understand this topic, try to answer the following questions. If you have problems, return to the corresponding place in the preceding text.

- Give the formulas for *macro*-averaging of *precision*, *recall*, and F_1.
- Give the formulas for *micro*-averaging of *precision*, *recall*, and F_1. Discuss the difference between *macro*-averaging and *micro*-averaging.
- What is meant by "averaging the performance over examples"?

13.8 Summary and Historical Remarks

- In some domains (such as text categorization), each example can be labeled with more than one class at the same time. These are so-called *multi-label* domains.
- In domains of this kind, classical machine-learning paradigms can sometimes be used. Unless special precautions have been taken, however, the results are rarely encouraging. For some paradigms, multi-label versions exists, but these are too advanced for an introductory text, especially in view of the fact that good results can be achieved with simpler means.
- The most common approach to *multi-label domains* induces a binary classifier for each class separately, and then submits the example to all these classifiers in parallel. This is called the *binary relevance* technique.
- What the basic version of *binary relevance* seems to neglect is the fact that the individual classes may not be independent of each other. The fact that an example has been identified as a representative of class C_A may strengthen or weaken its chances of belonging also to class C_B.
- The simplest mechanism for dealing with class interdependence in multi-label domains is the *classifier chain*. Here, the output of one binary classifier is used as an additional attribute describing the example to be presented to the next classifier in line.

- One weakness of classifier chains is that the user is expected to specify the sequence of classes (perhaps according to class subsumption). If the sequence is poorly designed, the results are disappointing.
- Another shortcoming is known as *error propagation*: an incorrect label given to an example by one classifier is passed on to the next classifier in the chain, potentially misleading it.
- A safer approach relies on the two-layered *stacking* principle. The upper-layer classifiers are induced from examples described by the original attribute vectors, and the lower-layer classifiers are induced from examples described by attribute vectors to which the class labels obtained in the upper layer have been added. When classifying an example, the outcomes of the lower-layer classifiers are used.
- Sometimes, it is possible to take advantage of known hierarchical order among the classes. Here, too, induction is carried out based on specially designed training sets. Again, the user has to be aware of the dangers of error propagation.
- Yet another possibility is to resort to *class aggregation* where each combination of classes is treated as a separate higher-level class. A special auxiliary training set is created for each of these higher-level classes.
- The engineer has to pay attention to ways of measuring the quality of the induced classifiers. Observing that each class may experience different classification performance, we need mechanisms for averaging over the classes (or examples). Two of them are currently popular: micro-averaging and macro-averaging.

Historical Remarks The problem of multi-label classification is relatively new. The first time it was encountered was in the field of text categorization—see McCallum [57]. The simplest approach, the *binary relevance* principle, was employed by Boutell et al. [7]. A successful application of classifier chains was reported by Read et al. [79], whereas Goldpole and Sarawagi [32] are credited with having developed the *stacking* approach. Apart from the approaches related to *binary relevance*, some authors have studied ways of modifying classical single-label paradigms. The ideas on nearest-neighbor classifiers in multi-label domains are borrowed from Zhang and Zhou [101] (whose technique, however, is much more sophisticated than the one described in this chapter). Induction of hierarchically ordered classes was first addressed by Koller and Sahami [47]. Multi-label decision trees were developed by Clare and King [15].

13.9 Solidify Your Knowledge

The exercises are to solidify the acquired knowledge. The suggested thought experiments will help the reader see this chapter's ideas in a different light and provoke independent thinking. Computer assignments will force the readers to pay attention to seemingly insignificant details they might otherwise overlook.

Table 13.5 An example of a multi-label domain

	True classes	Classifier's classes
ex_1	C_1	C_1, C_2
ex_2	C_1, C_2	C_1, C_2
ex_3	C_1, C_3	C_1
ex_4	C_2, C_3	C_2, C_3
ex_5	C_2	C_2
ex_6	C_1	C_1, C_2

Exercises

1. Consider the multi-label training set shown in the left part of Table 13.5. Show how the auxiliary training sets will be created when the principle of *binary relevance* is to be used.
2. For the same training set, create the auxiliary training sets for the approach known as *class aggregation*. How many such sets will we need?
3. Draw the schema showing how the problem from Table 13.5 would be addressed by *stacking*. Suppose the examples in the original training set are described by ten attributes. How many attributes will the lower-level classifiers have to use?
4. Suggest the classifier-chain schema for a domain with the following four classes: decision trees, machine learning, classification, pruning.
5. Returning to the set of examples from Table 13.5, suppose that a classifier has labeled them as indicated in the rightmost column. Calculate the macro- and micro-averages of *precision* and *recall*.

Give It Some Thought

1. Suggest a multi-label domain where the principle of *classifier chain* can be a reasonable strategy to follow. What would be the main requirement for such data?
2. Consider a domain where the majority of training examples are labeled each with only a single class, and only a small subset of the examples (say, 5%) are labeled with more than one class. Suggest a machine learning approach to induce reliable classifiers from such data.
3. Suppose that you have a reason to assume that a few classes are marked by strong interdependence while most of the remaining classes are mutually independent. You are thinking of using the *stacking* approach. What is the main problem that might compromise the performance of the induced classifiers? Can you suggest a mechanism that overcomes this pitfall?

4. Suppose you have been asked to develop machine-learning software for induction from multi-label examples. This chapter has described at least four approaches that you can choose from. Write down the main thoughts that would guide your decision.
5. Suggest a mechanism that would mitigate the problem of *error propagation* during multi-label induction with hierarchically ordered classes. Hint: after a testing run, consider "enriching" the training sets by "problematic" examples.

Computer Assignments

1. Write a program that accepts as input a training set of multi-label examples, and returns as output the set of auxiliary training sets needed for the *binary relevance* approach.
2. Write a program that converts the training set from the previous question into auxiliary training sets, following the principle of class aggregation.
3. Search the web for machine-learning benchmark domains that contain multi-label examples. Convert them using the data-processing program from the previous question, and then induce the classifiers by the *binary relevance* approach.
4. Write a program that first induces the classifiers using *binary relevance* as in the previous question. In the next step, the program redescribes the training examples by adding to their attribute vectors the class labels as required by the lower layer in the *classifier stacking* technique.
5. What data structures would you use for the input and output data when implementing the *classifier stacking* technique?
6. Write a program that takes as input the values of $N_{TP}, N_{TN}, N_{FP}, N_{FN}$ for each class, and returns micro- and macro-averaged *precision, recall*, and F_1.

Chapter 14
Unsupervised Learning

It would be a mistake to think that machine learning always requires examples with class labels. Far from it! Useful information can be gleaned even from examples whose classes are not known. This is sometimes called *unsupervised learning*, in contrast to the term *supervised learning* which is used when talking about induction from pre-classified examples.

While supervised learning focuses on induction of classifiers, unsupervised learning is interested in discovering useful properties of available data. Perhaps the most popular task looks for groups (called clusters) of similar examples. The centroids of these groups can then be used as gaussian centers for Bayesian or RBF classifiers, as predictors of unknown attribute values, and even as visualization tools for multidimensional data. Last but not least, techniques used in unsupervised learning can be used to create higher-level attributes from existing ones.

The chapter describes some practical techniques for unsupervised learning, explaining the basic algorithms, their behaviors in practical circumstances, and the benefits they offer.

14.1 Cluster Analysis

The fundamental task in unsupervised learning is *cluster analysis*. Here, the input is a set of examples, each described by a vector of attribute values—but no class labels. The output is a set of two or more clusters of examples.

Identifying Groups of Similar Examples Figure 14.1 shows a simple domain with a few examples described by two attributes: weight and height. An observer can easily see that the examples form three or four groups, depending on the subjective "level of resolution."

© Springer International Publishing AG 2017
M. Kubat, *An Introduction to Machine Learning*,
DOI 10.1007/978-3-319-63913-0_14

Fig. 14.1 A two-dimensional domain with clusters of examples

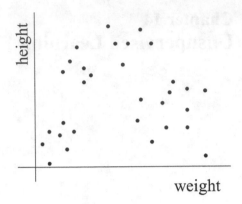

Visual identification of such groups in a two-dimensional space is easy, but in four or more dimensions, humans can neither visualize the data nor see the clusters. These can only be detected by cluster-analysis algorithms.

Representing Clusters by Centroids To begin with, we have to decide how the clusters are to be described. A few alternatives can be considered: it is possible to specify the clusters' locations, sizes, boundaries, and perhaps some other aspects. But the simplest approach relies on *centroids*.[1] If all attributes are numeric, the centroid is identified with the averages of the individual attributes. For instance, suppose a two-dimensional cluster consists of the following examples: $(2, 5), (1, 4), (3, 6)$. In this case, the centroid is described by vector $(2, 5)$ because the first attribute's average is $\frac{2+1+3}{3} = 2$ and the second attribute's average is $\frac{5+4+6}{3} = 5$.

The averages can be calculated even when the attributes are discrete if we know how to turn them into numeric ones. Here is a simple way of doing so. If the attribute can acquire three or more different values, we can replace each attribute-value pair with one boolean variable (say, `season=fall`, `season=winter`, etc.). The values of the boolean attributes are then represented by 0 or 1 instead of *false* and *true*, respectively.

What Should the Clusters Be Like? Clusters should not overlap each other: each example must belong to one and only one cluster. Within the same cluster, the examples should be relatively close to each other, certainly much closer than to the examples from the other clusters.

An important question will ask how many clusters the data contain. In Fig. 14.1, we noticed that the human observer discerns either three or four clusters. However, the scope of existing options is not limited to these two possibilities. At one extreme, the entire training set can be thought of as forming one big cluster; at the other,

---

[1]Machine learning professionals sometimes avoid the term "center" which might imply mathematical properties that are for the specific needs of cluster analysis largely irrelevant.

each example can be seen as representing its own single-example cluster. Practical implementations often side-step the problem by asking the user to supply the number of clusters by way of an input parameter. Sometimes, however, machine-learning software is expected to determine the number automatically.

Problems with Measuring Distances   Algorithms for cluster analysis usually need a mechanism to evaluate the distance between an example and a cluster. If the cluster is described by its centroid, the Euclidean distance between the two vectors seems to offer a good way of doing so—assuming that we are aware of situations where this can be misleading. This last statement calls for an explanation.

Euclidean distance may be inconvenient in the case of discrete attributes, but we already know how to deal with them. More importantly, we must not forget that each attribute is likely to represent a different quantity, which renders the use geometric distance rather arbitrary: a 4-year difference in age is hard to compare with a 4-foot difference in height. Also the problem of scaling plays its role: if we replace feet with miles, the distances will change considerably.

We have encountered these problems earlier, in Chap. 3. In the context of cluster analysis, however, these issues tend to be less serious than in k-NN classifiers. Most of the time, engineers get around the difficulties by normalizing all attribute values into the unit interval, $x_i \in [0, 1]$. We will return to normalization in Sect. 14.2.

A More General Formula for Distances   If the examples are described by a mixture of numeric and discrete attributes, we can rely on the sum of the squared distances along corresponding attributes. More specifically, the following expression is recommended (denoting by n the number of attributes):

$$d_M(\mathbf{x}, \mathbf{y}) = \sqrt{\Sigma_{i-1}^n d(x_i, y_i)} \tag{14.1}$$

In this formula, we will use $d(x_i, y_i) = (x_i - y_i)^2$ for continuous attributes. For discrete attributes (including boolean attributes), we put $d(x_i, y_i) = 0$ if $x_i = y_i$ and $d(x_i, y_i) = 1$ if $x_i \neq y_i$.

Which Cluster Should an Example Belong To?   Let us suppose that each example is described by an attribute vector, $\mathbf{x}$, and that each cluster is defined by its centroid—which, too, is an attribute vector.

Suppose there are N clusters whose centroids are denoted by $\mathbf{c}_i$, where $i \in (1, N)$. The example $\mathbf{x}$ has a certain distance $d(\mathbf{x}, \mathbf{c}_i)$, from each centroid. If $d(\mathbf{x}, \mathbf{c}_p)$ is the smallest of these distances, it is natural to expect that $\mathbf{x}$ be placed in cluster $\mathbf{c}_p$.

For instance, suppose that we use the Euclidean distance, and that there are three clusters. If the centroids are $\mathbf{c}_1 = (3, 4)$, $\mathbf{c}_2 = (4, 6)$ and $\mathbf{c}_3 = (5, 7)$, and if the example is $\mathbf{x} = (4, 4)$, then the Euclidean distances are $d(\mathbf{x}, \mathbf{c}_1) = 1$, $d(\mathbf{x}, \mathbf{c}_2) = 2$, and $d(\mathbf{x}, \mathbf{c}_3) = \sqrt{10}$. Since $d(\mathbf{x}, \mathbf{c}_1)$ is the smallest of the three values, we conclude that $\mathbf{x}$ should belong to $\mathbf{c}_1$.

Benefit 1: Estimating Missing Values   The knowledge of the clusters can help us estimate missing attribute values. Returning to Fig. 14.1, the reader will notice that if weight is low, the example is bound to belong to the bottom-left cluster. In this case, also height is likely to be low because it is low in all examples found in this cluster. This aspect tends to be even more strongly pronounced in realistic data described by multiple attributes.

In example descriptions, some attribute values are sometimes unknown. As a simple way of dealing with this issue, Sect. 10.4 suggested that the missing value be estimated as the average or the most frequent value encountered in the training set. However, an estimate of the missing value as the average or the most frequent value of the given cluster is sounder because it uses more information about the nature of the domain.

Benefit 2: Reducing the Size of RBF Networks and Bayesian Classifiers   Cluster analysis can assist such techniques as Bayesian learners and radial-basis-function networks. The reader will recall that these paradigms operate with centers.[2] In the simplest implementation, the centers are identified with the attribute vectors of the individual examples. In domains with millions of examples, however, this would lead to impractically big classifiers. The engineer then prefers to divide the training set into N clusters, and to identify the gaussian centers with the centroids of the clusters.

Benefit 3: A Simple Classifier   Finally, the knowledge of data clusters may be useful in supervised learning. It is quite common that all (or almost all) examples in a cluster belong to the same class. In that case, the developer of a supervised-learning software may decide first to identify the clusters, and then label each cluster with its dominant class.

What Have You Learned?

To make sure you understand the topic, try to answer the following questions. If needed, return to the appropriate place in the text.

- What is the main difference between unsupervised learning and supervised learning?
- What is a cluster of examples? Define some mechanisms to describe the clusters. How can we measure the distance between an example and a cluster?
- Summarize the main benefits of cluster analysis.

---

[2]For instance, the section on RBF networks denoted these centers by μ_i's.

14.2 A Simple Algorithm: k-Means

Perhaps the simplest algorithm to detect clusters of data is known under the name k-means. The "k" in the name denotes the requested number of clusters—a parameter whose value is supplied by the user.

The Outline of the Algorithm The pseudocode of the algorithm is provided in Table 14.1. The first step creates k initial clusters such that each example finds itself in one and only one cluster. After this, the coordinates of all centroids are calculated. Let us note that the problem of initialization is somewhat more complicated than that. We will return to this issue presently.

In the next step, k-means investigates one example at a time, calculating its distances from all centroids. The nearest centroid then defines the cluster to which the example should belong. If the example already *is* that cluster, nothing needs to be done; otherwise, the example is transferred from the current (wrong) cluster to the right one. After the relocation, the centroids of the two affected clusters (the one that lost the example, and the one that gained it) have to be recalculated. The procedure is graphically illustrated by the single-attribute domain from Fig. 14.2. Here, two examples find themselves in the wrong cluster and are therefore relocated. Note how, after the example relocation, the vertical bars (separating the clusters), and also the centroids, change their locations.

Termination The good thing about the algorithm described in Table 14.1 is that the process is guaranteed to reach a situation where each example finds itself in the nearest cluster so that, from this moment on, no further transfers are needed. The clusters do not overlap. Since this is usually achieved in a manageable number of steps, no sophisticated termination criterion is needed here.

Table 14.1 The clustering algorithm k-means

Input: a set of examples without class labels
 user-set constant k

---

1. Create k initial clusters. For each, calculate the coordinates of its centroid, C_i, as the numeric averages of the attribute values in the examples it contains.
2. Choose an example, $\mathbf{x}$, and find its distances from all centroids. Let j be the index of the *nearest centroid.*
3. If $\mathbf{x}$ already finds itself in the j-th cluster, do nothing. Otherwise, move $\mathbf{x}$ from its current cluster to the j-th cluster and recalculate the centroids.
4. Unless a stopping criterion has been satisfied, repeat the last two steps for another example.

Stopping criterion: each training example already finds itself in the nearest cluster.

Let us consider an almost trivial domain where 13 examples are described by a single numeric attribute. Suppose the examples have been initially divided into the three groups indicated here by the *vertical bars*. The following sequence shows how two examples (marked by *circles*) are moved from one cluster to another.

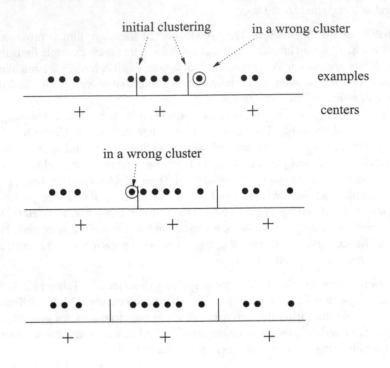

now: all examples in correct clusters

After the second transfer, the clusters are perfect and the calculations can stop.

Fig. 14.2 Illustration of the *k-means* procedure in a domain with one numeric attribute

Numeric Example In Table 14.2, a set of nine two-dimensional examples has been randomly divided into three groups (because the user specified $k = 3$), each containing the same number of examples. The table also provides the centroids for each group. *k-means* goes through these examples systematically, one by one—in this concrete case, starting with group-2. For each example, its distance from each centroid is calculated. It turns out that the first example from group-2 already finds itself in the right cluster. However, the second example is closer to group-1 than to group-2 and, for this reason, has to be transferred from its original cluster to group-1. After this, the affected centroids are recalculated.

Table 14.2 Illustration of the *k-means* procedure in a domain with two attributes

The table below contains three initial groups of vectors. The task is to find "ideal" clusters using the *k-means* ($k = 3$).

	Group-1	Group-2	Group-3
	(2, 5)	(4, 3)	(1, 5)
	(1, 4)	(3, 7)	(3, 1)
	(3, 6)	(2, 2)	(2, 3)
Centroids:	(2, 5)	(3, 4)	(2, 3)

Let us pick the first example in group-2. The Euclidean distances between this example, (4, 3), and the centroids of the three groups are $\sqrt{8}$, $\sqrt{2}$, and $\sqrt{4}$, respectively. This means that the centroid of group-2 is the one nearest to the example. Since this is where the example already is, *k-means* does not do anything.

Let us now proceed to the second example in group-2, (3, 7). In this case, the distances are $\sqrt{5}$, $\sqrt{9}$, and $\sqrt{17}$, respectively. Since the centroid of group-1 has the smallest distance, the example is moved from group-2 to group-1. After this, the averages of the two affected groups are recalculated.

Here are the new clusters:

	Group-1	Group-2	Group-3
	(2, 5)	(4, 3)	(1, 5)
	(1, 4)	(2, 2)	(3, 1)
	(3, 6)		(2, 3)
	(3, 7)		
Averages:	(2.25, 5.25)	(3, 2.5)	(2, 3)

The process continues as long as any example transfers are needed.

The Need for Normalization The reader will recall that, when discussing *k*-NN classifiers, Chap. 3 argued that inappropriate attribute scaling will distort the distances between attribute vectors. The same concern can be raised in cluster analysis. It is therefore always a good idea to normalize the vectors so that all numeric attributes have values from the same range, say, from 0 to 1.

The simplest way of doing so is to determine for the given attribute its maximum (*MAX*) and minimum (*MIN*) value in the training set. Then, each value of this attribute is re-calculated using the following formula:

$$x = \frac{x - MIN}{MAX - MIN} \tag{14.2}$$

As for boolean attributes, their values can simply be replaced with 1 and 0 for *true* and *false*, respectively. Finally, an attribute that acquires *n* discrete values (such as season, which has four different values) can be replaced with *n* boolean attributes, one for each value—and, again, for the values of these boolean attributes, 1 or 0 are used.

Computational Aspects of Initialization   To reach its goal, the *k-means* needs to go through a certain number of transfers of examples from wrong clusters to the right clusters. How many such transfers are needed depends on the contents of the initial clusters. At least in theory, it *can* happen that the randomly created initial clusters are already perfect, and not a single example needs to be moved. This, of course, is an extreme, but the reader surely understands that if the initialization is "lucky," fewer relocations have to be carried out than otherwise. Initialization matters in the sense that a better starting point ensures that the solution is found sooner.

How to Initialize   In some domains, we can take advantage of some background knowledge about the problem at hand. For instance, seeking to create initial clusters in a database of a company's employees, the data analyst may speculate that it makes sense to group them by their age, salary, or some other intuitive criterion, and that the groups thus obtained will be good initial clusters.

In other applications, however, no such guidelines exist. The simplest procedure then picks *k* random training examples and regards them as *code vectors* to define initial centroids. The initial clusters are then created by associating each of the examples with its nearest code vector.

A More Serious Problem with Initialization   There is another issue, and a more serious one than the computational costs. The thing is, also the *composition* of the resulting clusters (once the *k-means* has completed its work) may depend on initialization. Choose a different set of initial code vectors, and the technique may generate a different set of clusters.

The point is illustrated in Fig. 14.3. Suppose that the user wants two clusters ($k = 2$). If he chooses as code vectors the examples denoted by **x** and **y** then the initial clusters created with the help of these two examples are already perfect.

Fig. 14.3 Suppose $k = 2$. If the code vectors are [**x**,**y**], the initial clusters for *k*-means will be different than when the code vectors are [**x**,**z**]

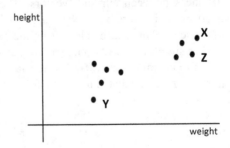

The situation changes when we choose for the code vectors examples **x** and **z**. In this event, the two initial clusters will have a very different composition, and *k-means* is likely to converge on a different set of clusters. The phenomenon will be more pronounced if there are "outliers," examples that do not apparently belong to any of the two clusters.

Summary of the Main Problems   The good thing about *k-means* is that it is easy to explain and easy to implement. Yet this simplicity comes at a price. The technique is sensitive to initialization; the user is expected to provide the number of clusters (though he may not know how many clusters there are); and, as we will see, some clusters can never be identified, in this manner. The next sections will take a look at some techniques to overcome these shortcomings.

What Have You Learned?

To make sure you understand the topic, try to answer the following questions. If needed, return to the appropriate place in the text.

- Describe the principle of the *k-means* algorithm. What is the termination criterion?
- What would be the consequence if we did not normalize the training set? Write down the simple normalization formula.
- Describe some methods for the initialization of *k-means*. What are the main consequences of good or bad initialization?

14.3   More Advanced Versions of *k*-Means

The previous section described the baseline version of *k-means* and explained its main weaknesses: sensitivity to initialization, and lack of flexibility in deciding about the number of clusters. Let us now take a look at some improvements meant to address these shortcomings.

Quality of the Clusters   An engineer always needs to be able to evaluate and compare alternative solutions. And thus also for the topic addressed by this section, we need a criterion to help us choose between different sets of clusters. Primarily, the criterion should reflect the fact that we want clusters that minimize the average distance between examples and the centroids of "their" clusters.

Let us denote by $d(\mathbf{x}, \mathbf{c})$ the distance between example **x** and the centroid, **c**, of the cluster to which **x** belongs. If all attributes are numeric, and if they all have been normalized, then $d(\mathbf{x}, \mathbf{c})$ can be evaluated either by the Euclidean distance or by the more general Eq. (14.1).

The following formula (in which SD stands for *summed distances*) sums up the distances of all examples from their clusters' centroids. Here, $\mathbf{x}_i^{(j)}$ denotes the i-th example in the j-th cluster, K is the number of clusters, n_j is the number of examples in the j-th cluster, and $\mathbf{c}_j$ is the j-th cluster's centroid.

$$SD = \sum_{j=1}^{K} \sum_{i=1}^{n_j} d(\mathbf{x}_i^{(j)}, \mathbf{c}_j) \qquad (14.3)$$

In cluster analysis, we seek to minimize SD. When calculating this quantity, we must not forget that the value obtained by Eq. (14.3) will go down if we increase the number of clusters (and thus decrease their average size), reaching $SD = 0$ in the extreme case when each cluster is identified with one and only one training example. The formula is therefore useful only if we compare solutions that have similar numbers of clusters.

Using Alternative Initializations Knowing that the composition of the resulting clusters depends on the algorithm's initialization, we can suggest a simple improvement. We will define two or more sets of initial code vectors, and apply *k-means* separately to each of them. After this, we will evaluate the quality of all the alternative data partitionings thus obtained, using the criterion defined by Eq. (14.3). The best solution is the one for which we get the lowest value. This solution is then retained, and the others discarded.

Experimenting with Different Values of k One obvious weakness of *k-means* is the requirement that the user should provide the value of k. This is easier said than done because, more often than not, the engineer has no idea into how many clusters the available data naturally divide. Unless more sophisticated techniques are used (about these, see later), the only way out is to try a few different values, and then pick the best according to an appropriate criterion (such as the one defined in Eq. (14.3)). As we already know, the shortcoming of this criterion is that it tends to give preference to small clusters. For this reason, data analysts often normalize the value of SD by k, the number of clusters.

Post-processing: Merging and Splitting Clusters The quality of the set of clusters created by *k-means* can often be improved by post-processing techniques that either increase the number of clusters by splitting, or decrease the number by merging.

As for merging, two neighboring clusters will be merged if their mutual distance is small. To find out whether the distance merits the merging, we simply calculate the distance of two centroids, and then compare it with the average cluster-to-cluster distance calculated by the following sum, where $\mathbf{c}_i$ and $\mathbf{c}_j$ are centroids:

$$S = \sum_{i \neq j} d(\mathbf{c}_i, \mathbf{c}_j) \qquad (14.4)$$

Conversely, splitting makes sense when the average example-to-example distance within some cluster is high. The concrete solution is not easy to formalize because once we have specified that cluster C is to be split into C_1 and C_2, we need to decide which of C's examples will go to the first cluster and which to the second. Very often, however, it is perfectly acceptable to identify in C two examples with the greatest mutual distance, and then treat them as the code vectors of newly created C_1 and C_2, respectively.

Hierarchical Application of k-Means Another modification relies on recursive calls. The technique begins with running k-means for $k = 2$, obtaining two clusters. After this, k-means is applied to each of these two clusters separately, again with $k = 2$. The process is continued until an appropriate termination criterion has been satisfied— for instance, the maximum number of clusters the user wants to obtain, or the minimum distance between neighboring clusters.

This hierarchical version side-steps one serious shortcoming of k-means: the necessity to provide the number of clusters because, in this case, this is found automatically. This is particularly useful when the goal is to identify the gaussian centers in Bayesian classifiers or in RBF networks.

What Have You Learned?

To make sure you understand the topic, try to answer the following questions. If needed, return to the appropriate place in the text.

- Discuss some shortcomings of the k-means algorithm, and describe the simple techniques for overcoming them.
- Specify how you would implement the cluster-splitting and cluster merging techniques described in this section.
- Explain the principle of hierarchical application of k-means.

14.4 Hierarchical Aggregation

Just as any other machine-learning technique, k-means has not only advantages, but also shortcomings. To avoid at least some of the latter, we need an alternative, an approach which is likely to prove useful in situations where k-means fails.

Another Serious Limitation of k-Means By minimizing the distance between the examples and the centroids of the clusters to which these examples belong, k-means essentially guarantees that it will discover "convex" clusters. Most of the time, this is indeed what we want. Such clusters can be useful in the context of Bayesian classifiers and the RBF networks where they reduce, sometimes very significantly, the number of employed gaussian centers.

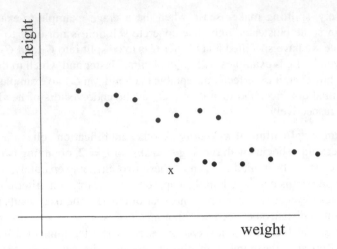

Fig. 14.4 Note that the leftmost examples in the "bottom" cluster are closer to the "upper" cluster's centroid than to its own. In a domain of this kind *k-means* will not find the best solution

This approach, however, will do a poor job if the clusters are of a different ("non-convex") nature. To see the point, consider the clusters in Fig. 14.4. Here, the leftmost example, **x**, in the bottom cluster is closer to the centroid of the upper cluster, and *k-means* would therefore relocate it accordingly—and yet we feel that this would not do justice to the nature of the two groups.

To deal with data of this kind, we need another technique, one capable of identifying clusters such as those in Fig. 14.4.

An Alternative Way of Measuring Inter-Cluster Distance In the previous section, the distance between two clusters was evaluated as the Euclidean distance between their centroids. However, for the needs of the approach described below, we will suggest another mechanism: we will measure the distances between all pairs of examples, [**x**,**y**], such that **x** comes from the first cluster and **y** from the second. The smallest value found among all these example-to-example distances then defines the distance between the two clusters.

Returning to Fig. 14.4, the reader will agree that, along this new distance metric, example **x** is closer to the bottom cluster than to the upper clusters. This means that the limitation mentioned in the previous paragraph has been in this particular case eliminated. The price for this improvement is increased computational costs: if N_A is the number of examples in the first cluster, and N_B the number of examples in the second, then $N_A \times N_B$ example-to-example distances have to be evaluated. Most of the time, however, the clusters are not going to so big as to make this an issue.

Numeric Example For the sake of illustration of how this new distance metric is calculated, consider the following two clusters, A and B.

A	B
$\mathbf{x}_1 = (1,0)$	$\mathbf{y}_1 = (3,3)$
$\mathbf{x}_2 = (2,2)$	$\mathbf{y}_2 = (4,4)$

Table 14.3 The basic algorithm of *hierarchical aggregation*

Input: a set of examples without labels

1. Let each example form one cluster. For N examples, this means creating N clusters, each containing a single example.
2. Find a pair of clusters with the smallest cluster-to-cluster distance. Merge the two clusters into one, thus reducing the total number of clusters to $N - 1$.
3. Unless a termination criterion is satisfied, repeat the previous step.

Using the Euclidean formula, we calculate the individual example-to-example distances as follows:

$$d(\mathbf{x}_1, \mathbf{y}_1) = \sqrt{13},$$
$$d(\mathbf{x}_1, \mathbf{y}_2) = \sqrt{25},$$
$$d(\mathbf{x}_2, \mathbf{y}_1) = \sqrt{2},$$
$$d(\mathbf{x}_2, \mathbf{y}_2) = \sqrt{8}.$$

Observing that the smallest of these values is $d(\mathbf{x}_2, \mathbf{y}_1) = \sqrt{2}$, we conclude that the distance between the two clusters is $d(A, B) = \sqrt{2}$.

Hierarchical Aggregation For domains such as the one in Fig. 14.4, the clustering technique known as *hierarchical aggregation* is recommended. The principle is summarized by the pseudocode in Table 14.3.

In the first step, each example defines its own cluster. This means that in a domain with N examples, we have N initial clusters. In a series of the subsequent steps, hierarchical aggregation always identifies a pair of clusters that have the smallest mutual distance along the distance metric from the previous paragraphs. These clusters are then merged. At an early stage, this typically amounts to merging pairs of neighboring examples. Later, this results either in adding an example to its nearest cluster, or in merging two neighboring clusters. The first few steps are illustrated in Fig. 14.5.

The process continues until an appropriate termination criterion is satisfied. One possibility is to stop when the number of clusters drops below a user-specified threshold. Alternatively, one can base the stopping criterion on the cluster-to-cluster distance (finish when the smallest of these distances exceeds a certain value).

What Have You Learned?

To make sure you understand the topic, try to answer the following questions. If needed, return to the appropriate place in the text.

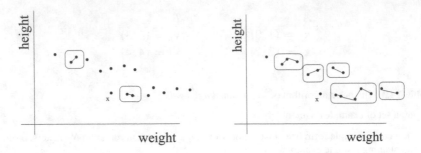

Fig. 14.5 Hierarchical aggregation after first two steps (*left*) and after first nine steps (*right*). Note how the clusters are gradually developed

- What kind of clusters cannot be detected by the *k-means* algorithm?
- What distance metric is used in hierarchical aggregation? What are the advantages and disadvantages of this metric?
- Describe the principle of the hierarchical-aggregation approach to clustering. For what kind of clusters is it particularly suited?

14.5   Self-Organizing Feature Maps: Introduction

Let us now introduce yet another approach to unsupervised learning, this time borrowing from the field of neural networks. The technique is known as a *self-organizing feature map, SOFM*.[3] Another name commonly used in this context is *Kohonen networks*, to honor its inventor.

The Idea Perhaps the best way to explain the nature of SOFM is to use, as a metaphor, the principle of physical attraction. A *code vector*, initially generated by a random-number generator, is subjected to the influence of a sequence of examples (attribute vectors), each "pulling" the vector in a different direction. In the long run, the code vector settles in a location that represents a compromise over all these conflicting forces.

The whole network consists of a set of *neurons* arranged in a two-dimensional matrix such as the one shown in Fig. 14.6. Each node (a neuron) in this matrix represents a code vector that has the same length (the same number of attributes) as the training examples. At the bottom is an input attribute vector that is connected to all neurons in parallel. The idea is to achieve by training a situation where neighboring neurons respond similarly to similar input vectors. For the sake of simplicity, the input vector in the picture only has two attributes, x_1 and x_2. In reality, it can have a great many.

---

[3]In statistics, and in neural networks, scientists often use the term *feature* instead of *attribute*.

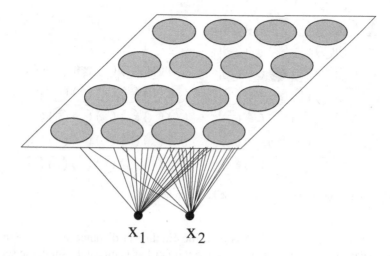

Fig. 14.6 General schema of a Kohonen network

How to Model Attraction Each neuron is described by a weight vector, $\mathbf{w} = (w_1, \ldots, w_n)$ where n is the number of attributes describing the examples. If $\mathbf{x} = (x_1 \ldots, x_n)$ is the example, and if $\eta \in (0, 1)$ is a user-specified learning rate, then the individual weights are modified according to the following formula:

$$w_i = w_i + \eta(x_i - w_i) \tag{14.5}$$

Note that the i-th weight is increased if $x_i > w_i$ because the term in the parentheses is then positive (and η is always positive). Conversely, the weight is decreased if $x_i < w_i$ because then the term is negative. It is in this sense that we say that the weight vector is attracted to $\mathbf{x}$. How strongly it is attracted is determined by the value of the learning rate.

Numeric Example Suppose that an example $\mathbf{x} = (0.2, 0.8)$ has been presented, and suppose that the winning neuron has the weights $\mathbf{w} = (0.3, 0.7)$. If the learning rate is $\eta = 0.1$, then the new weights are calculated as follows:

$w_1 = w_1 + \eta(x_1 - w_1) = 0.3 + 0.1(0.2 - 0.3) = 0.3 - 0.01 = 0.29$
$w_2 = w_2 + \eta(x_2 - w_2) = 0.7 + 0.1(0.8 - 0.7) = 0.7 + 0.01 = 0.71$

Note that the first weight originally had a greater value than the first attribute. By the force of the attribute's attraction, the weight has been reduced. Conversely, the second weight was originally smaller than the corresponding attribute, but the attribute's "pull" increases it.

Which Weight Vectors Are to Be Attracted by the Example Once an example, $\mathbf{x}$, has been presented to the neural matrix, a two-step process is launched. The task of the first step, a "competition," is to identify in the matrix the neuron whose weight

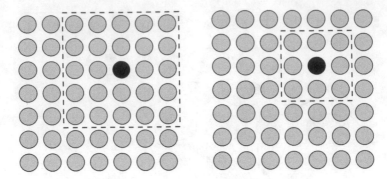

Fig. 14.7 The idea of "neighborhood" in the Kohonen network

vector is most similar to **x**. To this end, the Euclidean distance is used—smaller distance means greater similarity. Once the winner has been established, the second step updates the weights of this winning neuron as well as the weights of all neurons in the winner's physical neighborhood.

A Note on "Neighborhood"   Figure 14.7 illustrates what is meant by the neighborhood of the winning code vector, c_{winner}. Informally, the neighborhood consists of a set of neurons within a specific physical distance (in the matrix) from c_{winner}. Note that this results in a situation where the weights of all neurons in the neighborhood are modified in a like manner.

Usually, the size of the neighborhood is not fixed. Rather, it is a common practice to reduce it over time as indicated in the right part of Fig. 14.7. Ultimately, the neighborhood will degenerate to the single neuron, the one that has won the competition. The idea is to start with a coarse approximation that is later fine-tuned.

Why Does It Work?   The idea motivating the self-organizing feature map is to make sure that code vectors physically close to each other in the neural matrix respond to similar examples. This is why the same weight-updating formula is applied to all neurons in the winner's neighborhood.

What Have You Learned?

To make sure you understand the topic, try to answer the following questions. If needed, return to the appropriate place in the text.

- Describe the general architecture of self-organizing feature maps. Relate the notion of a *code vector* to that of a *neuron* in the matrix.
- Explain the two steps of the self-organizing algorithm: competition, and weight adjustment. Comment also on the role of the learning rate, η.
- What is meant by the *neighborhood* of the winning code vector? Is its size always constant?

14.6   Some Important Details

Having outlined the principle, let us now take a look at some details without which the technique would underperform.

Normalization The technique does not work well unless all vectors have been normalized to unit length (i.e., length equal to 1). This applies both to the vectors describing the examples, and to the weight vectors of the neurons. Fortunately, normalization to unit length is easy to carry out. Suppose an example is described as follows:

$$\mathbf{x} = (x_1, \ldots, x_n)$$

To obtain unit length, we divide each attribute's value by the length of the original attribute vector, $\mathbf{x}$:

$$x_i := \frac{x_i}{\sqrt{\Sigma_j x_j^2}} \tag{14.6}$$

Numeric Example Suppose we want to normalize the two-dimensional vector $\mathbf{x} = (5, 5)$. The length of this vector is $l(\mathbf{x}) = \sqrt{x_1^2 + x_2^2} = \sqrt{25 + 25} = \sqrt{50}$. Dividing the value of each attribute by this length results in the following normalized version of the vector:

$$\mathbf{x}' = (\frac{5}{\sqrt{50}}, \frac{5}{\sqrt{50}}) = (\sqrt{\frac{25}{50}}, \sqrt{\frac{25}{50}}) = (\frac{1}{\sqrt{2}}, \frac{1}{\sqrt{2}})$$

That the length of $\mathbf{x}'$ is equal to 1 is easy to verify: using the Pythagorean Theorem, we calculate it as $\sqrt{x_1^2 + x_2^2} = \sqrt{\frac{1}{2} + \frac{1}{2}} = 1$. We can see that the new attribute vector indeed has unit length.

Initialization The first step in SOFM is the initialization of the neurons' weights. Usually, this initialization is carried out by a random-number generator that chooses the values from an interval that spans equally the positive and negative domains, say, $[-1, 1]$. After this, the weight vector is normalized to unit length as explained above.

Another thing to be decided is the learning rate, η. Usually, a small number is used, say, $\eta = 0.1$. Sometimes, however, it is practical to start with a relatively high value such as $\eta = 0.9$, and then gradually decrease it. The point is to modify the weights more strongly at the beginning, during the period of early approximation. After this, smaller values are used so as to fine-tune the results.

The Algorithm The general principle of self-organizing feature maps is summarized by the pseudocode in Table 14.4. To begin with, all examples are normalized

Table 14.4 The basic algorithm of self-organizing feature maps

Input: set of examples without labels
     a learning rate, η.
     a set of randomly initialized neurons arranged in a matrix

1. Normalize all training examples to unit length.
2. Present a training example, and find its nearest code vector, c_{winner}
3. Modify the weights of c_{winner}, as well as the weights of the code vectors in the neighborhood of c_{winner}, using the formula $w_i = w_i + \eta(x_i - w_i)$. After this, re-normalize the weight vectors.
4. Unless a stopping criterion is met, present another training example, identify c_{winner}, and repeat the previous step.

Comments:
1. η usually begins with a relatively high value from $(0, 1)$, then gradually decreases.
2. Every now and then, the size of the neighborhood is reduced.

to unit length. Initial code vectors are created by a random-number generator and then normalized, too.

In the algorithm's main body, training examples are presented one by one. After the presentation of example x, the algorithm identifies a neuron whose weight vector, c_{winner}, is the closest to x according to the Euclidean distance. Then, the weights of c_{winner} as well as those of all neurons in its neighborhood in the matrix are modified using Eq. (14.5), and then re-normalized. The algorithm is usually run for a predefined number of epochs.[4]

In the course of this procedure, the value of the learning rate is gradually decreased. Occasionally, the size of the neighborhood is reduced, too.

What Have You Learned?

To make sure you understand the topic, try to answer the following questions. If needed, return to the appropriate place in the text.

- Which vectors have to be normalized? What is the goal of this normalization? Write down the formula that is used to this end.
- What is the subject of initialization? What values are used?
- Summarize the algorithm of self-organizing feature maps.

---

[4]Recall that one epoch means that all training examples have been presented once.

14.7   Why Feature Maps?

Let us now turn our attention to some practical benefits of self-organizing feature maps.

Reducing Dimensionality The technique introduced in the previous section essentially maps the original N-dimensional space of the original attribute vectors to the two-dimensional space of the neural matrix: each example has its winning neuron, and the winner can be described by its two coordinates in the matrix. These coordinates can be seen as new description of the original examples.

Creating Higher-Level Features Reducing the number of attributes is of vital importance in applications where the number of attributes is prohibitively high, especially if most of these attributes are either irrelevant or redundant. For instance, this may the case in the field of computer vision where each image (i.e., an example) can be described by hundreds of thousands of pixels.

   Having studied Chap. 7, the reader understands that a direct application of, say, perceptron learning to attribute vectors of extreme length is unlikely to succeed: in a domain of this kind, a classifier that does well on a training set tends to fail miserably on testing data. The mathematical explanation of this failure is that the countless attributes render the problem's VC-dimension so high as to prevent learnability unless the training set is unrealistically large. Advanced machine learning applications therefore seek to reduce the dimensionality either by attribute selection or, which is more relevant to this chapter, by way of mapping the multi-dimensional problem to a smaller space. This is accomplished by creating new features as functions of the original features.

What New Features Can Thus Be Created One might take this idea a step further. Here is a simple suggestion, just to offer inspiration. Suppose the training examples are described by 100 attributes, and suppose we have a reason to suspect that some attributes are mutually dependent whereas quite a few others are irrelevant.

   In such a domain, the dimensionality is likely to be unnecessarily high, and any attempt to reduce it is welcome. One way to do so is to extract, say, five different subsets of the attributes, and then redescribe the training examples five times, each time using a different attribute set. After this, each of these newly obtained training sets is subjected to SOFM which then maps each multi-dimensional space to two dimensions. Since this is done five times, we obtain $5 \times 2 = 10$ new attributes.

Visualization Human brain has no problem visualizing two- or three-dimensional data, and then develop an opinion about the similarity (or mutual distance) of concrete examples, about the examples' distribution, or about the groups they tend to form. However, this becomes impossible when there are more than three attributes.

   This is when self-organizing feature maps can be practical. By mapping each example onto a two-dimensional matrix, we may be able to visualize at least some of the relations inherent in the data. For instance, similar attribute vectors are likely to be mapped to the same neuron, or at least to neurons that are physically close to each other. Similar observations can be made about general data distribution. In this sense, feature maps can be regarded as a useful visualization tool.

Initializing k-Means Each of the weight vectors in the neural matrix can be treated as a code vector. The mechanism of SOFM makes it possible to find reasonably good values of these vectors—which can then be used to initialize such cluster-analysis methods as *k-means*. Whether this is a practical approach is another question because SOFM is a computationally expensive technique.

A Brief Mention of "Deep Learning" Let us also remark that methods for automated creation of higher-level features from very long attribute vectors are used in so-called *deep learning*. In essence, deep learning is a neural-networks technique that organizes the neurons in many layers, many more that we have seen in the context of multi-layer perceptrons in Chap. 5. It is in this sense the networks are "deep," and this is what gave the paradigm its name.

The top layers of these networks are trained by a supervised learning technique such as the backpropagation of error that the reader already knows. By contrast, the task for the lower layers is to create a reasonable set of features. It is conceivable that self-organizing feature maps be used to this end; in reality, more advanced techniques are usually preferred, but their detailed treatment is outside the scope of an introductory textbook.

One has to be cautions, though. The circumstance that a concrete paradigm is popular does not mean that it is a panacea that will solve all machine-learning problems. Far from it. If the number of features is manageable, and if the size of the training set is limited, then classical approaches will do just as well as deep learning, or even better.[5]

What Have You Learned?

To make sure you understand the topic, try to answer the following questions. If needed, return to the appropriate place in the text.

- In what way can the SOFM technique help us visualize the data?
- Explain how the SOFM technique can be used to reduce the number of attributes and to create new, higher-level features. Where can this be used?
- What is the principle of *deep learning*?

---

[5]A bulldozer is more powerful than a spade, and yet the gardener prefers the spade most of the time.

14.8   Summary and Historical Remarks

- In some machine-learning tasks, we have to deal with example vectors that do not have class labels. Still, useful knowledge can be induced for them by the techniques of *unsupervised learning*.
- The simplest task in unsupervised learning is *cluster analysis*. The goal is to find a way that naturally divides the training set into groups of similar examples. Each example should be more similar to examples in its group than to examples from any other group.
- One of the simplest *cluster analysis* techniques is known under the name of *k-means*. Once an initial clustering has been created, the algorithm accepts one example at a time, and evaluates its distance from the centroid of its own cluster as well as the distances from the centroids of the other clusters. If the example appears to find itself in a wrong cluster, it is transferred to a better one. The technique converges in a finite number of steps.
- The quality of the clusters discovered by *k-means* is sensitive to initialization and to the user-specified value of *k*. Methods to improve this quality by subsequent merging and splitting, and by alternative initializations sometimes have to be considered. Also hierarchical implementations of *k-means* are useful.
- In some domains, the shapes of the data clusters make it impossible for *k-means* to find them. For instance, this was the case of the training set shown in Fig. 14.4. In this event, the engineer will may give preference to some other clustering technique such as *hierarchical aggregation* that creates the clusters in a bottom-up manner, always merging the clusters with the smallest mutual distance.
- In the case of *hierarchical aggregation*, it is impractical to identify the distance between two clusters with the distance between their centroids. Instead, we use the minimum distance between [x,y] where x belongs to one cluster and y to the other.
- One of the problems facing the *k-means* algorithm is the question of how to define the initial *code vectors*. One way to address this issue is by the technique of *self-organizing feature maps*.
- As an added bonus, self-organizing feature maps are capable of converting a high-dimensional feature space onto only two attributes. They can also be used for the needs of data visualization.

Historical Remarks   The problems of cluster analysis have been studied since the 1960s. The *k-means* algorithm was described by McQueen [58] and hierarchical aggregation by Murty and Krishna [71]. The idea of merging and splitting clusters (not necessarily those obtained by *k-means* was studied by Ball and Hall [2]. The technique of SOFM, self-organizing feature maps, was developed by Kohonen [45]. In this book, however, only a very simple version of the technique was presented.

14.9   Solidify Your Knowledge

The exercises are to solidify the acquired knowledge. The suggested thought experiments will help the reader see this chapter's ideas in a different light and provoke independent thinking. Computer assignments will force the readers to pay attention to seemingly insignificant details they might otherwise overlook.

Exercises

1. Look at the three initial clusters of two-dimensional vectors in Table 14.5. Calculate the coordinates of their centroids.
2. Using the Euclidean distance, decide whether all of the examples from group 1 are where they belong. You will realize that one of them is not. Move it to a more appropriate group and recalculate the centroids.
3. Normalize the examples in Table 14.5 using Eq. (14.2).
4. Suppose the only information you have is the set of the nine training examples from Table 14.5, and suppose that you want to run the k-means algorithm for $k = 3$ clusters. What will be the composition of the three initial clusters if your code vectors are $(1, 4)$, $(3, 6)$, and $(3, 5)$?
5. Consider the three clusters from Table 14.5. Which pair of clusters will be merged by the hierarchical-aggregation technique?

Give It Some Thought

1. At the beginning of this chapter, we specified the benefits of cluster analysis. Among these was the possibility of identifying neurons in RBF networks with clusters instead of examples. In the case of k-means, this is straightforward: each gaussian center is identified with one cluster's centroid. However, how would you benefit (in RBF networks) from the clusters obtained by hierarchical aggregation?
2. Try to invent a machine-learning algorithm that first pre-processes the training examples using some cluster-analysis technique, and then uses them for classification purposes.

Table 14.5 An initial set of three clusters

Group 1	Group 2	Group 3
(1, 4)	(4, 3)	(4, 5)
(3, 6)	(6, 7)	(3, 1)
(3, 5)	(2, 2)	(2, 3)

3. Explain how self-organizing feature maps can be used to define the code vectors with which *k-means* sometimes starts. Will it be more meaningful to use the opposite approach (initialize SOFM) by *kmeans*)?

Computer Assignments

1. Write a program that accepts as input a training set of unlabeled examples, chooses among them k random code vectors, and creates the clusters using the *k-means* technique.
2. Write a program that decides whether a pair of clusters (obtained by *k-means*) should be merged. The easiest way of doing so is to compare the distance between the two clusters with the average cluster-to-cluster distance in the given clustering.
3. Write a program that creates the clusters using the *hierarchical aggregation* technique described in Sect. 14.4. Do not forget that the distance between clusters is evaluated differently than in the case of *k-means*.
4. Write a program that accepts a set of unlabeled training examples and subjects them to the technique of self-organizing feature maps.

Chapter 15
Classifiers in the Form of Rulesets

Some classifiers take the form of so-called *if-then* rules: if the conditions from the *if*-part are satisfied, the example is labeled with the class specified in the *then*-part. Typically, the classifier is represented not by a single rule, but by a set of rules, a *ruleset*. The paradigm has certain advantages. For one thing, the rules capture the underlying logic, and therefore facilitate explanations of why an example has to be labeled with the given class; for another, induction of rulesets is capable of discovering recursive definitions, something that is difficult to accomplish within other machine-learning paradigms.

In our search for techniques that induce rules or rulesets from data, we will rely on ideas borrowed from *Inductive Logic Programming*, a discipline that studies methods for automated creation and improvement of *Prolog* programs. Here, however, we are interested only in classifier induction.

15.1 A Class Described By Rules

To prepare the ground for simple rule-induction algorithms to be presented later, let us take a look at the nature of the rules we will want to use. After this, we will introduce some relevant terminology and define the specific machine-learning task.

The Essence of Rules Table 15.1 contains the training set of the "pies" domain we have encountered earlier. In Chap. 1, the following expression was given as one possible description of the positive class:

```
[ (shape=circle) AND (filling-shade=dark) ] OR
[ NOT(shape=circle) AND (crust-shade=dark) ]
```

When classifying example **x**, the classifier compares the example's attribute values with those in the expression. Thus if **x** is circular and its filling-shade happens to be dark, the expression is *true*, and the classifier therefore labels **x** with

© Springer International Publishing AG 2017
M. Kubat, *An Introduction to Machine Learning*,
DOI 10.1007/978-3-319-63913-0_15

Table 15.1 Twelve training
examples expressed in a
matrix form

		Crust		Filling		
Example	Shape	Size	Shade	Size	Shade	Class
ex1	Circle	Thick	Gray	Thick	Dark	pos
ex2	Circle	Thick	White	Thick	Dark	pos
ex3	Triangle	Thick	Dark	Thick	Gray	pos
ex4	Circle	Thin	White	Thin	Dark	pos
ex5	Square	Thick	Dark	Thin	White	pos
ex6	Circle	Thick	White	Thin	Dark	pos
ex7	Circle	Thick	Gray	Thick	White	neg
ex8	Square	Thick	White	Thick	Gray	neg
ex9	Triangle	Thin	Gray	Thin	Dark	neg
ex10	Circle	Thick	Dark	Thick	White	neg
ex11	Square	Thick	White	Thick	Dark	neg
ex12	Triangle	Thick	White	Thick	Gray	neg

the positive class. If the expression is *false*, the classifier labels the example with the negative class. Importantly, the expression can be converted into the following two *rules*:

R1: *if* [(shape=circle) AND (filling-shade=dark)] *then* **pos**.
R2: *if* [NOT(shape=circle) AND (crust-shade=dark)] *then* **pos**.
     else **neg**.

In the terminology of machine learning, each rule consists of an *antecedent* (the *if*-part), which in this context is a conjunction of attribute values, and a *consequent* (the *then*-part) which points to a concrete class label.

Note that the consequents of both rules indicate the positive class. For an example to be labeled as positive, it is necessary that the conditions in the antecedent of at least one rule be satisfied. Otherwise the classifier will label the example with the default class which, in this case, is **neg**. We will remember that when working with rulesets in domains of this kind, one must not forget to specify the default class.

Simplifying Assumptions Throughout this chapter, we will rely on the following simplifying assumptions:

1. All training examples are described by discrete-valued attributes.
2. The training set is noise-free.
3. The training set is consistent: examples described by the same attribute vectors must belong to the same class.

The Machine-Learning Task Our goal is an algorithm for the induction of rulesets from data that satisfy the simplifying assumptions from the previous paragraph. We will limit ourselves to rules whose consequents point to the positive class, the default always being the negative class.

Since the training set is supposed to be consistent and noise-free, we will be interested in classifiers that correctly classify all training examples. This means that for each positive example, the antecedent of at least one rule will be *true*. For any negative example, no rule's antecedent is *true*, and the example is labeled with the default (negative) class.

A Rule "Covers" An Example Let us introduce one useful term: an example either is or is not *covered* by a rule. A simple illustration will clarify the notion. Consider the following rule:

R: *if* (shape=circle) *then* **pos**.

If we apply this rule to the examples from Table 15.1, we will observe that the antecedent's condition, shape=circle, is satisfied by the following set of examples: $\{ex_1, ex_2, ex_4, ex_6, ex_7, ex_0\}$. We will say that **R** *covers* these six examples. Generally speaking, a rule covers an example if the expression in the rule's antecedent is *true* for this example. Note that four of the examples covered by this particular rule are positive and two are negative.

Rule Specialization Suppose we modify the above rule by adding to its antecedent another condition, filling-shade=dark, obtaining the following:

R1: *if* (shape=circle) AND (filling-shade=dark) *then* **pos**

Checking **R1** against the training set, we realize that it covers the following examples: $\{ex_1, ex_2, ex_4, ex_6\}$. We observe that this is a subset of the six examples originally covered by **R**. Conveniently, only positive (and no negative) examples are now covered.

This leads us to the definition of another useful term. If a modification of a rule's antecedent reduces the set of covered examples to a subset, we say that the modification has *specialized* the rule. In other words, specialization narrows the set of covered examples to a proper subset. A typical way of specializing a rule is to add a new condition to the rule's antecedent.

Rule Generalization Conversely, a rule is *generalized* if its modification enlarges the set of covered examples to a superset—if the new version covers all examples that were covered by the previous version, plus some additional ones. The easiest way to generalize a rule is by removing a condition from its antecedent. For instance, this happens when we drop from rule **R1** the condition (filling-shade=dark).

Specialization and Generalization of Rulesets We have said we are interested in induction of rulesets that label an example with the positive class if the antecedent of at least one rule is *true* for the example. For instance, this is the case of the ruleset consisting of the rules **R1** and **R2** above.

If we remove one rule from a ruleset, the ruleset may no longer cover some of the previously covered examples. This, we already know, is called specialization. Conversely, adding a new rule to the ruleset will generalize the ruleset because the new rule will add to the set of covered examples.

What Have You Learned?

To make sure you understand the topic, try to answer the following questions. If needed, return to the appropriate place in the text.

- Explain the nature of rule-based classifiers. What do we mean when we say that a rule *covers* an example? Using this term (*cover*), specify how the induced classifier should behave on a consistent and noise-free training set.
- Define the terms *generalization* and *specialization*. How will you specialize or generalize a rule? How will you specialize or generalize a ruleset?
- List the simplifying assumptions to be used throughout this chapter.

15.2 Inducing Rulesets by Sequential Covering

Let us now introduce a simple technique that induces rulesets from training data satisfying the simplifying assumptions from the previous section.

The Principle The goal is to find a ruleset such that each of its rules covers some positive examples, but no negative examples. Together, the rules should cover all positive examples and no negative ones. The procedure we will use creates one rule at a time, always starting with a very general initial version (covering also negative examples) that is then gradually specialized until all negative examples are excluded from coverage. The circumstance that the rules are created sequentially, and that each is supposed to cover those positive examples that were missed by previous rules, gives the technique its name: *sequential covering*.

Baseline Version of Sequential Covering Table 15.2 provides the pseudocode of a simple method for induction of rulesets. The main body contains the sequential covering algorithm. The idea is to find a rule that covers some positive examples, but no negative examples. Once the rule has been created, the examples it covers are removed from the training set. If no positive examples remain, the algorithm stops; otherwise, the algorithm is applied to the reduced training set.

The lower part describes induction of a single rule. The algorithm starts with the most general version of the antecedent that says, "all examples are positive." Assuming that the training set contains at least one negative example, this statement is obviously incorrect. The algorithm therefore seeks to rectify the situation by specialization, trying to exclude from coverage some negative examples, hopefully without losing the coverage of the positive examples. The specialization operator adds to the rule another conjunct in the form, $a_i = v_j$ (read: the value of attribute a_i is v_j).

A Concrete Example Let us "hand-simulate" the sequential-covering algorithm using the data from Table 15.1. The first rule, with the empty antecedent, covers all training examples. Adding to the empty antecedent the condition `shape=circle`

Table 15.2 The sequential covering algorithm

Input: training set T.

Sequential covering.

Create an empty ruleset.

While at least one positive example remains in T:

1. Create a rule using the algorithm below.
2. Remove from T all examples that satisfy the rule's antecedent.
3. Add the rule to the ruleset.

Create a single rule

Create an initial version of the rule, **R**: *if* () *then* **pos**

1. If **R** does not cover any negative example, stop.
2. Add to **R**'s antecedent a condition, $a_i = v_j$, and return to the previous step.

---

results in a rule that covers four positive and two negative examples. Adding one more condition, `filling-shade=dark`, specializes the rule so that, while still covering the four positive examples, it now no longer covers any negative example. We have obtained a rule that covers examples $\{\textbf{ex}_1, \textbf{ex}_2, \textbf{ex}_4, \textbf{ex}_6\}$. Note that is the rule **R1** from the previous section.

If we remove these four examples from the training set, we are left with only two positive examples, $\textbf{ex}_3$ and $\textbf{ex}_5$. The development of another rule again starts from the most general version (empty antecedent). Suppose that we then choose `shape=triangle` as the initial condition. This covers one positive and two negative examples. Adding to the antecedent the term `filling-shade=dark`, we succeed in excluding the negative examples while retaining the coverage of the positive example $\textbf{ex}_3$, which can now be removed from the training set. After the creation of this second rule, we are left with one positive example $\textbf{ex}_5$.

We therefore have to create yet another rule whose task will be to cover $\textbf{ex}_5$ without covering any negative example. Once we find such rule, $\textbf{ex}_5$ is removed from the training set. After this, we observe that there are no positive examples left, and the procedure can stop. We have created a ruleset consisting of three rules that cover all positive examples and no negative examples.

How to Identify the Best Attribute-Value Pair In the previous example, we always chose the condition to be added to the rule's antecedent more or less at random. But seeing that we could have selected it from quite a few alternatives, we realize that we need a mechanism capable of informing us about the quality of each choice. Perhaps the most natural criterion to be used here is based on information theory, a principle we have encountered in Chap. 6 where we used it in the course of induction of decision trees.

Let N_{old}^+ be the number of positive examples covered by the original version of the rule, and let N_{old}^- be number of negative examples covered by the original version of the rule. Likewise, the numbers of positive and negative examples covered by the new version of the rule will be denoted by N_{new}^+ and N_{new}^-, respectively.

Since the rule covers only positive examples, the information content of the message that a randomly picked example is labeled by it as positive is calculated as follows (for the old version and for the new version):

$$I_{old} = -\log(\frac{N_{old}^+}{N_{old}^+ + N_{old}^-})$$

$$I_{new} = -\log(\frac{N_{new}^+}{N_{new}^+ + N_{new}^-})$$

The difference between these two is the amount of information that has been gained by modifying the rule. Usually, machine-learning professionals normalize the information gain by the number, N_C, of covered examples so as to give preference rule modifications that optimize the number of covered examples. The quality of the rule-improvement is then calculated as follows:

$$Q = N_C \times |I_{new} - I_{old}| \tag{15.1}$$

When comparing alternative ways of modifying a rule, we choose the one with the highest value of Q.

What Have You Learned?

To make sure you understand the topic, try to answer the following questions. If needed, return to the appropriate place in the text.

- Summarize the principle of the *sequential covering* algorithm.
- Explain the mechanism of gradual specialization of a rule. What do we want to accomplish by this specialization?
- How will you use information gain when looking for the most promising way of specializing a rule?

15.3   Predicates and Recursion

The *sequential covering* algorithm has a much broader scope of applications than the previous section seems to have indicated. Perhaps most importantly, the technique can be employed for induction of concepts expressed in predicate calculus.

Predicates: Greater Expressive Power Than Attributes A serious limitation of attribute-value logic is that it is not sufficiently flexible to capture certain relations among data. For instance, the fact that **y** is located between **x** and **z** can be stated using the predicate between(x,y,z) —more accurately, the predicate is the term "between," whereas " (x,y,z) " is a list of the predicate's arguments.

The reader will agree that trying to express the same relation by means of attributes and their values would be difficult to say the least. An attribute can be seen as a special case of a one-argument predicate. For instance, the fact that, for a given example, x, the shape is circular can be written as circular(x). But the analogy is no longer as obvious in the case of predicates with more arguments.

Induction of Rules in Predicate Calculus Here is an example of a rule that says that if x is a parent of y, and at the same time x is a woman, then this parent is actually y's mother:

if parent(x,y) AND female(x) *then* mother(x,y)

We can see that this rule has the same structure as the rules **R1** and **R2** we have seen above: a list of conditions in the antecedent followed by a consequent. And indeed, the same sequential covering algorithm can be used here. There is one difference, though. When choosing among candidate predicates to be added to antecedent, we must not forget that the meaning of the predicate changes if we change the arguments. For instance, the previous rule's meaning will change if we replace parent(x,y) with parent(x,z) because, in this case, the fact that x is a parent of z surely does not guarantee that x is mother of some other subject, y.

Rulesets Allow Recursive Definitions The rules can be more interesting than the toy domain from Table 15.1 might lead us to believe. For one thing, they can be recursive—which is the case of the following two rules defining the term ancestor.

if parent(x,y) *then* ancestor(x,y).
if parent(x,z) AND ancestor(z,y) *then* ancestor(x,y).

The meaning of two rules is easy to see. Ancestor is a parent, or at least the parent's ancestor. For instance, a grandparent is the parent of a parent—and therefore an ancestor.

A Concrete Example of Induction Let us illustrate induction of rulesets using the problem from Table 15.3. Here, two concepts (classes), parent and ancestor, are characterized by a list of positive examples under the assumption that any example that is not in this list should be regarded as a negative example. Our goal is to induce the definition of ancestor, using the predicate parent.

We begin with the most-general rule, *if* () *then* ancestor(x,y). In the next step, we want to add a condition to the antecedent. To this end, we may consider various possibilities, but the simplest appears to be parent(x,y) —which will also be supported by the information-gain criterion. We have obtained the following rule:

Table 15.3 Illustration of induction from examples described using predicate logic

Consider the knowledge base consisting of the following positive examples of classes
parent and ancestor, defined using prolog-like facts (any other example will be
regarded as negative).

```
parent(eve,ted)      ancestor(eve,ted)ancestor(eve,ivy)
parent(tom,ted)      ancestor(tom,ted)ancestor(eve,ann)
parent(tom,liz)      ancestor(tom,ted)ancestor(eve,jim)
parent(ted,ivy)      ancestor(tom,ted)ancestor(tim,ivy)
parent(ted,ann)      ancestor(tom,ted)ancestor(eve,ann)
parent(ann,jim)      ancestor(tom,ted)ancestor(eve,jim)
                                       ancestor(ted,jim)
```

From the above examples, the algorithm creates the following first version of the rule. Note
that this rule does not cover any negative examples.

R3: *if* parent(x,y) *then* ancestor(x,y)

Removing all positive examples of this rule, the following set of positive examples of
ancestor(x,y) remains:

```
ancestor(eve,ivy)
ancestor(eve,ann)
ancestor(eve,jim)
ancestor(tim,ivy)
ancestor(eve,ann)
ancestor(eve,jim)
```

To cover these, another rule is created:

if parent(x,z) *then* ancestor(x,y)

After specialization, the second rule is turned into following:

R4: *if* parent(x,z) AND ancestor(z,y) *then* ancestor(x,y)

These two rules **R3** and **R4** now cover all positive examples and no negative examples.

R3: *if* parent(x,y) *then* ancestor(x,y)

Observing that the rule covers only positive examples and no negative examples,
we realize there is no need to specialize it.

However, the rule covers only the ancestor examples from the middle column,
and no examples from the rightmost column. Obviously, we need at least one more
rule. When considering the conditions to be added to the empty antecedent of the

next rule, we may consider the following (note that this is always the same predicate, but each time with a different set of arguments):

```
parent(x,z)
parent(z,y)
```

Suppose that the first leads to higher information gain. Seeing that the rule still covers some negative examples, we want to specialize it by adding another condition to its antecedent. Seeing that the `parent` predicate does not lead us anywhere, we try the predicate `ancestor`, again with various lists of arguments. Evaluating the information gain of all alternatives, we realize that the best option is `ancestor(z,y)`. This is how we obtain the second rule:

R4: *if* `parent(x,z)` AND `ancestor(z,y)` *then* `ancestor(x,y)`.

What Have You Learned?

To make sure you understand the topic, try to answer the following questions. If needed, return to the appropriate place in the text.

- How can a class be expressed using predicates? In what sense is the language of predicates richer than the language of attributes?
- Give an example of a recursively defined class. Can you think of a different example than `ancestor`?

15.4 More Advanced Search Operators

The technique described in the previous sections followed a simple strategy: in its attempts to find a good ruleset, the algorithm always sought to modify the rule(s) by specialization and generalization, evaluating alternative options by the information-gain criterion.

Operators for Ruleset Modification In reality, the search for rules can be more flexible than that. Other ruleset-modifying operators have been suggested. These, as we will see, do not necessarily represent specialization or generalization, but if we take a look at them, we realize they make sense. Let us mention in passing that these operators have been derived with the help of a well-known principle from logic, so-called *inverse resolution*. For our specific needs, however, the method of their derivation is unimportant.

In the following, we will simplify the formalism by writing a comma instead of AND, and using an arrow instead of the *if-then* construct. In all of the four cases, the operator converts the ruleset on the left into the ruleset on the right. The leftmost column gives the traditional names of these operators.

- identification:
$$\left\{ \begin{array}{l} b,x \to a \\ b,c,d \to a \end{array} \right\} \quad \Rightarrow \quad \left\{ \begin{array}{l} b,x \to a \\ c,d \to x \end{array} \right\}$$

- absorption:
$$\left\{ \begin{array}{l} c,d \to x \\ b,c,d \to a \end{array} \right\} \quad \Rightarrow \quad \left\{ \begin{array}{l} c,d \to x \\ b,x \to a \end{array} \right\}$$

- inter-construction:
$$\left\{ \begin{array}{l} v,b,c \to a \\ w,b,c \to a \end{array} \right\} \quad \Rightarrow \quad \left\{ \begin{array}{l} u,b,c \to a \\ v \to u \\ w \to u \end{array} \right\}$$

- intra-construction:
$$\left\{ \begin{array}{l} v,b,c \to a \\ w,b,c \to a \end{array} \right\} \quad \Rightarrow \quad \left\{ \begin{array}{l} v,u \to a \\ w,u \to a \\ b,c \to u \end{array} \right\}$$

Note that these replacements are not deductive: the rules on the right are never perfectly equivalent to those on the left. And yet, they do appear to make sense intuitively.

How to Improve Existing Rulesets? The operators from the previous paragraph can be used to improve rulesets that have been induced by the sequential covering algorithm. We can even consider a situation where not one, but several different classes were induced, which gave rise to several rulesets.

These rulesets can then be improved applying the hill-climbing search technique. The search operators are those listed in the previous paragraph. The evaluation function may give preference to more compact rules that classify correctly some auxiliary set of training examples meant to represent a concrete application domain.

What Have You Learned?

To make sure you understand the topic, try to answer the following questions. If needed, return to the appropriate place in the text.

- List the ruleset-modifying operators listing in this section. Which field of logic has helped derive them?
- Suggest how you might use these rules in an attempt to improve a given ruleset?

15.5   Summary and Historical Remarks

- Some classifiers have the form of rules. A rule consists of an antecedent (a list of conjuncted conditions) and a consequent (a class label). If the rule's antecedent is *true*, for the given example, then the example is labeled with the label pointed to by the consequent.

- If a rule's antecedent is *true*, for an example, we say that the rule *covers* the example.
- In the course of rule induction, we often rely on *specialization*. This reduces the set of covered examples to its subset. A rule is specialized if we add a condition to its antecedent. Conversely, *generalization* enlarges the set of covered examples to its superset.
- Usually, we induce a set of rules, a *ruleset*. The classifier then labels an example as positive if the antecedent of at least one of the rules is *true*. Adding a rule to a ruleset represents generalization. Removing a rule would represent specialization.
- The chapter introduced a simple algorithm for induction of rulesets from noise-free and consistent training data described by discrete attributes. The algorithm can so to some degree be optimized with the help of a criterion derived from information theory.
- The same algorithm can be used for induction of rules in domains where the examples are described using *predicate* calculus. Even recursive rules can thus be discovered.
- Some other "search operators" have been developed by the field of *inverse resolution*. They do not necessarily represent specialization or deduction.

Historical Remarks Induction of rules belongs to the oldest tasks of machine learning since the days when this discipline was seen as a means of inducing knowledge artificial-intelligence systems. The sequential-covering algorithm is a simplified version of an algorithm by Clark and Niblett [16]. Its use for induction of predicate-based rule was inspired by the FOIL algorithm developed by Quinlan [77]. The additional operators from Sect. 15.4 are based on the operators introduced by Muggleton and Buntine [70] in the framework of their work on *inverse resolution*.

15.6 Solidify Your Knowledge

The exercises are to solidify the acquired knowledge. The suggested thought experiments will help the reader see this chapter's ideas in a different light and provoke independent thinking. Computer assignments will force the readers to pay attention to seemingly insignificant details they might otherwise overlook.

Exercises

1. Hand-simulate the algorithm of *sequential covering* for the data from Table 15.1. Ignoring information gain, indicate how the first rule is created if we start from `crust-shade=gray`.

2. Show that, when we choose different ways of specializing a rule (adding different attribute-value pairs), we in the end obtain a different ruleset, often of a different size.

Give It Some Thought

1. Think of some other examples of classes (different from those discussed in this chapter) that are best defined recursively.
2. Think about how classes that are by nature recursive would be difficult to address in the framework of attribute-value logic. Demonstrate the superior power of the predicate calculus.
3. Suggest a learning procedure for "knowledge refinement." In this task, we assume that certain classes have already been defined in predicate calculus. When presented with another set of examples, the knowledge-refinement technique seeks to optimize the existing rules, either my making them more compact, or by making them more accurate in the presence of noise.

Computer Assignments

1. Write a computer program that implements the *sequential covering* algorithm. Use some simple criterion (not necessarily information gain) to choose which condition to add to a rule's antecedent.
2. In the UCI repository, find a domain satisfying the criteria specified in Sect. 15.1. Apply to it the program developed in the previous step.
3. How would you represent two-argument or three-argument predicates if you wanted to implement your machine-learning program in *C++*, *Java*? or some other programming language of a similar nature?
4. Write a program that applies the *sequential covering* algorithm to examples described in predicate calculus.

Chapter 16
The Genetic Algorithm

The essence of machine learning is the *search* for the best solution to our problem: to find a classifier which classifies as correctly as possible not only the training examples, but also future examples. Chapter 1 explained the principle of one of the most popular AI-based search techniques, the so-called *hill-climbing*, and showed how it can be used in classifier induction.

There is another approach to search: the *Genetic Algorithm*, inspired by the principles of Darwinian evolution. The reader needs to be acquainted with it because the technique can be very useful in dealing with various machine-learning problems. This chapter presents the baseline version, and then illustrates its use using certain typical issues from the field of k-NN classifiers.

16.1 The Baseline Genetic Algorithm

Let us first briefly describe the general principle of the genetic algorithm, relegating the details of implementation to the next section.

The Basic Philosophy In this section, the classifier will encode in the form of a *chromosome*, which most of the time will be a string of bits that are sometimes referred to as "genes." The genetic algorithm operates with a population of chromosomes, each describing one individual (a classifier). Each such individual is assigned a value by a *fitness function*; this value will usually depend on the classifier's performance. The fitness function plays a role analogous to that of the evaluation function in heuristic search.[1]

---

[1]This chapter will use the terms "evaluation function," "survival function", and "fitness function" interchangeably.

© Springer International Publishing AG 2017
M. Kubat, *An Introduction to Machine Learning*,
DOI 10.1007/978-3-319-63913-0_16

Fig. 16.1 The genetic algorithm's endless loop. Each individual in the population has its chance of survival. Recombination of the genetic information provided by mating partners creates new chromosomes that may be corrupted by mutation

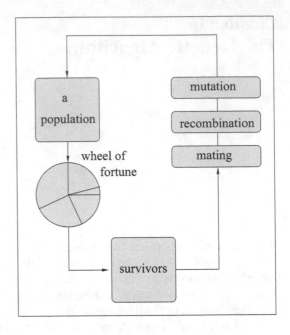

The Genetic Algorithm's Loop   The genetic algorithm operates in an endless loop depicted in Fig. 16.1. At each moment, there is a population of individuals, each with a certain value of the fitness function. This value then determines the size of the segment belonging to the individual in a "wheel of fortune" that determines the individual's chances of survival. It is important to understand the probabilistic nature of the process. While an individual with a larger segment enjoys a higher chance of survival, there is no guarantee of it because the survival game is non-deterministic. In the real world, too, a specimen with excellent genes may perish in a silly accident, while a weakling can make it by mere good luck. But in the long run, and in large populations, the laws of probability will favor genes that contribute to high fitness.

The surviving specimens will then choose "mating partners." In the process of mating, the chromosomes of the participating individuals are *recombined* (see below), which gives rise to a pair of new chromosomes. These new chromosomes may subsequently be subjected to *mutation*, which essentially adds noise to the strings of genes.

The whole principle is summarized by the pseudocode in Table 16.1.

How the Endless Loop Works   Once a new population has been created, the process enters a new cycle in which the individuals are subjected to the same wheel of fortune, followed by mating, recombination, and mutation, and the story goes on and on until stopped by an appropriate termination criterion. Note how occasional wrong turns are eliminated by the probabilistic nature of process. A low-quality chromosome may survive the wheel of fortune by a fluke; but if its children's fitness values remain low, the genes will perish in subsequent generations

Table 16.1 The principle of the genetic algorithm

---
initial state: a population of individual chromosomes

1. The fitness of each individual is evaluated. Based on its value, individuals are randomly selected for *survival*.
2. Survivors select mating partners.
3. New individuals are created by chromosome *recombination* of the mating partners.
4. Individual chromosomes are corrupted by random *mutation*.
5. Unless a termination criterion is satisfied, the algorithm returns to step 1.

---

anyway. Alternatively, some of an unpromising individual's genes may prove to be useful when embedded in different chromosomes which they may enter through recombination. By giving them an occasional second chance, the process offers flexibility that would be impossible in a more deterministic setting.

What Have You Learned?

To make sure you understand this topic, try to answer the following questions. If you have problems, return to the corresponding place in the preceding text.

- Explain the main principle of the genetic algorithm. How are the individuals described here? What is meant by their "survival chances"?
- Summarize the basic loop of the genetic algorithm.
- What is the advantage of the probabilistic implementation of the principle of survival as compared to a possible deterministic implementation?

16.2 Implementing the Individual Modules

Let us take a closer look at how to implement in a computer program the basic aspects of the genetic algorithm: the survival game, the mating process (partner selection), chromosome recombination, and mutation. To begin with, we will discuss only very simple solutions, relegating more advanced techniques to later sections.

For the sake of simplicity, we will assume that the chromosomes acquire the form of binary strings such as [1 1 0 1 1 0 0 1], where each bit represents a certain property that is either present, in which case the bit has value 1, or absent, in which case the bit is 0. Thus in a simplified version of the "pies" problem, the first bit may indicate whether or not the crust is thick, the second bit may indicate whether or not the filling is black, and so on.

Initial Population The most common approach to creating the initial population will employ a random-number generator. Sometimes, the engineer can rely on some knowledge that may help her create initial chromosomes known to outperform randomly generated individuals. In the "pies" domain, this role can be played by the descriptions of the positive examples. However, one has to make sure that the initial population is sufficiently large and has sufficient diversity.

The Survival Game The genetic algorithm assumes that there is a way to calculate for each specimen its survival chances. In some applications, these chances can be established by a practical experiment that lets the individual specimens to fight it out. In other domains, the fitness is calculated by a user-specified evaluation function whose value depends on the chromosome's properties. And if the chromosome represents a classifier, the fitness function can rely on the percentage of the training examples correctly labeled by the classifier.

An individual's survival is determined probabilistically. Here is how to implement this "wheel of fortune" in a computer program. Let F_i denote the i-th specimen's fitness and let $F = \Sigma_i F_i$ be the sum of all individual's fitness values that are then arranged along the interval $(0, F]$. The survival is modeled by a random-number generator that returns some $r \in (0, F]$: the sequential number of the subinterval that has been "hit" by r then points to the survivor. The principle is illustrated in Fig. 16.2 for a small population of four specimens and a random number that lands in the third interval so that individual 3 is selected. If the fate wants 20 specimens to survive, it has to generate 20 random numbers whose locations in the interval $(0, F]$ identify the survivors.

Whereas specimens with small fitness are likely to get eliminated, those with higher values can appear in the pool of survivors more than once. A biologist will wince at this "cloning" idea, but in the pragmatic world of computer programmers, the same individual can "survive" twice, three times, or even many times.

The Mating Operator The survival game is followed by mating. In nature, an individual judges a partner's suitability by strength, speed, or sharp teeth. Something similar is accomplished in a computer implementation by means of the fitness

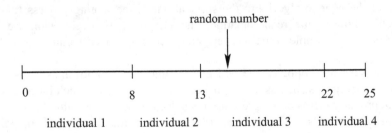

Fig. 16.2 The axis represents a population of four individuals whose fitness values are 8, 5, 9, and 3, respectively. Since the randomly generated number, 15, falls into the third subinterval, the third individual is selected

function. There is a difference, though: the notion of sex is usually ignored—any chromosome can mate with any other chromosome.

An almost trivial mating strategy will pair the individuals arbitrarily, perhaps generating random pairs of integers from the interval $[1, N_s]$, where N_s is the number of specimens in the population. However, this technique fails to do justice to the circumstance that specimens with high fitness are likely to be deemed more attractive than others. A simple way to reflect this in a computer program is to order the individuals in a descending order of their fitnesses, and then pair the neighbors.

Yet another strategy does it probabilistically. It takes the highest-ranking individual, then chooses its partner using the mechanism employed in the survival game—see Fig. 16.2. The same is done for the second highest-ranking individual, then for the third, and so on, until the new population has reached the required size. "Better" individuals are thus likely (though not guaranteed) to mate with other strong individuals. Sometimes, the partner will have low value (due to the probabilistic selection), but this gives rise to diversity that gives the system the opportunity to preserve valuable chromosome chunks that only have the bad luck of being currently incorporated in low-quality specimens.

Long-Living and Immortal Individuals One of the shortcomings of this algorithm is that a very good organism may be replaced by lower-valued children and useful genes may disappear. To prevent this from happening, some computer programs copy the best specimens into the new generation alongside their children. For instance, the program may directly insert in the new generation 20% best survivors, and then create the remaining 80% by applying the recombination and mutation operators to the best 95% individuals, totally ignoring the bottom 5%. In this way, not only will the best specimens live longer (even become "immortal"), but the program will also get rid of some very weak specimens that have survived by mere chance.

Chromosome Recombination: One-Point Crossover The simplest way to implement chromosome recombination is by the *one-point crossover*, an operator that swaps parts of the information in the parent chromosomes. The principle is simple. Suppose that each chromosome consists of a string of n bits and that a random-number generator has returned an integer $i \in [1, n]$. Then, the last i bits in the first chromosome (its i-bit tail) are replaced with the last i bits in the second chromosome and vice versa. A concrete implementation can permit the situation where $i = n$, in which case the two children are just replications of their parents. In the example below, the random integer is $i = 4$, which means that 4-bit tails are exchanged (the crossover point is indicated by a space).

$$
\begin{matrix}
1101\ 1001 \\
0010\ 0111
\end{matrix}
\Rightarrow
\begin{matrix}
1101\ 0111 \\
0010\ 1001
\end{matrix}
$$

The reader can see that the children tend to take after their parents, especially when the exchanged tails are short. The maximum distance between the children and the parents is achieved when $i = n - 1$.

In many applications, the recombination operator is applied only to a certain percentage of individuals. For instance, if 50 pairs have been selected for mating, and if the probability of recombination has been set by the user as 80%, then only 40 pairs will be subject to recombination, and the remaining 10 will just be copied into the next generation.

The Mutation Operator The task for mutation is to corrupt the inherited genetic information. Practically speaking, this is done by flip-flopping a small percentage of the bits in the sense that a bit's value 0 is changed to 1 or the other way round. The concrete percentage (the frequency of mutations) is a user-set parameter. Suppose that this parameter requires that $p = 0.001$ of the bits should on average be thus affected. The corresponding program module will then for each bit generate a random integer from the interval $[1, 1000]$. If the integer equals 1, then the bit's value is changed, otherwise it is left alone.

Let us give some thought to what frequency of mutations we need. At one extreme, very rare mutations will hardly have any effect at all. At the other extreme, very high mutation frequency would disrupt the genetic search by damaging too many chromosomes. If the frequency approaches 50%, then each new chromosome will behave as a randomly generated bit string; the genetic algorithm then degenerates to a random-number generator.

The mutation operator serves a different purpose than the crossover operator. In the one-point crossover, no new information is created, only existing substrings are swapped. Mutation introduces some new twist, previously absent in the population.

What Have You Learned?

To make sure you understand this topic, try to answer the following questions. If you have problems, return to the corresponding place in the preceding text.

- What is the main task of the survival game and how would you implement it in a computer program?
- Describe a simple mechanism to implement the selection of the mating partners. Describe the recombination operator, and the mutation operator.

16.3   Why It Works

Let us now offer an intuitive explanation of the genetic algorithm's performance.

Function Maximization The goal of the simple problem in Table 16.2 is to find the value of x for which the function $f(x) = x^2 - x$ is maximized. Each chromosome in the second column of the upper table is interpreted as a binary-encoded integer whose decadic value is given in the third column. The fourth column gives the

Table 16.2 Illustration of the genetic algorithm

Suppose we want the genetic algorithm to find the maximum of $f(x) = x^2 - x$. Let x be an integer represented by a binary string. The initial population consists of the four strings in the following table that for each of them gives the integer value, x, the corresponding $f(x)$, the survival chances (proportional to $f(x)$), and the number of times each exemplar was selected for the next generation.

No.	Initial population	x	$x^2 - x$	Survival chance	actual count
1	0 1 1 0 0	12	132	0.14	1
2	1 1 0 0 1	25	600	0.50	2
3	0 1 0 0 0	8	56	0.05	0
4	1 0 0 1 1	19	342	0.31	1
Average			282		
Maximum			600		

In the sample run reported here, the neighboring specimens mated, exchanging 1-bit tails and 3-bit tails, respectively, as dictated by the randomly generated tail lengths (the crossover sites indicated by spaces). No mutation is used here. The last two columns give the values of x and $f(x)$ for the new generation.

After reproduction	Mate with	Tail length	New population	x	$x^2 - x$
0 1 1 0 0	2	1	0 1 1 0 1	13	156
1 1 0 0 1	1	1	1 1 0 0 0	24	552
1 1 0 0 1	4	3	1 1 0 1 1	27	702
1 0 0 1 1	3	3	1 0 0 0 1	17	289
Average					425
Maximum					702

The reader can see that the value of the best specimen and the average value in the entire population have increased.

corresponding $f(x)$ whose relative value, shown in the fifth column, then determines for each individual its survival chances. For example, the first specimen has $f(x) = 12^2 - 12 = 132$ and the relative chances of survival (in this particular population) are 14% because $132/(132 + 600 + 56 + 342) = 0.14$. The rightmost column tells us how many times each individual has been selected for inclusion in the next generation.

In the next step, the survivors identify their mating partners. Let us assume that we have simply paired the neighboring specimens: the first with the second, and the third with the fourth. Then, the random selection of the crossover point dictates that 1-bit tails be exchanged in the first pair and 3-bit tails in the second. No mutation is applied. The result is shown in the bottom table where the last three columns

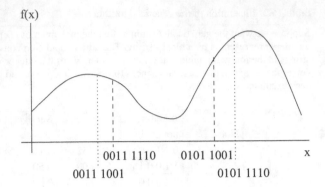

Fig. 16.3 After exchanging 4-bit tails, two parent chromosomes (upper strings) give rise to two children (lower strings). There is a chance that at least one child will "outperform" both parents

show, respectively, the new binary strings, their decadic values, and the values of $f(x)$. Note that both the average and the maximum value of the fitness function have increased.

Do Children Have to Outperform Their Parents? Let us ask what caused this improvement. An intuitive answer is illustrated in Fig. 16.3 that shows the location of two parents and the values of the survival function, $f(x)$, for each of them (the dashed vertical lines). When the two chromosomes swap their 4-bit tails, two children are created, each relatively close to one of the parents. The fact that each child finds itself in a region where the values of $f(x)$ are higher than those of the parents begs the question: are children always more fit than their parents? Far from that. All depends on the length of the exchanged tails and on the shape of the fitness function. Imagine that in the next generation the same two children get paired with each other and that the randomly generated crossover point is at the same location. Then, these children's children will be identical to the two original strings (their "grandparents"); this means that the survival chances decreased back to the original values. Sometimes, both children outperform their parents; in other cases, they are weaker than their parents; and quite often, we get a mixed bag. What matters is that in a sufficiently large population, most of the better specimens will survive because the selection process favors individuals with higher fitness, $f(x)$. Unfit specimens will occasionally make it, but they tend to lose in the long run.

If the exchanged string-tails are short, the children are close to their parent chromosomes. Long tails will give rise to children much less similar to their parents. As for mutation, its impact on the distance between the child and its parent depends on which bit is mutated. If it is the leftmost bit, the mutation will cause a big jump along the horizontal axis. If it is the rightmost bit, the jump is short. Either way, mutation complements recombination. Whereas the latter tends to explore the space in the vicinity of the parent chromosomes, the former may look elsewhere.

The Shape of the Fitness Function Some potential pitfalls inherent in the definition of the fitness functions are illustrated in Fig. 16.4. The function on the left is almost flat. The fact that different individuals have here virtually the same chances to survive defeats the purpose of the survival game. When the survivors are

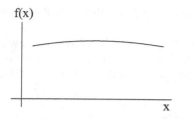
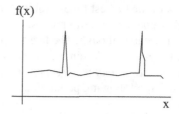

Fig. 16.4 Examples of two fitness functions that are poor guides for the genetic search. To be useful, the survival function should not be too flat and it should not contain isolated narrow peaks

chosen according to a near-uniform distribution, the qualities of the individuals will not give these individuals any perceptible competitive advantage. This drawback can be mitigated by making $f(x)$ less flat. There is an infinite number of ways this can be achieved, one possibility being to replace $f(x)$ with, say, $f(x) = f^2(x)$.

The right-hand part of Fig. 16.4 shows another pitfall: isolated narrow peaks. In comparison to the widths of the "peaks," children may find themselves too far from their parents. For instance, if the parent lies just at a hill's foot, the child may find itself on the opposite side, in which case the peak will go unnoticed. This problem is more difficult to prevent than the previous one.

What Have You Learned?

To make sure you understand this topic, try to answer the following questions. If you have problems, return to the corresponding place in the preceding text.

- Explain how the location of the crossover point determines how much the children will differ from their parents.
- Explain how the mutual interplay between recombination and mutation may affect the survival chances. Show how they also depend on the concrete shape of the survival function and on the location of the parents.

16.4   The Danger of Premature Degeneration

The fact that the genetic algorithm reached a value that does not seem to improve over a series of generation does not yet mean the search has been successful. The plateau may be explained by other circumstances.

Premature Degeneration A simple implementation of the genetic algorithm will stop after a predefined number of generations. A more sophisticated version will keep track of the highest fitness value achieved so far, and then terminate the search when this value no longer improves.

There is a catch, though. The fact that the fitness value has reached a plateau may not guarantee that a solution has been found. Rather, the search might have reached the stage called *premature degeneration*. Suppose that the search from Table 16.2 has reached the following population:

$$
\begin{array}{ccccc}
0 & 1 & 0 & 0 & 0 \\
0 & 1 & 0 & 0 & 1 \\
0 & 1 & 0 & 0 & 0 \\
0 & 1 & 0 & 0 & 0
\end{array}
$$

What are the chances of improving this population? Recombination will not get us anywhere. If the (identical) last two chromosomes mate, the children will only be copies of the parents. If the first two are paired, then 1-point crossover will only swap the rightmost bit, an operation that does not create a new chromosome, either. The only way to cause a change is to use mutation. By changing the appropriate bits, mutation can reignite the search. For instance, this will happen after the mutation of the third bit in the first chromosome and the fourth bit (from the left) of the last chromosome. Unfortunately, mutations are rare, and to wait for this to happen may be impractical. For all practical purposes, premature degeneration means the search got stuck.

Preventing Premature Degeneration   Premature degeneration has a lot to do with the population's *diversity*. The worst population is one in which all chromosomes have exactly the same bit string, something the engineer wants to avoid. Any computer implementation will therefore benefit from a module that monitors diversity and takes action whenever it drops below a certain level. A simple way to identify this situation is to calculate the average similarity between pairs of chromosomes, perhaps by counting the number of bits that have the same value in both strings. For instance, the similarity between [0 0 1 0 0] and 0 1 1 0 0] will be 4 (four bits are equal) and the similarity between [0 1 0 1 0] and [1 0 1 0 1] will be 0.

Once a drop in average chromosome-to-chromosome similarity has been detected, the system has to react. This is not yet a cause for alarm. Thus in the function-maximization example, advanced generations will be marked by populations where most specimens are already close to the maximum. This kind of "degeneration" will certainly *not* be deemed "premature." However, the situation is different if the best chromosome can be shown to be very different from the solution. In this event, we have to increase diversity.

Increasing Diversity   Several strategies can be used. The simplest will just insert in the current population one or more newly created random individuals. A more sophisticated approach will run the genetic algorithm on two or more populations in parallel, in isolation from each other. Then, either at random intervals, or whenever premature degeneration is suspected, a specimen from one population will be permitted to choose its mating partner in a different population. When implementing this technique, the programmer has to decide in which population to place the children.

The Impact of Population Size   Special attention has to be paid to the size of the population. Usually, though not always, the size is kept constant throughout the entire genetic search. The number of individuals in the population will be dictated by the concrete application. As a rule of thumb, smaller populations will need many generations to reach a good solution—unless they degenerated prematurely. Very large populations may be robust against degeneration, but they may incur impractical computational costs.

What Have You Learned?

To make sure you understand this topic, try to answer the following questions. If you have problems, return to the corresponding place in the preceding text.

- In what way does the success of the genetic algorithm depend on the definition of the fitness function? What are the two main pitfalls? How would you handle them?
- What criteria to terminate the genetic search would you recommend? What are their advantages and disadvantages?
- What is *premature degeneration?* How can it be detected and how can the situation be rectified? Why do we need diversity in the population?
- Discuss the impact of the population size.

16.5   Other Genetic Operators

We have introduced only a very simple version of the genetic algorithm and its operators. Now that the reader understands the principle, we take a look at some alternatives.

Two-Point Crossover   The one-point crossover introduced above is only a special case of the much more common *two-point crossover*. Here, the random-number generator is asked to return two integers that define two locations in the binary strings. The parents then swap the substrings between these two locations as illustrated below (the two crossover points are indicated by spaces).

$$
\begin{array}{c}
110\ 110\ 01 \\
001\ 001\ 11
\end{array}
\Rightarrow
\begin{array}{c}
110\ 001\ 01 \\
001\ 110\ 11
\end{array}
$$

The two crossover points can be different for each chromosome. In this event, each parent will "trade" a different substring of its chromosome as indicated below.

$$
\begin{array}{ll}
\text{1 101 1001} & \text{1 001 1001} \\
\text{001 001 11} & \text{001 101 11}
\end{array}
\Rightarrow
$$

Random Bit Exchange Yet another variation on the chromosome-recombination theme is the so-called *random bit exchange*. Here, the random-number generator selects a user-specified number of locations, and then swaps the bits at these locations as illustrated below.

$$
\begin{array}{ll}
\text{1 1 0 1 1 0 0 1} & \text{1 0 0 1 1 1 0 1} \\
\text{0 0 1 0 0 1 1 1} & \text{0 1 1 0 0 0 1 1}
\end{array}
\Rightarrow
$$

Here, the second and the sixth bits (counting from the left) were swapped. Note that nothing will happen if the leftmost bit is exchanged because it has the same value in both chromosomes. The number of exchanged bits can vary but most applications prefer the number to be much smaller than the chromosome's length.

A common practice in realistic applications is to combine two or more recombination operators. For instance, the selected pair of parents will with 50% probability be subjected to a 2-point chromosome, with 30% probability to a random bit exchange, and with 20% probability there will be no recombination at all.

Inversion Whereas the recombination operators act on pairs of chromosomes, other operators act on single specimens. One such operator is mutation; another is *inversion*. In a typical implementation, the random-number generator returns two integers that define two locations in the binary string (similarly as in the 2-point crossover). Then, the substring between the two positions is inverted as shown below.

$$110\ 110\ 01 \Rightarrow 110\ 011\ 01$$

Note that the order of the zeros and ones in the substring between the third and the seventh bit (counting from the left) was reversed. The location of the two points determines how much inversion impacts the chromosome. If the two integers are close to each other, say, 4 and 7, then only a small part of the chromosome is affected.

In advanced implementations, inversion is used to supplement mutation. For instance, the probability that a given bit is mutated can be set to 0.2% whereas each chromosome may have a 0.7% chance to see its random substring inverted. Similarly as with mutation, care has to be taken to make sure the inversion operator is used rarely. Excessive use may destroy the positive contribution of recombination.

Inversion and Premature Degeneration Much more than mutation, inversion is very good at extricating the genetic search from premature degeneration. To see why, take a look at the following degenerated population.

```
0 1 0 0 0
0 1 0 0 1
0 1 0 0 0
0 1 0 0 0
```

Inverting the middle three bits of the first chromosome, and the last three bits of the second chromosome will result in the following population:

```
0 0 0 1 0
0 1 1 0 0
0 1 0 0 0
0 1 0 0 0
```

The reader can see that the diversity has indeed increased. This observation suggests a simple way to handle premature degeneration: just increase, for a while, the frequency of inversions, and perhaps also that of mutations.

What Have You Learned?

To make sure you understand this topic, try to answer the following questions. If you have problems, return to the corresponding place in the preceding text.

- Explain the differences between one-point crossover, two-point crossover, and random bit exchange.
- What specific aspect makes the recombination operators different from the mutation and inversion operators?
- How does inversion affect the genetic search?

16.6 Some Advanced Versions

The genetic algorithm is a versatile general framework with almost infinite possibilities of variations. This section will introduce two interesting techniques.

A Note on the Lamarckian Alternative Computer programs are not constrained by the limitations of biology. Very often, the engineer discards some of these limitations, just as early aviators abandoned the idea of feathered wings. We have already encountered one such violation when making some specimens "immortal," copying them into the new generation to make sure they would not be destroyed by recombination and mutation. Let us now look at another deviation.

In the baseline genetic algorithm, new substrings come into being only as a result of random processes during such operators as recombination or mutation. After this, the genetic information remains unchanged throughout the specimen's entire life. One pre-Darwinian biologist, Jean-Baptiste Lamarck, suggested something more flexible: in his view, evolution might be driven by the individuals' needs. A giraffe that keeps trying to reach the topmost leaves will stretch his neck that will thus become longer. This longer neck is then passed on to the offspring. While the lamarckian hypothesis is untenable in the realm of biology, it is not totally irrational in other fields. For instance, by publishing a scientific paper, a researcher leaves to posterity the knowledge acquired during his lifetime.

Lamarckian evolution is much faster than that of the classical darwinian process, which is why we sometimes implement it in the genetic algorithm. The simplest way to incorporate this concept in the general loop from Fig. 16.1 is to place the "lamarckian" operator between the "wheel of fortune" and recombination. The task for the operator is to improve the chromosome by adaptation. For instance, one can ask what happens if a certain bit gets flipped-flopped by mutation. Whereas mutation by itself is irreversible, we can add flexibility by explicitly testing what happens when the i bit is flipped-flopped, and then choose the better version.

Multi-Population Search One motivation for multi-population search has to do with the many parameters the genetic algorithm depends on. Most of the time, the engineer has to rely only on her experience. Alternatively, we may choose to subject the same initial population to several parallel runs of the genetic algorithm, each with its own mutation frequency, with or without inversion, with a different mixture of recombination operators, or with a modified fitness function. Among the many alternatives, some will reach the solution faster than the others.

The reader will recall having encountered multi-population search in the section that discussed the threat of premature degeneration. In that particular context, the suggestion was to let two or more populations evolve in relative isolation that is disrupted by occasional interbreeding. Note that this interbreeding may not be easy to implement if each population uses a different way of chromosome definition as suggested in the previous paragraphs. In that case, the programmer has to implement a special program module for the conversion from one encoding to another.

Strings of Numbers, Strings of Symbols Chromosomes do not have to be binary strings; they can consist of numbers, or characters. The same recombination operators as before can then be used, though mutation may call for creativity. Perhaps the most common kind of mutation in numeric strings is to use "noise" superimposed on some (or all) of the chromosome's "genes." For instance, if all locations contain numbers from the interval [0, 100], then the noise can be modeled as a random number from $[-a, a]$ where a is a user-set parameter that plays here a role similar to that of mutation frequency in binary strings. Here is how it can work:

Before mutation	10	22	17	42	16
The "noise"		−3	1	−2	
After mutation	10	19	18	40	16

Fig. 16.5 A tree
representation of a candidate
expression from the "pies"
domain

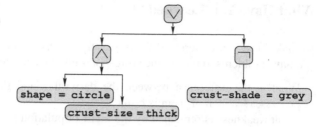

The situation is slightly different if the chromosomes have the form of strings
of symbols. Here, mutation can replace a randomly selected symbol in the chromosome with another symbol chosen by the random-number generator. For instance,
when applied to chromosome [d s r d w k l], the mutation can change from
r to s the third symbol from the left, the resulting chromosome being [d s s d
w k l].

Also possible are "mixed" chromosomes where some locations are binary, others
numeric, and yet others symbolic. Here, mutation is usually implemented as a
combination of the individual approaches. For instance, the program selects a
random location in the chromosome, determines whether the location is binary,
numeric, or symbolic, and then applies the appropriate type of mutation.

Chromosomes Implemented as Tree Structures In some applications, strings of
bits, numbers, or symbols are inadequate; a tree-structure may then be more flexible.
This, for instance, is the case of classifiers in the form of logical expressions—see
the example in Fig. 16.5 where the following expression is represented by a tree-like
chromosome.

```
(shape=circle ∧ crust-size=thick) ∨ ¬ crust-shade=gray
```

The expression consists of attributes, the values of these attributes, and the logical
operators of conjunction, disjunction, and negation. Note how naturally this is cast in
the tree structure. The internal nodes represent the logical operations and the leaves
contain the attribute-value pairs Recombination swaps random subtrees. Mutation
can affect the leaves: either attribute names or attribute values or both. Another
possibility for mutation is occasionally to replace ∧ with ∨ or the other way round.

Special attention has to be paid to the way the initial population is generated. The
programmer has to make sure that the population already contains some promising
expressions. One possibility is to create a group of random expressions and to
insert in it the descriptions of the positive examples. The survival function (to be
maximized) can be defined as the classification accuracy on the training set.

What Have You Learned?

To make sure you understand this topic, try to answer the following questions. If
you have problems, return to the corresponding place in the preceding text.

- What is the difference between the darwinian and the lamarckian evolution
  processes? Which of them is faster?
- What weakness is remedied by the multi-population genetic algorithm? In what
  way do multiple populations address this problem?
- How would you implement the mutation operator if the chromosome is a "mixed"
  string of bits, numeric values, and symbols?
- How would you implement the recombination and mutation operators in domains
  where chromosomes have the form of tree data structures?

16.7   Selections in k-NN Classifiers

Let us now illustrate a possible application of the genetic algorithm on a realistic
problem from the field of machine learning.

Attribute Selection and Example Selection   The reader knows that the success of
the k-NN classifier depends on the quality of the stored examples and also on the
choice of the attributes to describe these examples. The problem of choosing the
right examples and attributes is easily cast in the search paradigm. For instance, the
initial state can be defined as the complete set of examples, and the complete set
of attributes; the search operators will remove examples and/or attributes; and the
evaluation function (whose value is to be minimized) will be defined as the error rate
reached by the 1-NN rule as measured on an independent set of testing examples.

Another possibility is to employ the genetic algorithm. In essence, we have to
decide how to represent the problem in terms of chromosomes, how to define the
fitness function, and the recombination and mutation operators. Then, we have to be
clear about how to interpret (and utilize) the result of the search.

Chromosomes to Encode the Problem   A very simple approach will divide the
binary chromosome into two parts: each location in the first part corresponds to one
training example, and each location in the second part corresponds to one attribute.
If the value of a certain bit is 0, the corresponding example or attribute is ignored,
otherwise it is kept. The fitness function will be designed in a way that seeks to
minimize the number of 1s.

This solution may lead to impractically long chromosomes in domains where
the training set contains many examples: if the training set has ten thousand
examples, ten thousand bits would be needed. A better solution will then opt for
the more flexible variable-length scheme where each element in the chromosome
contains an integer that points to a training example or an attribute. The length of

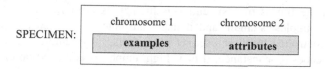

Fig. 16.6 Each specimen is described by two chromosomes, one representing examples and the other representing attributes. Recombination is applied to each of them separately

the chromosome would be the number of relevant attributes plus the number of representative examples. This mechanism is known as *value encoding*.

Interpreting the Chromosomes We must be sure to interpret the pairs of chromosomes properly. For instance, the specimen [3,14,39],[2,4] represents a training subset consisting of the third, the fourteenth, and the thirty-ninth training example, described by the second and the fourth attribute. When such specimen is used as a classifier, the system selects the examples determined by the first chromosome and describes them by the attributes determined by the second chromosome (Fig. 16.6). The distances between vectors $\mathbf{x} = (x_1, \ldots x_n)$ and $\mathbf{y} = (y_1, \ldots, y_n)$ are calculated using the formula:

$$D(\mathbf{x}, \mathbf{y}) = \sqrt{\Sigma_{i=1}^{n} d(x_i, y_i)}$$ (16.1)

where $d(x_i, y_i)$ is the contribution of the ith dimension. For numeric attributes, this contribution can be calculated by the usual formula for Euclidean distance, $d(x_i, y_i) = (x_i - y_i)^2$; for boolean attributes and for discrete attributes, we may define $d(x_i, y_i) = 0$ if $x_i = y_i$ and $d(x_i, y_i) = 1$ if $x_i \neq y_i$.

The Fitness Function The next problem is how to quantify each individual's survival chances. Recall that we want to reduce the number of examples and the number of attributes without compromising classification accuracy. These requirements may contradict each other because, in noise-free domains, the entire training set tends to give higher classification performance than a reduced set. Likewise, removing attributes is hardly beneficial if each of them provides relevant information.

The involved trade-offs therefore should be reflected in fitness-function parameters that give the user the chance to specify the concrete preferences. The fitness function should make it possible to place emphasis either on maximizing the classification accuracy or on minimizing the number of the retained training examples and attributes. This requirement is expressed by the following formula where E_R is the number of training examples misclassified by the given specimen, N_E is the number of retained examples, and N_A is the number of retained attributes:

$$f = 1/(c_1 * E_R + c_2 * N_E + c_3 * N_A)$$ (16.2)

Note that the fitness of a specimen is high if its error rate is low, if the set of retained examples is small, and if many attributes have been eliminated. The function is controlled by three user-set parameters, c_1, c_2, and c_3, that weigh the user's preferences. For instance, if c_1 is high, emphasis is placed on classification accuracy. If c_2 or c_3 are high, emphasis is placed on minimizing the number of retained examples and on minimizing the number of retained attributes, respectively.

Genetic Operators for This Application Parents are selected probabilistically. In particular, the following formula is used to calculate the probability that the specimen S' will be chosen:

$$Prob(S') = \frac{f(S')}{\sum f(S)} \tag{16.3}$$

Here, $f(S)$ is the fitness of specimen S as calculated by Eq. (16.2). The denominator sums up the values of the fitness functions of all specimens in the population—this makes the probabilities sum up to 1.

Once the pair of parents have been chosen, their chromosomes are recombined by the two-point crossover. Since each specimen is defined by a pair of chromosomes, each with a different meaning, we apply the recombination operator to each of them separately. Let the length of one parent's chromosome be denoted by N_1 and let the length of the other parent's chromosome be denoted by N_2. Using the uniform distribution, the algorithm selects one pair of integers from the closed interval $[1, N_1]$ and another pair of integers from the closed interval $[1, N_2]$. Each of these pairs then defines a substring in the respective chromosome (the first and the last locations are included in the substring). The crossover operator then exchanges the substrings from one of the parent chromosomes with the substrings of the other parent. Note that, as each of these substrings can have a different size, the children's lengths are likely to be different from the parents' lengths.

Graphical Illustration The principle is illustrated in Fig. 16.7 where the middle parts of chromosomes A and B have been exchanged. Note how the lengths of A and B are affected. The engineer has to decide whether to permit the situation where the exchanged segments have size 0; in the other extreme, a segment can represent the entire parent.

The *mutation* operator should prevent premature degeneration of the population and make sure the population represents a representative part of the search space.

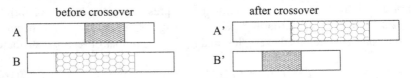

Fig. 16.7 The two-point crossover operator creates the children by exchanging randomly selected substrings in the parent chromosomes

One possibility is to select, randomly, a pre-specified percentage of the locations in the newly created population and to add to each of them a random integer generated separately for the location. The result is then taken modulo the number of examples/attributes. Let the original number of examples/attributes be 100 and let the location selected for mutation contains be 95. If the randomly generated integer is 22, then the value after mutation is $(95 + 22) \bmod 100 = 17$.

What Have You Learned?

To make sure you understand this topic, try to answer the following questions. If you have problems, return to the corresponding place in the preceding text.

- What can be accomplished by choosing the best attributes and the most representative examples?
- What are the advantages of using two chromosomes instead of just one?
- How does the chosen fitness function reflect the competing requirements of small sets of attributes and examples versus high classification accuracy?
- Why did we use a recombination operator that exchanges substrings of different lengths? How was mutation carried out?

16.8 Summary and Historical Remarks

- The genetic algorithm, inspired by the Darwinian evolution, is a popular alternative to classical artificial-intelligence search techniques. The simplest implementation works with binary strings.
- The algorithm subjects a population of individuals to three essential operations: fitness-function based survival, recombinations of pairs of chromosomes, and mutation. Also inversion of a substring is sometimes used.
- One of the frequently encountered problems in practical applications of the genetic algorithm is a population's premature degeneration. One way of detecting it is to consider the diversity of the chromosomes in the population. One solution will add artificially created chromosomes to the population. Also the inversion operator is useful, here.
- Alternative implementations of the genetic algorithm use strings of numbers, symbols, mixed strings, or even tree structures.
- The chapter illustrated the practical use of the genetic algorithm using a simple problem from the field of nearest-neighbor classifiers.

Historical Remarks The idea to cast the principle of biological evolution in the form of the genetic algorithm is due to Holland [37], although some other authors suggested something similar a little earlier. Among these, perhaps Rechenberg [80] deserves to be mentioned, while Fogel et al. [29] should be credited with pioneering the idea of genetic programming. The concrete way of applying the genetic algorithm to selections in the k-classifier is from Rozsypal and Kubat [82].

16.9   Solidify Your Knowledge

The exercises are to solidify the acquired knowledge. The suggested thought experiments will help the reader see this chapter's ideas in a different light and provoke independent thinking. Computer assignments will force the readers to pay attention to seemingly insignificant details they might otherwise overlook.

Exercises

1. Hand-simulate the genetic algorithm with a pencil and paper in a similar way as in Table 16.2. Use a fitness function of your own choice, a different initial population, and the random points for a one-point crossover. Then repeat the exercise with the two-point crossover.

Give It Some Thought

1. Explain how different population sizes may affect the number of generations needed to reach a good solution. Elaborate on the relation of population size to the problem of premature degeneration. Discuss also the effect of the shape of the fitness function.
2. What types of search problems are likely to be more efficiently addressed by the genetic algorithm than by classical search algorithms?
3. Identify concrete engineering problems (other than those in the previous text) appropriate for the genetic algorithm. Suggest problems where the chromosomes are best represented by binary or numeric strings, and suggest problems where trees are more appropriate.
4. Name some differences between natural evolution and its computer model. Speculate on whether more inspiration can be taken from nature. Where do you think are the advantages of the computer programs as compared to biological evolution?

Computer Assignments

1. Implement the baseline genetic algorithm to operate on binary-string chromo-somes. Make sure you have separate modules for the survival function, the wheel of fortune, recombination, and mutation, and that these modules are sufficiently general to enable easy modifications.
2. Create the initial populations for the "pies" and "circles" domains from Chap. 1 and use them as input to the program developed in the previous task. Note that, in the case of the "circles" domain, you might have to consider a slight modification of the original program so that it can handle numeric-string chromosomes.
3. For a domain of your choice, implement a few alternative mating strategies. Run systematic experiments to find out which strategy will most quickly find the solution. The speed can be measured by the number of chromosomes whose fitness values have to be evaluated before the solution is found.
4. For a domain of your choice, experiment with alternative "cocktails" of different recombination operators, and with different frequencies of recombinations, mutations, and inversions. Plot graphs that show how the speed of search (measured as in the previous task) depends on the concrete settings of these parameters.

Chapter 17
Reinforcement Learning

The fundamental problem addressed by this book is how to induce a classifier capable of determining the class of an object. We have seen quite a few techniques that have been developed with this in mind. In *reinforcement learning*, though, the task is different. Instead of induction from a set of pre-classified examples, the agent "experiments" with a system, and the system responds to this experimentation with rewards or punishments. The agent then optimizes its behavior, its goal being to maximize the rewards and to minimize the punishments.

This alternative paradigm differs from the classifier-induction task to such an extent that a critic might suggest that reinforcement learning should perhaps be relegated to a different book, perhaps a sequel to this one. The wealth of available material would certainly merit such decision. And yet, the author feels that this textbook would be incomplete without at least a cursory introduction of the basic ideas. Hence this last chapter.

17.1 How to Choose the Most Rewarding Action

To establish the terminology, and to convey some early understanding of what reinforcement learning is all about, let us begin with a simplified version of the task at hand.

N-Armed Bandit Figure 17.1 shows five slot machines. Each gives a different average return, but we do not know how big these average returns are. If we want to maximize our gains, we need to find out what these average returns are, and then stick with the most promising machine. This is the essence of what machine learning calls the problem of an *N-armed bandit*, alluding to the notorious tendency of the slot machines to rob you of your money.

© Springer International Publishing AG 2017
M. Kubat, *An Introduction to Machine Learning*,
DOI 10.1007/978-3-319-63913-0_17

Fig. 17.1 The generic problem: which of the slot machines offers the highest average return?

In theory, this should be easy. Why not simply try each machine many times, observe the returns, and then choose the one where these returns have been highest? In reality, though, this is not a good idea. Too many coins may have to be wasted before a reliable decision about the best machine can be made.

A Simple Strategy   Mindful of the incurred costs, the practically minded engineer will limit the experimentation, and make an initial choice based on just a few trials. Knowing that this early decision is unreliable, she will not be dogmatic. She will occasionally experiment with the other machines: what if some of them might indeed be better? If yes, it will be quite reasonable to replace the "previously best" with this new one. The strategy is quite natural. One does not have to be machine-learning scientist to come up with something of this kind.

This then is the behavior that the *reinforcement learning* paradigm seeks to emulate. In the specific case from Fig. 17.1, there are five actions to choose from. The principle described above combines *exploitation* of the machine currently believed to be the best, and the *exploration* of alternatives. Exploitation dominates; exploration is rare. In the simplest implementation, the frequency of the exploration steps is controlled by a user-specified parameter, ϵ. For instance, $\epsilon = 0.1$ means that the "best" machine (the one that appears best in view of previous trials) is chosen 90% of the time; in the remaining 10% cases, a chance is given to a randomly selected other machine.

Keeping a Tally of the Rewards   The "best action" is defined as the one that has led to the highest average return.[1] For each action, the learner keeps a tally of the previous returns; and the average of these returns is regarded as this action's *quality*. For instance, let us refer to the machines in Fig. 17.1 by integers, $1, 2, 3, 4,$ and 5. Action a_i then represents the choice of the i-th machine. Suppose the leftmost machine was chosen three times, and these choices resulted in the following returns $r_1 = 0, r_2 = 9,$ and $r_3 = 3$. The quality of this particular choice is then $Q(a_1) = (r_1 + r_2 + r_3)/3 = (0 + 9 + 3)/3 = 4$.

To avoid the necessity to store the rewards of all previously taken actions, the engineer implementing the procedure can take advantage of the following formula where $Q_k(a)$ is the quality of action a as calculated from k rewards, and r_{k+1} is the $(k + 1)$st reward.

---

[1] At this point, let us remark that the returns can be negative—"punishments," rather.

Table 17.1 The algorithm for the ϵ-greedy reinforcement learning strategy

Input: user-specified parameter ϵ, e.g., $\epsilon = 0.1$;
     a set of actions, a_i, and their initial value-estimates, $Q_0(a_i)$;
     for each action, a_i, let $k_i = 0$ (the number of times the action has been taken);

1. Generate a random number, $p \in (0, 1)$, from the uniform distribution.
2. If $p \geq \epsilon$, choose the action with the highest value (*exploitation*).
   Otherwise, choose a randomly selected other action (*exploration*).
3. Denote the action chosen in the previous step by a_i.
   Observe the reward, r_i.
4. Update the value of a_i using the following formula:

$$Q(a_i) = Q(a_i) + \frac{1}{k_i + 1}[r_i - Q(a_i)]$$

5. Set $k_i = k_i + 1$ and return to 1.

$$Q_{k+1}(a) = Q_k(a) + \frac{1}{k+1}[r_{k+1} - Q_k(a)] \qquad (17.1)$$

Thanks to this formula, it is enough to "remember" for each action only the values of k and $Q_k(a)$—these are all that is needed, together with the latest reward, to update the action's value at the $(k + 1)$st step.

The procedure just described is sometimes called the ϵ-greedy strategy. For the user's convenience, Table 17.1 summarizes the algorithm in a pseudocode.

Initializing the Process To be able to use Formula (17.1), we need to start somewhere: we need to set for each action its initial value, $Q_0(a_i)$. An elegant possibility is to choose a value well above any realistic single return. For instance, if all returns are known to come from the interval $[0, 10]$, the following will be reasonable initial values: $Q_0(a_i) = 50$.

At each moment, the system chooses, with $(1 - \epsilon)$ probability, the action with the highest value, breaking ties randomly. At the beginning, all actions have the same chance of being taken. Suppose that a_i is picked. In consequence of the received reward, this action's quality is then reduced using Formula (17.1). Therefore, when the next action is to be selected, it will (if *exploitation* is to be used) have to be some other action—whose value will then get reduced, too. Long story short, the reader can see that initialization of all action values to the same big number makes sure that, in the early stages of the game, all actions will be systematically experimented with.

What Have You Learned?

To make sure you understand this topic, try to answer the following questions. If you have problems, return to the corresponding place in the preceding text.

- Describe the ϵ-*greedy* strategy to be used when searching for the best machine in the N-armed bandit problem. Explain the meaning of actions and their values. What is meant by *exploitation* and *exploration*?
- Describe the simple mechanism for maintaining the average rewards. How does this mechanism update the action's values?
- Why did this section recommend that the initial values, $Q_0(a_i)$, of all actions should be set to a multiple of the typical reward?

17.2   States and Actions in a Game

The example with slot machines is a simplification that has made it easy to explain the basic terminology. Its main limitation is the existence of only one *state* in which an appropriate action is to be selected.

In reality, the situation is more complicated than that. Usually, there are many states, each with several actions to choose from. The essence can be illustrated on the tic-tac-toe game.

The Tic-Tac-Toe Game   The principle is shown in Fig. 17.2 for the elementary case where the size of the playing board is three by three squares. Two players are taking turns, one placing crosses on the board, the other one circles. The goal is to achieve a line of three crosses or circles—either in a column or in a row or diagonally. Who succeeds first, wins. If, in the situation on the left, it is the turn of the player that plays with crosses, he wins by putting his cross in the bottom left corner. If, conversely, it were his opponent's turn, the opponent would prevent this by putting there a circle.

States and Actions   Each board-position represents a *state*. At each state, the player is to choose a concrete *action*. Thus in the state depicted on the left, there are three empty squares, and thus three actions to choose from (one of them winning). The whole situation can be represented by a look-up table in which each state-action

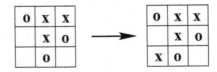

Fig. 17.2 In tic-tac-toe, two players took turns at placing their *crosses* and *circles*. The winner is the one who obtains a triplet in a line (vertical, horizontal, or diagonal)

pair has a certain value, $Q(s, a)$. Based on these values, the ϵ-*greedy* policy decides which action should be taken in the particular state. The action results in a reward, r, and this reward is then used to update the value of the state-action pair by means of Formula (17.1).

The most typical way of implementing the learning scenario is to let the program play a long series of games with itself, starting with ad hoc choices for actions (based on only the initial values of $Q(s, a)$), then gradually improving them until it achieves very high playing strength.

The main problem is how to determine the rewards of the concrete actions. In principle, three alternatives can be considered.

Episodic Formulation This is perhaps the simplest way of dealing with the reward-assignment problem. A whole game is played. If it is won, then all state-actions pairs encountered throughout the game by the learning agent are treated as if they received reward 1. If the game is lost, they are treated as if they all received reward -1.

The main weakness of this method is that it ignores the circumstance that not all actions taken in a game have equally contributed to the final outcome. A player may have lost only because of a single blunder that followed a long series of excellent moves. In this case, it would of course be unfair, even impractical, to punish the good moves. The same goes for the opposite: weak actions might actually receive the reward only because the game happened to be eventually won thanks to the opponent's unexpected blunder. One can argue, however, that, in the long run, these little "injustices" get averaged out because, most of the time, the winner's actions will be good.

The advantage of the episodic formulation is its simplicity.

Continuing Formulation The aforementioned problem with the episodic formulation (the fact that it may punish a series of good moves on account of a single blunder) might be removed under the assumption that we know how to determine the reward right after each action. This is indeed sometimes possible; and even in domain where this is *not* possible, one can often at least make an estimate.

Most of the time, however, an attempt to determine the reward for a given action before the game ends is speculative—and thus misleading. This is why this approach is rarely used.

Compromise: Discounted Returns This is essentially an episodic formulation improved in a way that determines the rewards based on the length of the game. For instance, the longer it has taken to win a tic-tac-toe game, the smaller the reward should be. There is some logic, in this approach: stronger moves are likely to win sooner. The way to implement this strategy is to discount the final reward by the number of steps taken before the victory.

Here is how to formulate the idea more technically: Let r_k denote the reward obtained at the k-th trial and let $\gamma \in (0, 1)$ is a user-set discounting constant. The *discounted return* R is then calculated as follows:

Fig. 17.3 The task: keep the pole upright by moving the cart left or right

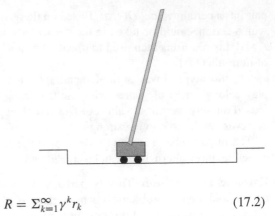

$$R = \Sigma_{k=1}^{\infty} \gamma^k r_k \qquad (17.2)$$

Note how the growing value of k decreases the coefficient by which r_k is multiplied. If the ultimate reward comes at the 10th step, and if "1" is the reward for the winning game, then the discounted reward for $\gamma = 0.9$ is $R = 0.9^{10} \cdot 1 = 0.35$.

Illustration: Pole Balancing A good illustration of when the discounted return may be a good idea is the pole-balancing problem shown in Fig. 17.3. Here, each *state* is defined by such attributes as the cart location, the cart's velocity, the pole's angle, and the velocity of the change in the pole's angle. There are essentially two *actions* to choose from: (1) apply force in the left-right direction or (2) apply force in the right-left direction. However, a different amount of force may be used. The simplest version of this task assumes that the actions can only be taken at regular intervals, say, 0.2 s.

In this game, the longer the time that has elapsed before the pole falls, the greater the perceived success, and this is why longer games should be rewarded more than short games. A simple way to implement this circumstance is to reward each state during the game with a 0, and the final fall with, say, $r = -10$. The discounted return will then be $R = -10\gamma^N$ where N is the number of steps before the pole has fallen.

What Have You Learned?

To make sure you understand this topic, try to answer the following questions. If you have problems, return to the corresponding place in the preceding text.

- Explain the difference between *states* and *actions*. What is the meaning of the "value of the state-action pair"?
- When it comes to reward-assignment, what is the difference between the episodic formulation and the continuing formulation?
- Discuss the motivation behind the idea of *discounted returns*. Give the precise formula, and illustrate its use on the pole-balancing game.

17.3 The SARSA Approach

The previous two sections introduced only a very simplified mechanism to deal with the reinforcement-learning problem. Without going into details, let us describe here a more popular approach that is known under the name of SARSA. The pseudocode summarizing the algorithm is provided in Table 17.2.

Essentially, the episodic formulation with discounting is used. The episode begins with selecting an initial state, s (in some domains, this initial state is randomly generated). In a series of successive steps, actions are taken according to the ϵ-greedy policy. Each such action results in a new state, s', being reached, and reward, r, being received. The same ϵ-greedy policy is then used to choose the next action, a' (to be taken in state s'). After this, the quality, $Q(s, a)$, of the given state-action pair is updated by the following formula:

$$Q(s, a) = Q(s, a) + \alpha[r + \gamma Q(s', a') - Q(s, a)] \qquad (17.3)$$

Here, α is a user-set constant and γ is the discounting factor.

Note that the update of the state-action pair's quality is based on the quintuple (s, a, r, s', a'). This is how the technique got its name.

What Have You Learned?

To make sure you understand this topic, try to answer the following questions. If you have problems, return to the corresponding place in the preceding text.

- Describe the principle of the SARSA approach to reinforcement learning. Where did this name come from?

Table 17.2 The SARSA algorithm—using the ϵ-greedy strategy and the episodic formulation of the task

---

Input: user-specified parameters ϵ, α, γ
  Initialized values of all action-value pairs, $Q_0(s_i, a_j)$;
  for each state-action pair, s_i, a_j, initialize $k_{ij} = 0$;

1. Choose an initial state, s.
2. Choose action a using the ϵ-greedy strategy from Table 17.1.
3. Take action a. This results in a new state, s', and reward, r.
4. In state s', choose action a' using the ϵ-greedy strategy.
   Update $Q(s, a) = Q(s, a) + \alpha[r + \gamma Q(s', a') - Q(s, a)]$
5. Let $s = s'$ and $a = a'$.

   If s is a terminal state, start a new episode by going to 1; otherwise, go to 3.

17.4   Summary and Historical Remarks

- Unlike the classifier-induction problems from the previous chapters, reinforcement learning assumes that an agent learns from direct experimentation with a system it is trying to control.
- In the greatly simplified formalism of the N-armed bandit, the agent seeks to identify the most promising action—the one that offers the highest average returns. The simplest practical implementation relies on the so-called ϵ-greedy policy.
- More realistic implementations of the task assume the existence of a set of states. For each state, the agent is to choose from a set of alternative actions. The choice can be made by the ϵ-greedy policy that relies on the qualities of the state-action pairs, $Q(s, a)$.
- The problem of assigning the rewards to the state-action pairs can be addressed by its episodic formulation, by continuing formulation, or by episodic formulation with discounting.
- Of the more advanced approaches to reinforcement learning, the chapter briefly mentioned the SARSA method.

Historical Remarks   One of the first systematic treatments of the "bandit" problem was offered by Bellman [3] who, in turn, was building on some earlier work still. Importantly, the same author later developed the principle of *dynamic programming* that can be regarded as a direct precursor to reinforcement learning [4]. The basic principles of reinforcement learning probably owe most for their development to Sutton [87].

17.5   Solidify Your Knowledge

The exercises are to solidify the acquired knowledge. The suggested thought experiments will help the reader see this chapter's ideas in a different light and provoke independent thinking. Computer assignments will force the readers to pay attention to seemingly insignificant details they might otherwise overlook.

Exercises

1. Calculate the number of state-action pairs in the tic-tac-toe example from Fig. 17.2.

Give It Some Thought

1. This chapter is all built around the idea of using the ϵ-*greedy* policy. What do you think are the limitations of this policy? Can you suggest how to overcome them?
2. The principles of reinforcement learning have been explained using some very simple toy domains. Can you think of an interesting real-world application? The main difficulty will be how to cast the concrete problem into the reinforcement-learning formalism.
3. How many episodes might be needed to solve the simple version of the tic-tac-toe game shown in Fig. 17.2?

Computer Assignments

1. Write a computer program that implements the N-armed bandit as described in Sect. 17.1.
2. Consider the maze-problem illustrated in Fig. 17.4. The task is to find the shortest path from the starting point, S, to the goal, G. A computer can use the principles of reinforcement learning to learn this shortest path based on great many training runs.

   Suggest the data structures to capture the states and actions of this game. Write a computer program that relies on the episodic formulation and the ϵ-*greedy* policy when addressing this task.

Fig. 17.4 The agent starts at S; the task is to find the shortest path to G

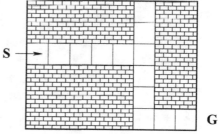

Bibliography

1. Ash, T. (1989). Dynamic node creation in backpropagation neural networks. *Connection Science: Journal of Neural Computing, Artificial Intelligence, and Cognitive Research, 1*, 365–375.
2. Ball, G. H. & Hall, D. J. (1965). ISODATA, a novel method of data analysis and clasification. *Technical Report of the Standford University*, Stanford, CA
3. Bellman, R. E. (1956). A problem in the sequential design of experiments. *Sankhya, 16*, 221–229.
4. Bellman, R. E. (1957). *Dynamic programming*. Princeton: Princeton University Press.
5. Blake, C. L. & Merz, C. J. (1998). Repository of machine learning databases. Department of Information and Computer Science, University of California at Irvine. www.ics.uci.edu/~mlearn/MLRepository.html.
6. Blumer, W., Ehrenfeucht, A., Haussler, D., & Warmuth, M. K. (1989). Learnability and the Vapnik-Chervonenkis dimension. *Journals of the ACM, 36*, 929–965.
7. Boutell, M. R., Luo, J., Shen, X., & Brown, C. M. (2004). Learning multi-label scene classification. *Pattern Recognition, 37*, 1757–1771
8. Bower, G. H. & Hilgard, E. R. (1981). *Theories of learning*. Englewood Cliffs. Prentice Hall.
9. Breiman, L. (1996). Bagging predictors. *Machine Learning, 24*, 123–140.
10. Breiman, L. (2001). Random forests. *Machine Learning, 45*, 5–32.
11. Breiman, L., Friedman, J., Olshen, R., & Stone, C. J. (1984). *Classification and regression trees*, Belmont: Wadsworth International Group.
12. Broomhead, D. S. & Lowe, D. (1988). Multivariable functional interpolation and adaptive networks. *Complex Systems, 2*, 321–355.
13. Bryson, A. E. & Ho, Y.-C. (1969). *Applied optimal control*. New York: Blaisdell.
14. Chow, C. K. (1957). An optimum character recognition system using decision functions. *IRE Transactions on Computers, EC-6*, 247–254.
15. Clare, A. & King, R. D. (2001). Knowledge discovery in multi-label phenotype data. In *Proceedings of the 5th European conference on principles of data mining and knowledge discovery, PKDD'01*, Freiburg, Germany (pp. 42–53)
16. Clark, P. & Niblett, R. (1989). The CN2 induction algorithm. *Machine Learning, 3*, 261–284.
17. Coppin, B. (2004). *Artificial intelligence illuminated*. Sudbury: Jones and Bartlett.
18. Cover, T. M. (1965). Geometrical and statistical properties of systems of linear inequalities with applications in pattern recognition. *IEEE Transactions on Electronic Computers, EC-14*, 326–334.
19. Cover, T. M. (1968). Estimation by the nearest neighbor rule. *IEEE Transactions on Information Theory, IT-14*, 50–55.

© Springer International Publishing AG 2017
M. Kubat, *An Introduction to Machine Learning*,
DOI 10.1007/978-3-319-63913-0

20. Cover, T. M. & Hart, P. E. (1967). Nearest neighbor pattern classification. *IEEE Transactions on Information Theory, IT-13*, 21–27.

21. Dasarathy, B. V. (1991). *Nearest-neighbor classification techniques*. Los Alomitos: IEEE Computer Society Press.

22. Dietterich, T. (1998). Approximate statistical tests for comparing supervised classification learning algorithms. *Neural Computation, 10*, 1895–1923.

23. Dudani, S. A. (1975). The distance-weighted *k*-nearest-neighbor rule. *IEEE Transactions on Systems, Man, and Cybernetics, SMC-6*, 325–327.

24. Fayyad, U. M. & Irani, K. B. (1992). On the handling of continuous-valued attributes in decision tree generation. *Machine Learning, 8*, 87–102.

25. Fisher, R. A. (1936). The use of multiple measurement in taxonomic problems. *Annals of Eugenics, 7*, 111–132.

26. Fisher, D. (1987). Knowledge acquisition via incremental conceptual clustering. *Machine Learning, 2*, 139–172.

27. Fix, E. & Hodges, J. L. (1951). Discriminatory analysis, non-parametric discrimination. USAF School of Aviation Medicine, Randolph Field, TX, Project 21-49-004, Report 4, Contract AF41(128)-3

28. Freund, Y. & Schapire, R. E. (1996). Experiments with a new boosting algorithm. In *Machine Learning: Proceedings of the Thirteenth International Conference*, Bari (pp. 148–156).

29. Fogel, L. J., Owens, A. J., & Walsh, M. J. (1966). *Artificial intelligence through simulated evolution*. New York: Wiley.

30. Friedman, J. H., Bentley, J. L., & Finkel, R. A. (1977). An algorithm for finding best matches in logarithmic expected time. *ACM Transactions on Mathematical Software, 3*(3), 209–226.

31. Gennari, J. H., Langley, P., & Fisher, D. (1990). Models of incremental concept formation. *Artificial Intelligence, 40*, 11–61.

32. Godbole, S. & Sarawagi, S. (2004). Discriminative methods for multi-label classification. In H. Dai, R. Srikant, & C. Zhang (Eds.), *Lecture Notes in Artificial Intelligence* (Vol. 3056, pp. 22–30). Berlin/Heidelberg: Springer.

33. Good, I. J. (1965). *The estimation of probabilities: An essay on modern Bayesian methods*. Cambridge: MIT.

34. Gordon, D. F. & desJardin, M. (1995). Evaluation and selection of biases in machine learning. *Machine Learning, 20*, 5–22.

35. Hart, P. E. (1968). The condensed nearest neighbor rule. *IEEE Transactions on Information Theory, IT-14*, 515–516.

36. Hellman, M. E. (1970). The nearest neighbor classification rule with the reject option. *IEEE Transactions on Systems Science and Cybernetics, 6*, 179–185.

37. Holland, J. H. (1975). *Adaptation in natural and artificial systems*. Ann Arbor: University of Michigan Press.

38. Holte, R. C. (1993). Very simple classification rules perform well on most commonly used databases. *Machine Learning, 11*, 63–90.

39. Hunt, E. B., Marin, J., & Stone, P. J. (1966). *Experiments in induction*. New York: Academic Press.

40. Katz, A. J., Gately, M. T., & Collins, D. R. (1990). Robust classifiers without robust features. *Neural Computation, 2*, 472–479.

41. Kearns, M. J. & Vazirani, U. V. (1994). *An introduction to computational learning theory*. Cambridge, MA: MIT Press.

42. Kodratoff, Y. (1988). *Introduction to machine learning*. London: Pitman.

43. Kodratoff, Y. & Michalski, R. S. (1990). *Machine learning: An artificial intelligence approach* (Vol. 3). San Mateo: Morgan Kaufmann.

44. Kohavi, R. (1997). Wrappers for feature selection. *Artificial Intelligence, 97*(1–2), 273–324.

45. Kohonen, T. (1982). Self-organized formation of topologically correct feature maps. *Biological Cybernetics, 43*, 59–69

46. Kohonen, T. (1990). The self-organizing map. *Proceedings of the IEEE, 78*(9), 1464–1480.

47. Koller, D. & Sahami, M. (1997). Hierarchically classifying documents using very few words. In *Proceedings of the 14th International conference on machine learning, ICML'07*, San Francisco, USA (pp. 170–178)

48. Kononenko, I., Bratko, I., & Kukar, M. (1998). Application of machine learning to medical diagnosis. In R. Michalski, I. Bratko, & M. Kubat (Eds.), *Machine learning and data mining: Methods and applications*. Chichester: Wiley.

49. Kubat, M. (1989). Floating approximation in time-varying knowledge bases. *Pattern Recognition Letters, 10*, 223–227.

50. Kubat, M., Pfurtscheller, G., & Flotzinger D. (1994). AI-based approach to automatic sleep classification. *Biological Cybernetics, 79*, 443–448.

51. Kubat, M., Holte, R., & Matwin, S. (1997). Learning when negatives examples abound. In *Proceedings of the European conference on machine learning (ECML'97)*, Apr 1997, Prague (pp. 146–153).

52. Kubat, M., Holte, R., & Matwin, S. (1998). Detection of oil-spills in radar images of sea surface. *Machine Learning, 30*, 195–215.

53. Kubat, M., Koprinska, I., & Pfurtscheller, G. (1998). Learning to classify medical signals. In R. Michalski, I. Bratko, & M. Kubat (Eds.), *Machine learning and data mining: Methods and applications*. Chichester: Wiley.

54. Littlestone, N. (1987). Learning quickly when irrelevant attributes abound: A new linear threshold algorithm. *Machine Learning, 2*, 285–318.

55. Lewis, D. D. & Gale, W. A. (1994). A sequential algorithm for training text classifiers. In *Proceedings of the 17th annual international ACM SIGIR conference on research and development in information retrieval (SIGIR'94)*, Dublin (pp. 3–12).

56. Louizou, G. & Maybank, S. J. (1987). The nearest neighbor and the bayes error rates. *IEEE Transactions on Pattern Analysis and Machine Intelligence, 9*, 254–262.

57. McCallum, A. (1999). Multi-label text classification with a mixture model trained by EM. In *Proceedings of the workshop on text learning (AAAI'99)* (pp. 1–7).

58. McQueen, J. (1967). Some methods for classification and analysis of multivariate observations. In *Proceedings of the 5th Berkeley symposium on mathematical statistics and probability*, Berkeley (pp. 281–297).

59. Michalski, R. S. (1969). On the quasi-minimal solution of the general covering problem. In *Proceedings of the 5th international symposium on information processing (FCIP'69)*, Bled, Yugoslavia (Vol. A3, pp. 125–128).

60. Michalski, R. S. & Tecuci, G. (1994). *Machine learning: A multistrategy approach*. Palo Alto: Morgan Kaufmann.

61. Michalski, R. S., Carbonell, J. G., & Mitchell, T. M. (1983). *Machine learning: An artificial intelligence approach*. Palo Alto: Tioga Publishing Company.

62. Michalski, R. S., Carbonell, J. G., & Mitchell, T. M. (1986). *Machine learning: An artificial intelligence approach* (Vol. 2). Palo Alto: Tioga Publishing Company.

63. Michalski, R., Bratko, I., & Kubat, M. (1998). *Machine learning and data mining: Methods and applications*. New York: Wiley.

64. Michell, M. (1998). *An introduction to genetic algorithm*. Cambridge, MA: MIT.

65. Mill, J. S. (1865). *A system of logic*. London: Longmans.

66. Minsky, M. & Papert, S. (1969). *Perceptrons*. Cambridge, MA: MIT.

67. Mitchell, T. M. (1982). Generalization as search. *Artificial Intelligence, 18*, 203–226.

68. Mitchell, T. M. (1997). *Machine learning*. New York: McGraw-Hill.

69. Mori, S, Suen, C. Y., & Yamamoto, K. (1992). Historical overview of OCR research and development. *Proceedings of IEEE, 80*, 1029–1058.

70. Muggleton, S. & Buntine, W. (1988). Machine invention of first-order predicates by inverting resolution. In *Proceedings of the 5th international machine learning conference*, Ann Arbor, Michigan (pp. 339–352)

71. Murty, M. N. & Krishna, G. (1980). A computationally efficient technique for data clustering. *Pattern Recognition, 12*, 153–158.

72. Neyman, J. & Pearson E. S. (1928). On the use and interpretation of certain test criteria for purposes of statistical inference. *Biometrica, 20A*, 175–240.

73. Ogden, C. K. & Richards, I. A. (1923). *The meaning of meaning* (8th ed., 1946). New York: Harcourt, Brace, and World.

74. Parzen E. (1962). On estimation of a probability density function and mode. *Annals of Mathematical Statistics, 33*, 1065–1076.

75. Quinlan, J. R. (1979). Discovering rules by induction from large collections of examples. In D. Michie (Ed.) *Expert systems in the micro electronic age*. Edinburgh: Edinburgh University Press.

76. Quinlan, J. R. (1986). Induction of decision trees. *Machine Learning, 1*, 81–106.

77. Quinlan, R. (1990). Learning logical definitions from relations. *Machine Learning, 5*, 239–266

78. Quinlan, J. R. (1993). *C4.5: Programms for machine learning*. San Mateo: Morgan Kaufmann.

79. Read, J., Pfahringer, B., Holmes, G., & Frank, E. (2011). Classifier chains for multi-label classification. *Machine Learning, 85*, 333–359.

80. Rechenberg, I. (1973). *Evolutionsstrategie: Optimierung technischer Systeme nach Principien der biologischen Evolution*. Stuttgart: Frommann-Holzboog.

81. Rosenblatt, M. (1958). The perceptron: A probabilistic model for information storage and organization in the brain. *Psychological Review, 65*, 386–408.

82. Rozsypal, A. & Kubat, M. (2001). Using the genetic algorithm to reduce the size of a nearest-neighbor classifier and to select relevant attributes. In *Proceedings of the 18th international conference on machine learning*, Williamstown (pp. 449–456).

83. Rumelhart, D. E. & McClelland, J. L. (1986). *Parallel distributed processing*. Cambridge: MIT Bradford Press.

84. Russell, S. & Norvig, P. (2003). *Artificial intelligence, a modern approach* (2nd ed.). Englewood Cliffs: Prentice Hall.

85. Schapire, R. E. (1990). The strength of weak learnability. *Machine Learning, 5*, 197–227.

86. Shawe-Taylor, J., Anthony, M., & Biggs, N. (1993). Bounding sample size with the Vapnik-Chervonenkis dimension. *Discrete Applied Mthematics, 42*(1), 65–73.

87. Sutton, R. S. (1984). *Temporal credit assignment in reinforcement learning*. PhD Dissertation, University of Massachusetts, Amherst.

88. Thrun, S. B. & Mitchell, T. M. (1995). Lifelong robot learning. *Robotics and Automonous Systems, 15*, pp. 24–46.

89. Tomek, I. (1976). Two modifications of CNN. *IEEE Transactions on Systems, Man and Communications, SMC-6*, 769–772.

90. Turney, P. D. (1993). Robust classification with context-sensitive features. In *Proceedings of the sixth international conference of industrial and engineering applications of artificial intelligence and expert systems*, Edinburgh (pp. 268–276).

91. Valiant, L. G. (1984). A theory of the learnable. *Communications of the ACM, 27*, 1134–1142.

92. Vapnik, V. N. (1992). *Estimation of dependences based on empirical data*. New York: Springer.

93. Vapnik, V. N. (1995). *The nature of statistical learning theory*. New York: Springer.

94. Vapnik, V. N. & Chervonenkis, A. Y. (1971). On the uniform convergence of relative frequencies of events to their probabilities. *Theory of Probability and its Applications, 16*, 264–280.

95. Werbos, P. (1974). *Beyond regression: New tools for prediction and analysis in the behavioral sciences*. PhD thesis, Harvard University.

96. Whewel, W. (1858). *History of scientific ideas*. London: J.W. Parker.

97. Widmer, G. & Kubat, M. (1996). Learning in the presence of concept drift and hidden contexts. *Machine Learning, 23*, 69–101.

98. Widrow, B. & Hoff, M. E. (1960). Adaptive switching circuits. In *IRE WESCON convention record*, New York (pp. 96–104).
99. Wolpert, D. (1992). Stacked generalization. *Neural Networks, 5*, 241–259.
100. Wolpert, D. (1996). The lack of a priori distinctions between learning algorithms. *Neural Computation, 8*, 1341–1390.
101. Zhang, M.-L. & Zhou, Z.-H. (2007). ML-KNN: A lazy learing approach to multi-label learning. *Pattern Recognition, 40*, 2038–2048.

Index

A
applications, 151, 167
attributes
    continuous, 30, 38, 45–47, 84, 122, 145
    discrete, 8, 22, 44, 137
    irrelevant, 13, 49, 74, 75, 118, 144, 156
    redundant, 13, 59, 118, 144, 156
    selection, 204, 205, 324
    unknown, 202

B
backpropagation, 98
bias, 67, 143, 191, 193

C
clustering
    hierarchical aggregation, 283
    intercluster distance, 275, 284
    k-means, 277, 281
    normalization, 278
    principle, 273
    SOFM, 286
context, 191, 199, 201

D
decision trees
    as classifiers, 113
    converted to rules, 130
    induction, 117
    numeric, 122
    pruning, 126
distance, 275

G
gaussian function
    in Bayes, 33
    in RBF networks, 107

I
imbalanced classes, 78, 194, 215, 225, 255
interpretability, 114, 126

L
linear classifiers
    in RBF networks, 108
    perceptron, 69
    WINNOW, 73
linearly ordered classes, 207

M
multi-label classification
    binary relevance, 254
    class aggregation, 263
    classifier chains, 256
    nearest-neigbor classifiers, 253
    neural networks, 252
    stacking, 258

N
nearest neighbor
    dangerous examples, 57
    weighted, 55

© Springer International Publishing AG 2017
M. Kubat, *An Introduction to Machine Learning*,
DOI 10.1007/978-3-319-63913-0

neural networks
    backpropagation, 97
    MLP architecture, 100
    MLP as classifiers, 91
    RBF networks, 106
noise
    in attributes, 14, 45, 57
    in class labels, 14
normalization, 51, 278, 279, 289

P
performance criteria
    F_β, 219
    error rate, 15
    macro-averaging, 265
    micro-averaging, 266
    precision, 215
    recall, 215
    sensitivity, 220
    specificity, 220
polynomial classifiers, 79
predicates
    alternative search operators, 305
    informal definition, 303
    recursive rules, 303
probability, 19
pruning
    decision tree, 127, 128, 184, 192
    rules, 130, 131

R
regression, 207

reinforcement
    sarsa, 337
    states and actions, 334
rule induction
    predicates, 303
    recursion, 303
    rulesets, 130, 298
    sequential covering, 300

S
search
    genetic, 309
    hill-climbing, 1, 5, 6, 8, 97, 309
similarity, 43
statistical evaluation, 241
statistical significance
    margin or error, 239
    type I error, 243
support vector machines
    linear, 84
    RBF-based, 108

T
time-varying classes, 200

V
voting
    plain, 174, 176
    weighted majority, 179, 181

Printed in the United States
By Bookmasters